# Emerging Separation and Separative Reaction Technologies for Process Waste Reduction

ADSORPTION AND MEMBRANE SYSTEMS

# Emerging Separation and Separative Reaction Technologies for Process Waste Reduction

ADSORPTION AND MEMBRANE SYSTEMS

**EDITORS**

Peter P. Radecki
John C. Crittenden
David R. Shonnard
John L. Bulloch

Michigan Technological University/
National Center for Clean Industrial and Treatment Technologies
Houghton, Michigan

Center for Waste Reduction Technologies
American Institute of Chemical Engineers
3 Park Avenue
New York, NY 10016-5901

Copyright © 1999
American Institute of Chemical Engineers
3 Park Avenue
New York, New York 10016-5901

All rights reserved. No part of this publication may be reproduced, stored in a retrieval system, or transmitted in any form or by any means, electronic, mechanical, photocopying, recording, or otherwise without the prior permission of the copyright owner.

**Library of Congress Cataloging-in-Publication Data**
Emerging separation and separative reaction technologies for process
  waste reduction : adsorption and membrane systems / editors, Peter
  P. Radecki . . . [et al.].
      p.     cm.
   Includes bibliographical references and index.
   ISBN 0-8169-0789-7
   1. Waste minimization.  2. Adsorption.  3. Membrane separation.
I. Radecki, Peter P.
TD793.8.E44  1998                       98-33101
628.4—dc21                               CIP

PRINTED IN THE UNITED STATES OF AMERICA
10  9  8  7  6  5  4  3  2  1

# Contents

Contributors      xi

Preface      xiii

## 1
## Adsorption, Membrane, and Separative Reactor Processes

| | | |
|---|---|---|
| **1.1.** | **Summary** | **1** |
| 1.1.1. | Process Modifications to Produce Less Pollution | 2 |
| 1.1.2. | Recovery and Recycle of Potential Contaminants for Reuse within Production Unit Boundaries | 3 |
| 1.1.3. | Research Needs | 3 |
| **1.2.** | **Adsorption Processes** | **4** |
| 1.2.1. | Adsorption Process Fundamentals | 4 |
| 1.2.2. | Adsorbents | 5 |
| 1.2.3. | Regeneration Cycles | 6 |
| 1.2.4. | Adsorption Process Configurations | 7 |
| 1.2.5. | Advantages/Disadvantages of Adsorption | 11 |
| 1.2.6. | Factors Favoring Adsorption | 11 |
| 1.2.7. | Applications of Adsorption for Pollution Prevention | 12 |
| 1.2.8. | Economics of Adsorption versus Competing Processes for Clean Air Applications | 12 |
| 1.2.9. | Future Directions in Adsorption Technology | 15 |
| | 1.2.9.1. Clean Regeneration Configurations | 15 |
| | 1.2.9.2. Simulation Models | 17 |
| **1.3.** | **Membrane Processes** | **17** |
| 1.3.1. | Membrane Fundamentals | 17 |

v

|  |  |  |
|---|---|---|
| 1.3.2. | Advantages/Disadvantages of Membranes | 18 |
| 1.3.3. | Factors Favoring Membrane Processes | 19 |
| 1.3.4. | Membrane Applications for Pollution Prevention | 19 |
| 1.3.5. | Membrane Phase Contactors in Pollution Prevention | 22 |
| 1.3.6. | Future Directions in Membrane Technology | 22 |

**1.4. Separative Reactor Processes** — 24

|  |  |  |
|---|---|---|
| 1.4.1. | Factors Favoring Separative Reactor Processes | 24 |
| 1.4.2. | Reactive Distillation | 24 |
| 1.4.3. | Absorption Reactors | 25 |
| 1.4.4. | Adsorption Reactors | 26 |
| 1.4.5. | Membrane Reactors | 26 |
| 1.4.6. | Future Directions in Separative Reactors | 29 |

**1.5. References** — 29

# 2

# Adsorption Technologies

**2.1. Adsorption Technology Overview** — 34

|  |  |  |
|---|---|---|
| 2.1.1. | Introduction | 34 |
| 2.1.2. | Current Adsorption Processes | 35 |
| 2.1.3. | Adsorption Fundamentals | 36 |
| 2.1.4. | Adsorption Equilibrium Description | 38 |
|  | 2.1.4.1. Heat of Adsorption | 43 |
| 2.1.5. | Kinetics of Adsorption | 44 |
| 2.1.6. | Molecular Simulation of Adsorption | 46 |
| 2.1.7. | Introduction of Different Adsorbents and Their Usage | 48 |
|  | 2.1.7.1. Carbon Adsorbents | 48 |
|  | 2.1.7.2. Zeolites | 49 |
|  | 2.1.7.3. Polymeric Adsorbents and Ion Exchange Resins | 51 |
|  | 2.1.7.4. Activated Alumina | 51 |
|  | 2.1.7.5. Silica Gel | 52 |
|  | 2.1.7.6. Clays | 52 |
| 2.1.8. | Selection of an Adsorbent | 53 |

**2.2. Adsorption Process Design—Engineering, Economic, Environmental, and Energy Considerations** — 54

|  |  |  |
|---|---|---|
| 2.2.1. | Mechanisms of Adsorptive Separations | 54 |
| 2.2.2. | Process Configuration | 55 |
|  | 2.2.2.1. Nonregenerative Adsorption Processes | 55 |

## Contents

|  |  |  |
|---|---|---|
| | 2.2.2.2. Regenerative Processes | 56 |
| | 2.2.2.4. Moving Bed Adsorbers | 63 |
| | 2.2.2.5. Pressure Drop / Cost Issues—Packed Bed Problems | 66 |
| 2.2.3. | **Equilibrium and Mass Transfer Parameters for Adsorber Design** | **69** |
| 2.2.4. | **Economic Viability of Adsorptive Separations** | **71** |
| | 2.2.4.1. Capital Cost and Operating Cost vs. Other Separation Technologies | 71 |
| 2.2.5. | **Environmental Benefits and Challenges** | **72** |

### 2.3. Progress toward Implementation: Incentives and Impediments, Research Needs — 74

- 2.3.1. Adsorbent Material Development — 75
- 2.3.2. Adsorption Process Improvements — 76
- 2.3.3. Advances in Engineering Design Information — 77

### 2.4. Selected Emerging and Proven Nonreactive Uses of Adsorption — 79

- 2.4.1. Gases/Bulk Separations — 81
- 2.4.2. Gases/Purifications — 82
- 2.4.3. Adsorption for BTU Adjustment in Natural Gas — 83
  - 2.4.3.1. Technology Description — 83
  - 2.4.3.2. Engineering, Economic, Environmental, and Energy Considerations — 84
  - 2.4.3.3. Contacts, References, and Suggested Vendors — 85
- 2.4.4. $H_2$ Separation from Low-Grade Refinery Gases — 85
  - 2.4.4.1. Technology Description — 85
  - 2.4.4.2. Engineering, Economic, Environmental, and Energy Considerations — 85
  - 2.4.4.3. Contacts, References, and Suggested Vendors — 86

### 2.5. General Listing of Adsorbent and Adsorption Process Suppliers — 86

### 2.6. References for Nonreactive Adsorption — 88

### 2.7. Adsorptive Chemical Reactors — 90

- 2.7.1. Introduction — 90
- 2.7.2. Reaction Chromatography — 92
- 2.7.3. The Rotating Cylindrical Annulus Chromatographic Reactor — 93
- 2.7.4. The Countercurrent Moving Bed Chromatographic Reactor — 94
- 2.7.5. The Simulated Countercurrent Moving Bed Chromatographic Reactor — 97
  - 2.7.5.1. Equilibrium Stage Model — 98
  - 2.7.5.2. Multiple Column Configuration — 101
  - 2.7.5.3. Esterification of Acetic Acid — 104
  - 2.7.5.4. Reactor Dynamics — 105
- 2.7.6. Natural Gas Utilization — 108

| | | |
|---|---|---|
| | 2.7.6.1. Oxidative Coupling of Methane | 109 |
| | 2.7.6.2. Partial Oxidation of Methane to Methanol | 112 |
| | 2.7.6.3. Methanol from Synthesis Gas | 114 |
| 2.7.7. | The Pressure Swing Reactor | 114 |
| 2.7.8. | The Gas–Solid–Solid Trickle Flow Reactor | 120 |
| 2.7.9. | The Temperature Swing Reactor | 122 |
| **2.8.** | **References for Adsorptive Reactors** | **124** |
| **2.9.** | **General Texts on Adsorption and Adsorption Processes** | **128** |

# 3

# Membrane Technologies

| | | |
|---|---|---|
| **3.1.** | **Membrane Technology Overview** | **132** |
| 3.1.1. | General Overview | 132 |
| | 3.1.1.1. Existing Membrane Separation Processes | 136 |
| | 3.1.1.2. Membrane Materials | 161 |
| | 3.1.1.3. Membrane Modules and Systems | 172 |
| **3.2.** | **Selected Emerging Nonreactive in-Process Source Reduction Membrane Applications** | **180** |
| 3.2.1. | Membrane Gas Separation Opportunities in the Control of Greenhouse Effect | 180 |
| | 3.2.1.1. Technology Description | 183 |
| | 3.2.1.2. Commercial Applications and Their Economics | 187 |
| 3.2.2. | Solvent Vapor Recovery from Gas Streams | 192 |
| | 3.2.2.1. Technology Description | 192 |
| | 3.2.2.2. Engineering, Economic, Environmental, and Energy Considerations | 193 |
| 3.2.3. | Metal Ion Recovery from Aqueous Waste Streams | 199 |
| 3.2.3.1. | Technology Description | 204 |
| | 3.2.3.2. Engineering, Economic, Environmental, and Energy Considerations | 210 |
| | 3.2.3.3. Vendors and Contacts | 212 |
| 3.2.5. | Pervaporation/Aqueous Streams | 212 |
| | 3.2.5.1. Technology Description | 212 |
| | 3.2.5.2. Engineering, Economic, Environmental, and Energy Considerations | 216 |
| **3.3.** | **Selected Emerging Separative Reactors Using Membranes** | **220** |
| 3.3.1. | Overview: Membranes in Reactors | 220 |

| | |
|---|---|
| 3.3.2. Hydrocarbon-Selective Oxidation | 227 |
|    3.3.2.1. Technology Description | 228 |
|    3.3.2.2. Engineering, Economic, Environmental, and Energy | 230 |
|    3.3.2.3. Contacts and Suggested Vendors | 232 |
| 3.3.3. Dehydrogenation Reactions | 234 |
|    3.3.3.1. Technology Description | 234 |
|    3.3.3.2. Engineering, Economic, Environmental, and Energy | 235 |
| 3.3.4. Methane Reforming Reactions | 237 |
|    3.3.4.1. Technology Description | 238 |
|    3.3.4.2. Engineering, Economic, Environmental, and Energy | 242 |
| 3.3.5 Industry Implementation Considerations | 243 |
|    Economics and Productivity Issues | 243 |
| **References** | **248** |

# 4

# Findings of the National Workshop on Process Waste Reduction via Separation Technologies and Separative Reactors

| | |
|---|---|
| **4.1. Overview of Workshop** | **261** |
| **4.2. Workshop Sponsors** | **261** |
| **4.3. Summaries of Technical Presentations** | **262** |
|    4.3.1. Introduction to the Center for Waste Reduction Technologies | 262 |
|    4.3.2. Sustainability—The Future of Pollution Prevention | 264 |
|    4.3.3. An Overview of Emerging Adsorption, Membrane, and Separative Reactor Technologies for Process Pollution Prevention | 266 |
|    4.3.4. Introduction to Adsorption Technology | 267 |
|    4.3.5. Adsorption Technology Panel Presentations and Discussion | 270 |
|       4.3.5.1. Adsorption's Role in Achieving In-Process Pollution Prevention | 270 |
|       4.3.5.2. Suggestions: New Areas of Waste Reduction via Adsorption | 271 |
|       4.3.5.3. Perspectives for the Development of New Sorption Based Technologies | 273 |
|    4.3.6. Design and Optimization of Waste Reduction by Adsorption with ADSIM™ | 275 |
|    4.3.7. Process Integration for Pollution Prevention | 277 |
|    4.3.8. Membrane Basics | 281 |
|    4.3.9. Membrane Technology Panel Presentations and Discussion | 282 |

|  |  |
|---|---|
| 4.3.9.1. Membrane Technologies: Applications—Present and Future | 283 |
| 4.3.9.2. Applications of Membrane Technologies for Process Waste Reductions in Gaseous Streams | 284 |
| 4.3.9.3. Environmental/Health Implications | 285 |
| 4.3.9.4. Process Waste Reduction via Membranes | 286 |
| 4.3.10. Introduction to Separative Reactor Technology | 289 |
| **4.3.11. Separative Reactor Technology Panel Presentations and Discussion** | **293** |
| 4.3.11.1 Adsorption-Based Separative Reactor Technology | 293 |
| 4.3.11.2. Membrane-Based Separative Reactor Technology | 294 |
| 4.3.11.3. Synthesis of an Integrated Separative Reactor Process | 294 |
| 4.3.12. Conclusions of CWRT Separation and Separative Reactor Monograph | 297 |
| **4.4. Processes and Process Stream Recommendations** | **298** |
| **4.5. List of Workshop Participants** | **309** |
| **References** | **312** |

*Index*     *313*

# Contributors

## Primary Authors

Robert W. Carr, *University of Minnesota, St. Paul, Minnesota*
Jimmy L. Humphrey, *J.L. Humphrey & Associates, Austin, Texas*
Kent Knaebel, *Adsorption Research Inc., Dublin, Ohio*
Pushpinder Puri, *Air Products and Chemicals, Inc., Allentown, Pennsylvania*
Douglas M. Ruthven, *University of Maine, Orono, Maine*
Kamalesh K. Sirkar, *New Jersey Institute of Technology, Newark, New Jersey*
Anna Lee Y. Tonkovich, *Pacific Northwest National Laboratory, Richland, Washington*

## Contributing Authors

Richard W. Baker, *Membrane Technology and Research, Inc., Menlo Park, California*
Dibakar Bhattacharyya, *University of Kentucky, Lexington, Kentucky*
John L. Bulloch, *Michigan Technological University/CenCITT, Houghton, Michigan*
John C. Crittenden, *Michigan Technological University/CenCITT, Houghton, Michigan*
Jeffrey R. Hufton, *Air Products and Chemicals, Inc., Allentown, Pennsylvania*
Alan L. Myers, *University of Pennsylvania, Philadelphia, Pennsylvania*
Peter P. Radecki, *Michigan Technological University/CenCITT, Houghton, Michigan*
David R. Shonnard, *Michigan Technological University, Houghton, Michigan*

## Other Reviewers and Contributors

Earl Beaver, *Monsanto, St. Louis, Missouri*
Edward L. Cussler, *University of Minnesota, St. Paul, Minnesota*
Russell F. Dunn, *Solutia, Inc.*
Douglas E. Fain, *Lockheed Martin Energy Systems*
Nick Hankins, *Aspen Technologies, Inc.*
Mike Harold, *E.I. du Pont de Nemours and Company, Wilmington, Delaware*
Darryl W. Hertz, *The M.W. Kellogg Company, Houston, Texas*
Jamie Hestekin, *University of Kentucky, Lexington, Kentucky*
Kerry Irons, *The Dow Chemical Company, Midland, Michigan*
Felix Jegede, *Aspen Technologies, Inc.*
Gregory Keeports, *Rohm & Haas*
George Keller II, *Consultant*
Richard Kiesel, *Michigan Technological University, Houghton, Michigan*
Jerry Lin, *University of Cincinnati, Cincinnati, Ohio*
Richard Noble, *University of Colorado, Boulder, Colorado*
Eric D. Sall, *Monsanto Corporate Research*
B.K. Srinivas, *General Electric Research & Development*
Jack Weaver, *Center for Waste Reduction Technologies, New York, New York*
Hans Wijmans, *Membrane Technology and Research, Inc., Menlo Park, California*

# Preface

This monograph is part of the Center for Waste Reduction Technologies' (CWRT) long-term objective of developing technologies and tools useful for promoting waste minimization practices in industry. The monograph, which focuses primarily on adsorption and membrane separation technologies, also contains information on the emerging science of reactor technologies using membranes, adsorption, and reactive distillation. The development of this monograph was directed by a task force of CWRT member companies led by Darryl W. Hertz of the M.W. Kellogg Company.

Information contained in this monograph comes from two sources: (1) nationally recognized technical experts in the technologies, and (2) the findings from the National Workshop on Process Waste Reduction via Separation Technologies and Separative Reactors which was held in New Orleans, Louisiana, February 1998. Chapters 1, 2, and 3 outline the fundamental concepts important to each of the technologies, describe state-of-the-art processes using the technologies, and highlight broad areas of need for each technology. Chapter 4 contains the workshop findings which include technology-specific recommendations and a compilation of a number of processes and process streams which workshop participants felt would be useful targets for process waste reduction research and new separation technology demonstrations.

The workshop was sponsored by CWRT, with co-sponsorship from U.S. Department of Energy/Office of Industrial Technologies (DOE-OIT), Industrial and Engineering Chemistry Division of the American Chemical Society, the Council for Chemical Research, the American Institute of Chemical Engineers Separations Division, and the National Center for Clean Industrial and Treatment Technologies (CenCITT). The workshop was attended by eighty-nine participants representing twenty-eight companies, eleven government agencies and laboratories, and eight universities. Immediately following the workshop, a one-day brainstorming exercise was conducted in which participants were asked to define the present challenges faced by industries producing and using chemicals and to identify the technical barriers and the research needs required to overcome those barriers so that the technologies would play important roles in improving future processing economics. The *Vision 2020: 1998 Separations Roadmap* produced by CWRT includes the results from this brainstorming exercise as well as the output of a second workshop focusing on distillation, extraction, and crystallization.

Together, this monograph and the roadmap are an effort to address the often-cited need to have closer coordination between industry, academia, and government in their individual efforts relating to sustainable development and ecoefficiency. A clear picture of industry needs will help government and academia channel their R&D resources toward well-defined and meaningful goals. In contrast, a lot of present day work in nonindustrial sectors is directed toward areas of intellectual interest with little regard for commercial applications or for the challenging and costly development work needed to bring new technologies to commercial fruition.

It is hoped that both this monograph and the roadmap will act as a stimulus for the R&D communities in government and academia and allow funding agencies to focus support on "real world" needs which could have a significant impact on the chemical industry's drive toward greater ecoefficiency. Interim findings of this project have already been used by DOE-OIT, the National Institute for Standards and Technology, and CenCITT to identify research opportunities and to develop research programs.

It is also hoped that the process put in place by the creation of the monograph and roadmap will be part of an ongoing effort between industry and government to create and implement a shared vision in the development and commercialization of new technologies. The monograph and roadmap are just first steps in this journey.

As we seek to become more sustainable in all of our operations we should recognize that future problems may have very different boundary conditions than we have today. In his keynote speech at the workshop, Earl Beaver of Monsanto challenged all to consider the magnitude of the challenge we face in making our processes and societies sustainable. We must find more than incremental improvements in separation technologies to meet future global demands for products and the environment. We must find step changes, for example, by combining technologies into single process units such as separative reactors. Step changes can also result from more intelligent selection and more efficient use of materials and energy. As proposed by one of the workshop participants, Bruce Cranford of the Department of Energy, the future may be one in which we will need to reduce drastically our consumption of materials and our emissions of

## Preface

$CO_2$, and do so with considerably higher oil prices in place. We need to maintain a focus on the characterization and solution of tomorrow's problems, because focusing on only today's problems may eventually result in "solutions" of yesterday's problems.

We wish to thank the many authors and contributors to this monograph. It is a collective work built on scientific and engineering expertise, and organizational leadership. We also wish to thank June Hansen with CenCITT for all her efforts in support of this project and the U.S. Environmental Protection Agency's support of CenCITT. Finally, our sincere gratitude goes to the members of CWRT and to DOE-OIT, for without their support, this project would not have been possible.

PETER P. RADECKI<br>
JOHN C. CRITTENDEN<br>
DAVID R. SHONNARD<br>
JOHN L. BULLOCH

# 1

# Adsorption, Membrane, and Separative Reactor Processes

## NEW DEVELOPMENTS OFFER OPPORTUNITIES FOR PROCESS WASTE REDUCTION

**AUTHOR**

Jimmy L. Humphrey
J. L. Humphrey & Associates, Austin, TX

**EDITORS**

John L. Bulloch
John C. Crittenden
Peter P. Radecki
CenCITT/Michigan Technological University, Houghton, Michigan

## 1.1. Summary

Adsorption, membrane, and separative reactor processes may be used for process waste reduction through process modifications which produce less pollution and through recovery of potential contaminants for reuse within production unit boundaries.

This chapter provides characterizations of adsorption, membrane, and separative reactor processes with respect to their applications for pollution prevention. It includes descriptions of factors which affect efficiency, covers technology status and new directions, and identifies research needs. A summary of applications for pollution prevention together with research needs follows. In this chapter, the terms "process waste reduction" and "pollution prevention" are used interchangeably. This may not be rigorously correct, but it follows from the assumption that process wastes comprise the bulk of pollution sources.

### 1.1.1. Process Modifications to Produce Less Pollution

- ***Elimination of Solvents:*** Both adsorption and membrane processes may be used to dehydrate azeotropes and hydrocarbon gases, thereby eliminating the use of solvents which themselves become waste sources. In the dehydration of azeotropes, azeotropic agents such as benzene and cyclohexane are eliminated, whereas for natural gas dehydration, solvents such as triethylene glycol are eliminated.
- ***Elimination of Purge Streams:*** Many industrial processes involve recycle streams that build up with undesirable contaminants. To control contaminant levels, a portion of the recycle stream is purged. Adsorbents designed to selectively remove contaminants would eliminate purging, thereby reducing costs while eliminating a waste source.
- ***Recovery of Catalysts:*** Homogeneous catalysts are gaining popularity in the process industries, but they are thermally sensitive and cannot be recovered by distillation. Membranes appear to be an ideal process for recovering such catalysts, though current membrane materials are not durable enough to withstand the corrosive nature of reaction mixtures. The highly durable reverse osmosis membranes which are needed for this application are not commercially available.
- ***Absorptive Reactors:*** Possible applications for absorption reactor processes include bioventing, biological scrubbing, biostripping. These processes are described in more detail in the body of this paper. These processes are in the conceptual stage and additional data and evaluations to determine their potential are needed.
- ***Adsorptive Reactors:*** In this process the adsorbent, which is normally the mass separating agent, is also the catalyst or is mixed with a catalyst. This technology is still largely untapped though it has significant potential. The challenge here is one of materials development involving matching of catalysts/adsorbents to process conditions and regeneration cycles to achieve optimum performance.
- ***Membrane Reactors:*** Dehydrogenation is one of several possible applications for membrane reactors. In this application, one could achieve higher conversions per pass by selective removal of hydrogen while reducing capital investment.

## 1.1.2. Recovery and Recycle of Potential Contaminants for Reuse within Production Unit Boundaries

- *Recovery from Plant Leak-off Streams:* Many plants have effluents containing valuable organics. For example, polyethylene plants have intermittent leak-off streams that contain significant concentrations of ethylene. Recovery of the ethylene would result in significant credit for ethylene while eliminating a waste source. New adsorbents may be needed for quantitative recoveries of ethylene.
- *Recovery of Organics from Air:* Membranes may be used to recover valuable organics, such as ethylene and propylene, from nitrogen streams used for purging polyethylene and polypropylene pellets, respectively. Valuable monomers can be recovered while allowing recycle of the nitrogen.
- *Wastewater Recovery:* There are a number of ways that membranes can be used for recovery of water and important by-products. Examples include the recovery of automobile paint from spent bath solutions, and the removal of particulates and oil emulsions from wastewaters. Reverse osmosis can be used for wastewaters contaminated with dissolved nickel and other catalyst species. In one application, reverse osmosis has saved $300,000 per year by eliminating reagents used to precipitate dissolved nickel species while recovering water for use as boiler feed water.
- *Recovery of Water from By-Product Streams:* Ceramic membranes can be used to concentrate corn-stillage streams. Unlike the centrifuge process it replaces, a recycle stream is obtained that is free of suspended particles and allows recycle of 90% of the feed water while preventing accumulation of solids.

## 1.1.3. Research Needs

- *More Adsorbents:* Only a dozen or so classes of adsorbents (e.g., activated carbon, synthetic polymeric resins, or zeolites) are available for commercial applications. More adsorbents are needed, including "designer" adsorbents which have the appropriate properties for separation of adsorbates for specific applications.
- *Clean Regeneration Configurations:* Regeneration configurations are needed that reduce costs while minimizing sources of pollution. For example, traditional ion exchange processes used to deionize water are regenerated with sulfuric acid and sodium hydroxide. Regenerant wastes later become sources of pollution. New regeneration cycles are needed which do not require such regenerants. Possibilities include use of "clean" fluids or

sources of external energy for regeneration. Examples of external energy sources include electric energy, electromagnetic fields, and microwaves.
- ***Better Membranes:*** Only a limited number of organic liquids can be processed with conventional polymeric membranes. This limitation has severely impeded the application of membranes across the chemical process industries. Thus, more durable polymer and inorganic membranes are needed. And to better compete with other processes economically, higher performance membrane materials are needed which provide higher flux rates and separation factors.
- ***Long Range Research Goals:*** In both adsorption and membranes, longer range research objectives should include the development of nanoscale models to predict megascale performance. Adsorption models should include

—simulation of molecular structure of adsorbents,
—prediction of interactions of adsorbates at surfaces of adsorbents,
—multicomponent phase equilibria, and
—rates of heat and mass transfer.

Models and simulations for membranes are need to predict flux and separation factors from the properties of the membrane material, properties of the fluids involved, and operating conditions. Accomplishing these goals will not be easy nor will they come quickly, but they should be included in the road maps for longer term research programs.

## 1.2. Adsorption Processes

### 1.2.1 Adsorption Process Fundamentals

In adsorption, the separation mechanism involves attracting one or more solutes in a liquid or gas mixture to the surface of a solid adsorbent where they are held by intermolecular forces. If the solute(s), called adsorbates, are attracted selectively to the adsorbent, then a separation of the components in the mixture will occur. Desorption, or regeneration, is the reverse process where adsorbate is removed from the surface of the adsorbent.

Since it is desirable to regenerate adsorbents once they are loaded with adsorbate, and it is difficult to move solids while the process is operating, there must be at least two fixed beds of adsorbent for continuous operation. Thus in a two-bed process one bed is operated in the adsorption mode while the other is being regenerated.

# 1. Adsorption, Membrane, and Separative Reactor Processes

Adsorption processes must deal with heat releases during adsorption and heat demands during desorption. The heat released upon adsorption will heat the adsorbent and vessel and some will leave with the fluid leaving the bed. As a rule-of-thumb, if the concentration of adsorbate(s) in the feed is greater than about 10 wt% (bulk mixture) then significant amounts of heat will remain trapped and raise the temperature of the bed. In this case, the rise in temperature will limit the amount of adsorbate that can be removed because as the bed heats up the tendency to adsorb decreases. If the concentration of the adsorbate(s) in the feed is low, of say 3–5% or less (dilute mixture), then the heat of adsorption mostly departs with the effluent, and the bed remains essentially isothermal. Adsorption can take place by three different mechanisms:

- Selectively binding one or more components to the surface of the adsorbent
- Excluding components based on geometric incompatibilities (molecular sieving)
- Taking advantage of differences of component intra-particle diffusion rates

Most adsorption processes rely on the first and second mechanisms. The only industrial application based on the last mechanism is the separation of air to produce nitrogen using carbon molecular sieves.

## 1.2.2 Adsorbents

An adsorbent must have three characteristics to have commercial potential.

- It must selectively attract adsorbate(s) from the feed mixture.
- It must have the ability to release the adsorbate so it can be reused.
- It must have a high working capacity—the change of weight of adsorbate per unit weight of adsorbent between adsorption and regeneration modes. Loadings of adsorbates range from less than 1% to greater than 10% of the weight of the adsorbent.

Because adsorbents incorporate at most a very few layers of molecules on their surfaces, it is generally advantageous for adsorbent surface areas to be as large as possible. Surface areas of commercial adsorbents commonly range from a few hundred square meters per gram to about 3000 square meters per gram. Examples of commercially important adsorbents are:

- Activated Carbon
  - Generally hydrophobic, separates approximately by boiling point
  - Low cost ($1/lb to $3/lb)

- Silica Gel
  - Generally hydrophilic, with strong affinity for water
  - Can sometimes be used for adsorption of hydrocarbons from water or gases
- Activated Alumina
  - Hydrophilic, used for dehydration of gas streams and gas sweetening
- Zeolite Molecular Sieves
  - Separates based on polarity and size
  - Wide variety of uses
- Carbon Molecular Sieves
  - Separates based on relative differences in intraparticle diffusion rates
- Hydrophobic Molecular Sieves
  - Separates like activated carbon
  - Can be easier to regenerate than activated carbon
  - Expensive (5–10 times the cost of activated carbon)
- Polymer-Based Adsorbents
  - Separates like activated carbon
  - Several types available
  - Mechanically durable for use in fluid/moving bed processes
  - Easier to regenerate than activated carbon
  - Expensive (10–50 times the cost of activated carbon)
- Biosorbents
  - Biofilms growing on supports (e.g., volcanic rock, peat, wood chips, and soil)
  - Removes organics from gas streams
  - Very sensitive to upsets and contaminants
  - Process has a very large "footprint"
- Irreversible Adsorbents
  - Are "regenerated" offsite
  - Removal of low levels of formaldehyde, $H_2S$, $SO_2$, etc. from gases and mercury, silver, etc. from water

## 1.2.3. Regeneration Cycles

Three types of cycles are commonly used for regeneration:

- ***Thermal-Swing Adsorption (TSA):*** Increasing the temperature of the adsorbent to achieve regeneration or desorption is called thermal-swing or temperature-swing adsorption. Frequently a hot fluid, such as steam or nitrogen, is used to heat the bed. Rotary wheel adsorption, described in detail later, is an example of a TSA process.

- **Pressure Swing Adsorption (PSA):** Altering the partial pressure of one adsorbate in the gas phase or the adsorbate concentration in the liquid phase. This is called pressure swing adsorption in the gas phase, and concentration-swing adsorption (CSA) in the liquid phase.
- **Displacement Purge Adsorption (DPA):** In displacement purge adsorption, a component is added which competes with adsorbate for adsorbent surface sites and limits surface area that is available to adsorbate. This type of regeneration cycle is used mostly in liquid separations. Typical ranges of feed concentration for TSA and PSA regeneration cycles are given in Figure 1.1. However, the applicability of different regeneration cycles is strongly case specific depending on the recovery objective, process flow rate, type and concentration of compound, etc.

### 1.2.4. Adsorption Process Configurations

**Fixed Beds**

Adsorption processes are normally carried out in fixed beds with adsorbent shapes in the form of particulates, pellets, fibers, or spheres. As described earlier, beds are regenerated using TSA, PSA, or DPA cycles. Of note is the fixed bed PSA processes which have been commercialized in recent years to remove organics from air. Depending on the application, activated carbon or polymer resins are used as adsorbents. When pressure drop is a significant consideration,

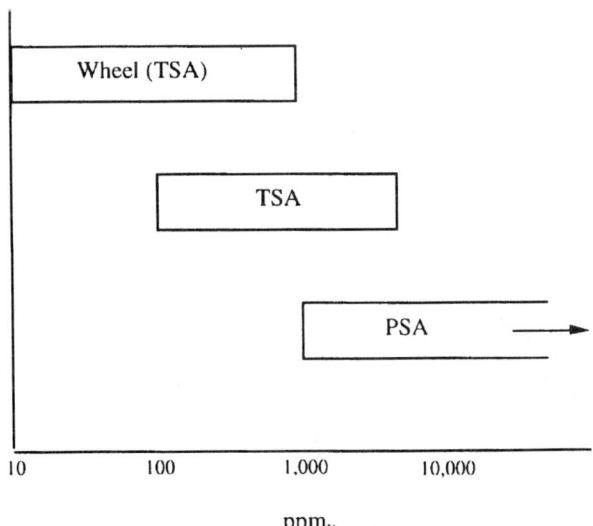

**FIGURE 1.1** Approximate ranges of applicability for regeneration cycles for gas streams separations.

novel fixed bed configurations are available based on radially feeding into a concentric opening in the middle of the bed. Regeneration is accomplished by reversing flow from the inside to the outside. This bed configuration is available using carbon fibers as the adsorbent.

**Moving/Fluid Beds**

Moving/fluid bed processes have been developed based on the use of mechanically stable polymer bead adsorbents. One commercially available process is called Polyad™ which is offered by Weatherly, Inc. (Atlanta, Georgia). There are five units in commercial operation in the US and about ten units are in operation in Europe. As shown in Figure 1.2, the Polyad™ design is a tower process which contains trays which support the adsorbent. The spherical shape of polymer bead adsorbents allows the design of the tower to mimic a distillation operation, with polymer beads overflowing each tray onto subsequent trays while the air stream passes countercurrently through the adsorbent on the trays. The trays are similar to sieve trays used in distillation.

In the Polyad™ process, fresh adsorbent enters the uppermost adsorber tray and overflows to subsequent trays. The VOC-laden gas stream first contacts the adsorbent in the lowest tray of the adsorption tower. The tower includes an entrainment zone above the uppermost tray to minimize loss of adsorbent in the off-gas. The adsorbent in the lowest tray also overflows, and is conveyed to the top of a separate unit that is the desorber tower. The desorber operates with a counter-current heated purge gas that also flows through trays designed similarly to distillation sieve trays. During desorption, polymer beads overflow consecu-

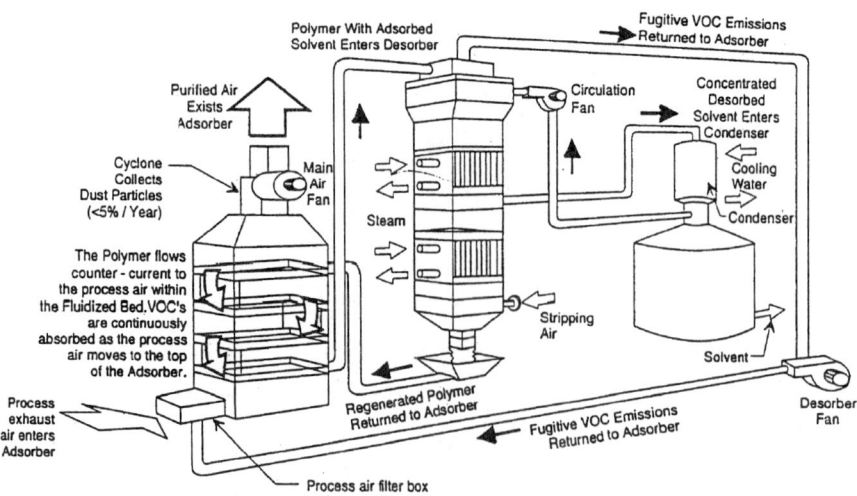

**Figure 1.2** Polyad™ process.

tive trays and are finally returned to the adsorber tower. Indirect heat may be applied to assist the purge during desorption. The goal of this process is to obtain a reject stream that will have a concentration of 10–50 times that of the original adsorber inlet. Such streams may be suitable for recovery or destruction of the organics (Abrahim et al., 1997; Niezgodski, 1997).

## Rotary Wheels

Adsorbents configured in the form of monolithic rotary wheels are rapidly gaining favor as a means of removal of concentrations of organics up to about 1000 ppm from large volumes (5000 scfm–30,000 scfm) of low pressure air streams. These wheels can reduce contaminated air volumes by 90–95% and thereby concentrate organics by a factor of 10–20 times. The concentrate from wheels is generally fed to thermal oxidizers or other types of destruction units. Figure 1.3, which is simplified, shows that as the wheel rotates, the adsorbent is alternately contacted by the feed gas and by a regeneration gas at an elevated temperature. Rotary wheels rotate slowly, about two revolutions per hour. In a typical applica-

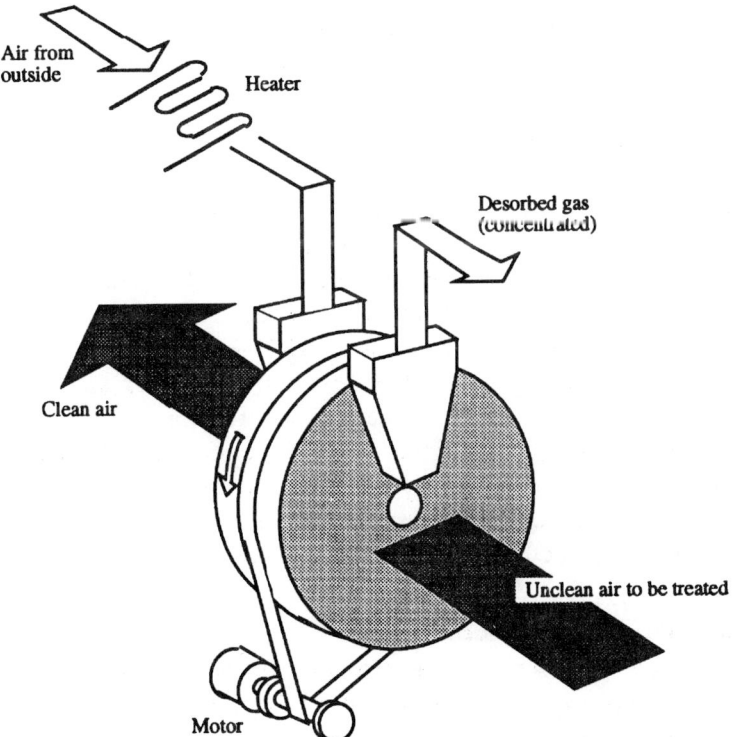

**Figure 1.3.** Adsorbent wheel with monolithic adsorbent (simplified). (After Humphrey & Keller, 1997.)

tion, the volume of regeneration gas is about 10% of the feed and thus feed components are concentrated by a factor of 10. About 75–90% of the wheel is used for adsorption while the remainder is regeneration. The desorption section is isolated from the rest of the rotor by precision seals.

Not only do wheels offer a degree of process simplicity through the use of only one unit, but also considerably shorten cycle times as compared to traditional fixed-bed TSA processes. Instead of cycle times of several hours or more for traditional processes, wheels can have regeneration times of only 5–10 minutes. Wheels are made with diameters in excess of 13 feet and are available with carbon, silicalite, alumina, and silica gel adsorbents. In some cases, two adsorbents, such as carbon and silicalite, are used on the same wheel. The carbon adsorbs hydrocarbon solvents such as xylene, while silicalite adsorbs low molecular weight oxygenates such as methanol (Hussey and Gupta, 1997).

A complication associated with rotary wheels is that there can be leakage of process gas into the regeneration section and of regeneration gas into the adsorption section. This cross contamination can limit practical removals below what can be achieved with fixed-bed processes. In addition, the mass-transfer rate per unit length of bed traversed is not as high as that in a fixed-bed of adsorbent. These factors combine to limit wheel-based removals to 95–97%.

### Simulated Moving Beds

Bulk liquid separations via adsorption are limited for the most part to DPA approaches involving simulated moving beds (SMBs). This type of process is used because of the difficulty in dealing with the problems of heat of adsorption and the high density of liquid feeds. In less than one bed volume, liquid feeds can fully load adsorbent even if there are no heat effects.

An example of the SMB process is the UOP Sorbex process. As shown in Figure 1.4, the Sorbex (and other SMB processes) are complex, and nearly always lead to one or more adsorption beds and two distillation columns. As a result, bulk-liquid separations via adsorption are relatively rare and are limited to separations which cannot be handled by distillation. Other companies such as US Filter/IWT have also commercialized SMB processes, and their applications are similar in nature to UOP's.

### Physical Movement of Beds

In one process, adsorbent beds are physically transported between adsorption and desorption zones. Separate beds of adsorbent move in a circular manner on a carousel, encountering fresh feed, feed from other beds, and desorbent purge fluid as the carousel rotates. This process is also used with ion-exchange operations.

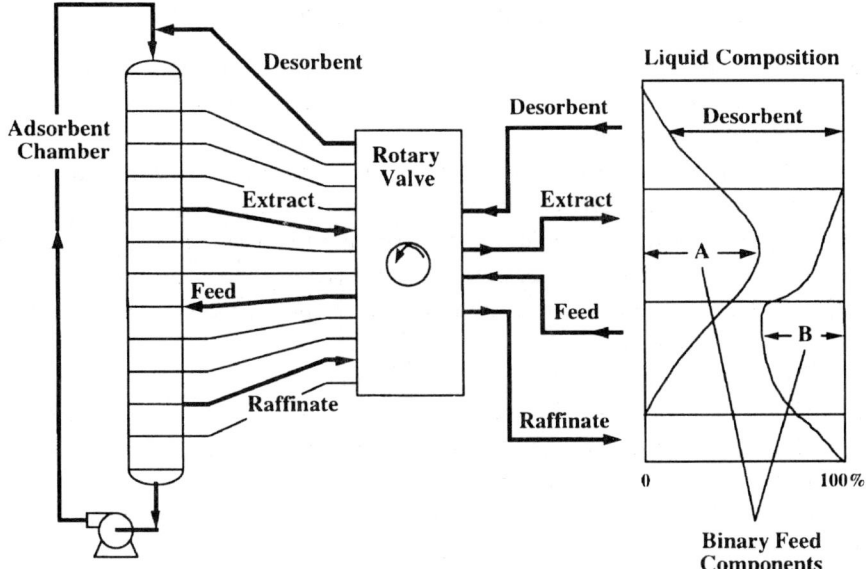

**Figure 1.4.** Sorbex simulated moving bed process for separations of bulk liquid mixtures.

### 1.2.5. Advantages/Disadvantages of Adsorption

*Advantages:* Adsorption processes can selectively remove adsorbates present in small concentrations to obtain high purity "product" streams. Additionally, adsorption processes can operate at low temperatures, which is important in applications where thermally sensitive products are involved.

*Disadvantages:* Adsorption is an unsteady-state process and suffers from the disadvantages associated with cyclic operations. Performance of adsorbents can be affected significantly by the presence of poisons or solids in the feed.

### 1.2.6. Factors Favoring Adsorption

Processes with the following characteristics favor the use of adsorption:

- A high degree of product purity is needed
- Loaded bed(s) can be regenerated easily
- Low-to-moderate operating temperatures are desired
- Adsorbent is not susceptible to fouling or attack by feed components.

## 1.2.7. Applications of Adsorption for Pollution Prevention

Adsorption processes play key roles in pollution prevention with applications for product recovery and purifications, and removal (and recycle) of contaminants from air and water streams.

Adsorption processes using 3A and 4A molecular sieves are used to dehydrate natural gas, thereby eliminating the use of solvents such as triethylene glycol. The implication for pollution prevention is that triethylene glycol solvent becomes contaminated with organics and subsequently becomes a source of pollution. When adsorption can be used as a replacement process, the solvent and the source of pollution are eliminated.

For dehydration of azeotropes such as ethanol/water and isopropanol/water, adsorption processes using 3A molecular sieves can take the place of traditional azeotropic distillation. The result is that the use of azeotropic solvents, such as cyclohexane and benzene, are eliminated.

Industrial plants have effluents which contain valuable organics. Recovery and recycle of these organics could reduce costs while eliminating sources of pollution. For example, polyethylene plants have compressor leak-off in streams which contain ethylene. Recovery credit of ethylene would be significant, but new adsorbents may be needed to achieve quantitative recoveries of ethylene.

Many industrial processes involve recycle streams which contain undesirable contaminants. To control concentrations of these contaminants a portion of the recycle stream is purged. Adsorbents designed to selectively remove such contaminants would alleviate the need for the purge, thereby reducing costs while eliminating a source of pollution. For example, higher molecular weight organic contaminants build up in ethylene and propylene recycle streams in polyethylene plants and polypropylene plants, respectively. To control concentrations of contaminants, these streams are "purged" to nearby olefin plants where the contaminants are removed by costly distillation processes. Use of selective adsorption processes would eliminate these purges.

## 1.2.8. Economics of Adsorption versus Competing Processes for Clean Air Applications

Recovery of volatile organic compounds (VOCs) from air is an application for which a number of processes compete including membranes, absorption, pressure swing adsorption, fixed-bed adsorption, rotary wheel adsorption, condensation, and freezing. Except for freezing, capital and operating costs have been determined for each of these processes to assist in developing a guide for process selection of the various processes (Humphrey et al., 1996). Results for the removal of acetone from air are shown in Figure 1.5. Capital and operating costs

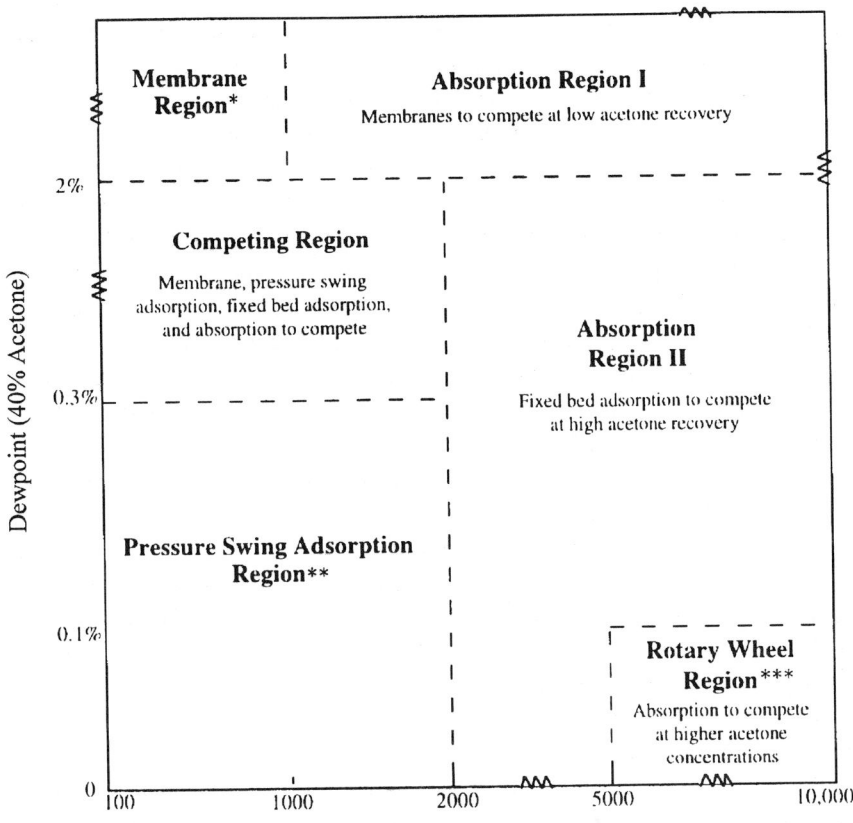

**Figure 1.5.** Process selection map—clean air technologies
- Giving regions of cost-effective applicability for clean air technologies
- For 95% acetone recovery from acetone/air mixture

(After Humphrey and Keller 1997.)

were based on removal of acetone from air at feed rates of 100 standard cubic feed per minute (scfm) and 10,000 scfm, acetone concentrations of 0.3 vol% and 2.0 vol%, and at 95% acetone removal. Lower capital and operating costs were the criteria used to determine the regions in which each process is considered to be the favored process. Differences in capital costs between the processes were found to be more significant than operating costs. Although Figure 1.5 is unique for 95% recovery of acetone from air, it does illustrate the important trends generally observed in choosing between available separation technologies.

As shown in Figure 1.5, there are six regions which can be distinguished. The first is the *membrane region*. Because the membrane process has lower capital costs than other processes, it is the process of choice at acetone concentrations above 2% and for gas feed rates up to about 1000 scfm. The membrane process can also compare favorably with absorption at rates higher than 1000 scfm, but applicability is limited to lower acetone recoveries of approximately 50%. For removals down to ppm and ppb levels, membrane surface areas become excessive, and pressure swing adsorption would likely become the process of choice. Higher membrane flux rates would extend the membrane rate range, whereas higher separation factors would allow membrane processes to be cost-effective at lower values of feed composition.

The second region is called *absorption region I*. Absorption is the favored process for acetone concentrations greater than 2% and for gas feed rates greater than 1000 scfm. Because absorption is a tower-type process, capital costs increase with rates approximately equal to the six-tenths power rule and capital costs of absorption tend to be lower than other processes at high rates. Though attractive for acetone removal, absorption quickly loses favor if the VOC forms an azeotrope with water because simple stripping could no longer be used for water recovery. Although acetone does not form an azeotrope with water, a number of organics such as benzene and xylene do form azeotropes.

The third region is the *competing region*. Membrane, adsorption and absorption processes all compete for acetone concentrations in the range of 0.3% to 2%, and for air feed rates less than about 2000 scfm. The process selected in this region will depend on acetone concentration and air feed rate. Within this region, the factors favoring each process are

- Membranes are favored by high acetone concentrations and low rates,
- PSA is favored at flow low rates and high acetone recovery,
- Adsorption is favored at low acetone concentrations and high acetone recovery, and
- Absorption is favored by high rates and low acetone recovery.

The fourth region is *absorption region II*. Absorption is favored for acetone concentrations of up to 2% and for gas feed rages above 2000 scfm. However, when organic concentrations must be reduced to low levels, such as those required for clean-air applications, adsorption becomes the favored process.

In the *pressure swing adsorption region*, the pressure swing process is favored with acetone concentrations below 0.2% and gas feed rates less than 2000 scfm. For clean air applications and rates less than 2000 scfm, PSA may compete favorably over all ranges of acetone concentration in the feed but should be restricted to rates less than about 2000 scfm because of significant costs of vacuum pump systems at higher rates. There is a small area of opportunity, at very low rates of less than approximately 100–200 scfm, for membranes within this region.

The sixth and last region is the *rotary wheel region*. Rotary wheels are cost-effective at low concentrations of up to 1000 ppm of acetone and at rates of 5000 scfm to 30,000 scfm and higher. Absorption processes can compete with rotary wheels in this region and would become favored at higher acetone concentrations in the feed.

Though capital and operating costs were determined for *condensation*, this process is not shown in Figure 1.5 because it does not compete economically with the other processes at any conditions included in the analysis. It was found that expensive compression steps were required to condense acetone, which is a highly volatile VOC.

In cases where acetone concentrations are to be reduced to parts per million and lower levels, membrane surface area becomes excessive, as would the number of trays required in an absorption column. At such low concentrations adsorption processes usually become the methods of choice. Fixed bed adsorption processes can be used to achieve 99.99% removals of VOCs from air. Pressure swing adsorption is favored for higher concentrations and lower flow rates, whereas thermal swing adsorption is favored at lower concentrations and higher flow rates.

### 1.2.9. Future Directions in Adsorption Technology

The need for higher efficiency processes for removal of solutes from air and water effluents has stimulated major developments in new adsorbents and regeneration configurations. New adsorbents include hydrophobic molecular sieves, polymer adsorbents, and biosorbents. New regeneration cycles include pressure swing using carbon "clean" and moving bed configurations using polymer beads for removal of organics from air. Rotary wheels, which have become extremely popular in the last five years, are effective at removing low concentrations of organic contaminants of (up to 1000 ppm) from large volumes of air (up to 30,000 scfm).

#### 1.2.9.1. Clean Regeneration Configurations

R&D programs in adsorption should include developing less costly and pollution-free regeneration cycles. Some candidates are described below.

*Regeneration with Natural Gas:* Preheated natural gas can be used in the place of nitrogen as a displacement fluid to regenerate adsorption beds. The advantage of this regeneration cycle is that the regenerant gas stream, containing natural gas and a hazardous adsorbate such as benzene, could be used as boiler fuel where the fuel value of the natural gas would be realized and the hazardous material destroyed (Humphrey, 1989). This application is most attractive when the concentration of hazardous material is low in the feed gas and it is one which must be thermally destroyed.

*Use of Mosaic Resins:* One approach to removing solutes selectively from dilute solutions is through the use of composite, charge-mosaic ion-exchange resins. Processes built around these materials, which are being investigated in Australia, use mosaic resins for removing electrolytes from aqueous solutions. The advantage of these resins is that regeneration can be accomplished by a temperature swing rather than requiring displacement purge cycles based on the use of strong acids and bases which can later become pollutants.

*Electrodesorption:* Adsorption at an electrode and solution interface can be controlled by the electrode potential. Reducing the potential across the bed reduces the amount of adsorbate and thus, this process affords a possible new method to regenerate fixed bed systems without the use of a potentially polluting purge stream. Based on removal of ethyleneamine from brine, Union Carbide has completed lab scale measurements, designed a commercial process, and completed an economic evaluation. Results show high capital investment requirements for the electrodes (Eisinger and Keller, 1989).

*Electromagnetic Desorption:* Magnetic polymer resins capable of radioactive species removal, including actinides and heavy metals from water, have been developed together with methods for making, using, and regenerating them. The bead resins are polyamine-epichlorohydrin beads with ferrites attached to their surfaces. Water decontamination has been demonstrated using these magnetic polymer resins in the presence of a magnetic field, as compared with water decontamination methods employing ordinary ion exchange resins or ferrites taken separately. Regeneration is accomplished by reducing the magnetic field and changing the pH (Kochen and Navratil, 1997).

*Microwave Desorption:* Spent carbon beds may be heated and regenerated with microwaves. Contaminants with high dielectric constants will be removed faster than compounds with lower constants. Some purge gas may be needed to flush contaminants off adsorbents. The primary advantage of using microwaves is that the bed could be regenerated much faster than with conventional heating, which would have the effect of minimizing the number of beds required. The disadvantages are the amount of electric energy required and safety concerns centering around possible ignition of activated carbon and organics.

### 1.2.9.2. Simulation Models

Long range research objectives should include developing nanoscale models to predict megascale performance. In adsorption this would involve simulations of molecular structure of adsorbents, interactions of adsorbates at surfaces of adsorbents, multicomponent phase equilibria, and predicting rates of heat and mass transfer. This will not be easy, nor will it come fast, but should remain a primary goal for longer term research programs.

## 1.3. Membrane Processes

### 1.3.1. Membrane Fundamentals

The two most important separation mechanisms used in industrial membrane processes are hydrodynamic sieving and sorption diffusion. Hydrodynamic-sieving membranes rely on the separation of molecules based on their relative sizes and rates of diffusion. Simply knowing typical fluxes and solute sizes that can be rejected by membranes based on sieving is adequate for most engineering designs. In sorption–diffusion membranes, both size (diffusivity) and condensibility (solubility) selectivity factors interact to determine which component passes the fastest.

Developments in both polymer and inorganic membrane materials have resulted in new applications of membrane processes. An example of a newer family of membrane materials is the polyimides, which are finding new commercial applications for the gas phase separations of nitrogen/oxygen and alcohols/water (Ninomiya et al., 1991). Some representative membrane materials and applications are presented in Table 1.1.

**Table 1.1.** Membrane Materials and Applications

| Membrane Material | Example Application | Membrane Processes |
|---|---|---|
| Cellulose acetate | Desalination of seawater | Microfiltration; ultrafiltration; reverse osmosis; gas |
| Polyimide | Production of nitrogen from air | Ultrafiltration; reverse osmosis; gas |
| Polyarylether sulfone (Polysulfone) | Recovery of hydrogen from hydrocarbon gas streams | Microfiltration; ultrafiltration; distillation; gas |
| Polydimethylsiloxane (Silicone rubber) | Removal of VOCs from air | Gas |
| Ceramic (gamma alumina) | Concentration of fermentation broth | Microfiltration; ultrafiltration |

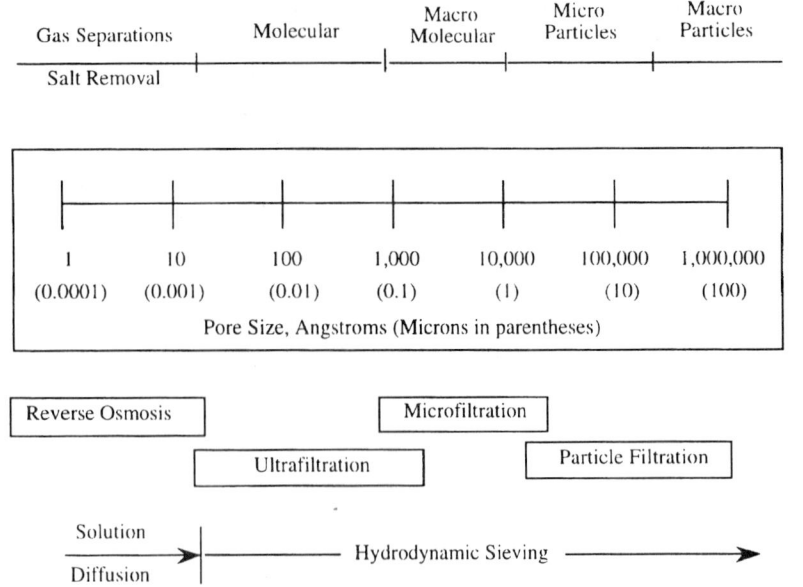

**Figure 1.6.** Classification of membrane separations.

Membranes are housed in modules having various configurations, which are discussed in more detail in the membrane section of this monograph. The three primary types of modules are *hollow-fiber*, *spiral-wound*, and *plate-and-frame*. Hollow-fiber and spiral-wound modules are more common than *plate-and-frame* flat-sheet modules, because these former configurations maximize area per unit of module volume. However, in some applications, the need to minimize fouling can dictate the use of a plate-and-frame configuration with its ability to more precisely control fluid dynamics.

Membrane processes are classified by the size of molecules or particles being separated as shown in Figure 1.6. Although its not shown in Figure 1.6, nanofiltration (NF) is a relatively new category; it includes the upper molecular weight range of the reverse osmosis (RO) domain and the lower molecular weight range of the ultrafiltration (UF) domain.

### 1.3.2. Advantages/Disadvantages of Membranes

*Advantages:* One advantage of membrane separation processes is energy efficiency. Membrane processes also have simple flow sheets and are easy to scale up. With membranes, only electric energy is required, an important factor at sites where steam and other forms of thermal energy are not available.

***Disadvantages***: With current materials, membrane processes are generally limited to bulk rather than precise separations. The disadvantages of membrane processes include fouling and possible lack of durability of membrane material. Only a limited number of organic liquids can be processed with conventional polymeric membranes, and this limitation has impeded the application of membranes across the chemical process industries. Additionally, capital costs for membrane units are often too high to compete with other processes at higher volumes.

### 1.3.3. Factors Favoring Membrane Processes

The process characteristics that favor membranes are as follows:

- Bulk rather than precise separations are sufficient
- Low to moderate processing rates
- Membrane is resistant to fouling by system components
- Energy cost is a key consideration

### 1.3.4. Membrane Applications for Pollution Prevention

***Gas Separations:*** Gas and vapor separations are based on sorption–diffusion as their primary separation mechanisms. Industrial applications include the separations of oxygen/nitrogen, hydrogen/methane, carbon dioxide/methane, hydrogen sulfide/methane, hydrogen/nitrogen, hydrogen/carbon monoxide, water/methane, and organics/air.

Membranes may be used to recover hydrogen from purge streams. They can also be used to recover ethylene and propylene from nitrogen streams used to purge vessels containing polyethylene and polypropylene pellets. This application saves resources while minimizing pollution.

A relatively new and important commercial application for membranes is the dehydration of natural gas. The implication for pollution prevention with this application is that triethylene glycol and other solvents used for dehydration of natural gas are eliminated instead of being wasted after contamination.

***Reverse Osmosis:*** The drive to lower costs has spurred an increasing interest in the use of homogeneous catalysts, such as metal–ligand complexes. These catalysts are expensive and thermally unstable, requiring recovery below approximately 100°C. Such conditions rule out the use of distillation unless the products boil at relatively low temperatures. Membranes appear to be ideal candidates for recoveries of homogeneous catalysts. A hypothetical flowsheet is shown in Figure 1.7. However, it must be recognized that many reaction mixtures are hostile to tradi-

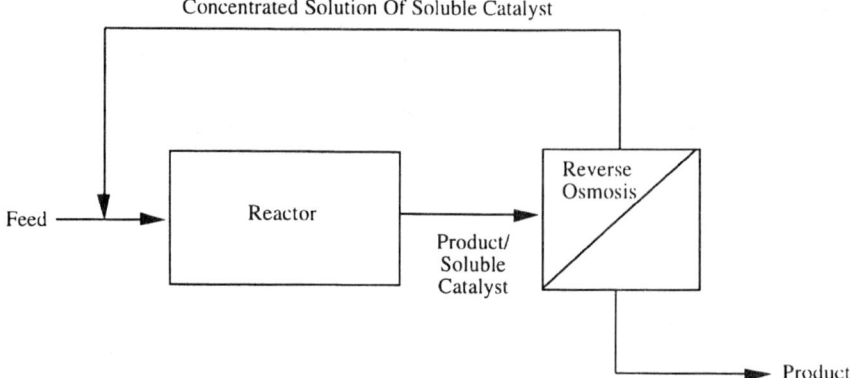

**Figure 1.7.** Reverse osmosis for recovery of soluble catalysts.

tional membrane materials. New and rugged polymer or inorganic reverse osmosis membranes are needed to take advantage of this application (Humphrey and Keller, 1997).

Reverse osmosis can be used to recover water from wastewaters contaminated by dissolved catalyst species. In one application about 200 million gallons/year of wastewater contaminated by dissolved nickel catalyst species is processed in an RO unit which recovers water for use as boiler feed water, while reducing costs. The process, which is illustrated in Figure 1.8, has saved

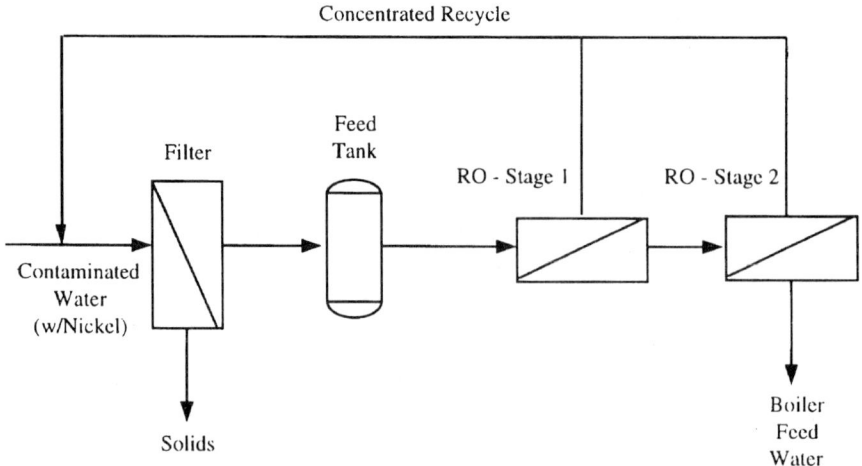

**Figure 1.8.** Reverse osmosis reduces costs and protects environment while converting wastewater to boiler feed water.

$300,000/yr by eliminating the use of reagents previously used to precipitate the dissolved nickel species as a sludge (EPRI TechApplication, 1997).

There are significant new opportunities for using reverse osmosis for recovery and recycle of water from landfill leachates and laundry effluents. Laundry effluents alone represent 38,000 cubic meters per year of wastewaters (Mulder, 1996).

*Ultrafiltration and Microfiltration:* High surface area ceramic membrane modules are available in monolithic configurations that afford more surface area per unit volume of module than traditional tubular module configurations. This type of module allows higher volumes to be processed than in previous ceramic membrane configurations.

Ultrafiltration (UF), based primarily on a sieving mechanism, is used for recovery of automobile paints from spent paint bath solutions. This process recovers expensive coating materials while preventing pollution. Painting 300 small cars per day requires roughly 100 $m^2$ of membrane area, which amounts to 40 conventional UF modules (Koros, 1995).

Removal of particulates and oil emulsions is becoming an increasingly popular application for microfiltration (MF) and UF membranes. In this case, water permeates are produced that in some cases may be suitable for discharge to a sewer with no post treatment. The concentrated retentate, which is 3–5% of the original volume, can be incinerated.

MF processes using ceramic membranes are used for concentration of corn-stillage streams. Unlike the centrifuge process it replaces, MF produces a recycle stream that is free of suspended particles, which allows recycle of 90% of the feed water while preventing accumulation of solids (Cheryan, 1995).

To combat fouling in UF, MF, and filtration applications, a new plate-and-frame module is commercially available which is mechanically oscillated at 60 cycles/sec to minimize fouling. The oscillation creates a shear on the surface of the membrane, thereby eliminating build-up on the membrane surface. This configuration is capable of concentrating solutions to 50% solids without fouling the membrane. The process operates at steady state and thus can take the place of cyclic filtration processes. This technology has become extremely popular in recent years.

*Pervaporation:* The pervaporation process, which is based on a sorption–diffusion mechanism, can be used to dewater azeotropes. The pollution prevention implication is that it may be used in place of azeotropic distillation which is based on the use of cyclohexane, benzene, and other azeotropic solvents. The use of pervaporation eliminates the use of these solvents. Pervaporation is used a great deal in Europe and Japan and to a lesser extent in the United States.

*Electrodialysis:* Electrochemical membrane processes are built around the use of ion-exchange membranes in which the driving force for transport is an electrical potential. An example of this process is electrodialysis. Used for water desalination, electrodialysis is based on an electrical potential to provide a driving force for cations to migrate toward the cathode and for anions to migrate toward the anode. Electrodialysis is often the process of choice when osmotic pressures are so high that reverse osmosis cannot be used. Japan has long used electrodialysis to produce table salt from sea water. Japanese equipment suppliers offer the most efficient electrodialysis technologies to industry.

### 1.3.5. Membrane Phase Contactors in Pollution Prevention

Similar to trays and packings, membranes can be used as phase contactors in stripping, absorption, and extraction processes (Humphrey and Keller, 1997). Though early commercial units produce high purity water for the microelectronics industry, this process can be extended to remove contaminants from air and water streams, through the absorption and stripping modes, respectively. The flowsheet included in Figure 1.9 shows a commercial membrane phase contactor unit being used for production of high purity water for the microelectronics industry. When operated in the extraction mode, it has been estimated that membrane phase contactors can offer a 25% lower capital investments than traditional Karr extractors for removal of phenol from wastewaters.

### 1.3.6. Future Directions in Membrane Technology

There are expectations that membrane processes will play increasingly important roles in future plants. Improved membrane materials are needed which give higher separation factor and fluxes. High separation factors are critical to offset the cost of staging membrane modules and high fluxes are needed to process the high rates which are characteristic of the process industries. Development of robust membrane materials, such as inorganic membranes, which are more tolerant of chemicals, would spur applications in the chemical and petroleum refinery industries. Additionally, new ways must be found to combat fouling, which is perhaps the single biggest complaint industry has about membrane processes.

In the near term, gas separations and the removal of contaminants from air and water streams probably represent the best opportunities for membranes. New developments in polymers and inorganic materials, such as ceramics, carbon, and molecular sieves, plus new module configurations, will expand the horizon of applications for clean air and water applications.

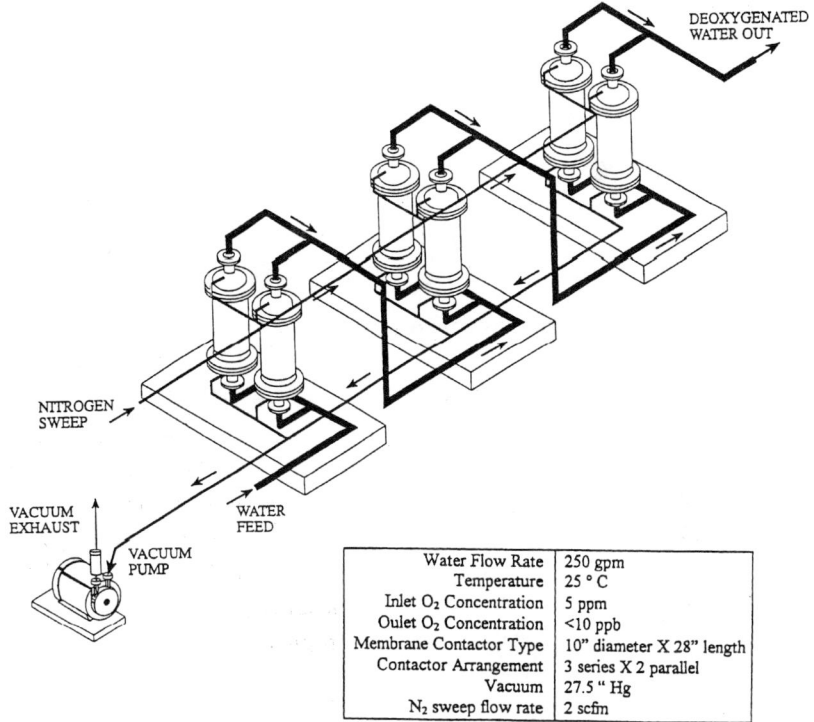

**Figure 1.9.** Membrane phase contactor unit to produce ultra high purity water.

Research is being done to expand inorganic membrane technology to include reverse osmosis and gas separations. Such applications require membrane pore sizes of a few Angstroms, but quality control problems during synthesis of some inorganic materials, such as the ceramics, have slowed the availability of commercial modules. Development of durable RO membranes would allow recovery and recycle of soluble catalysts.

A relatively new commercial development is the high-pressure reverse osmosis modules capable of operating at pressures up to 2000–3000 psig (Miller and Stanford, 1994), and testing is underway on modules which will operate at pressures of up to 4000 psig (Johnston, 1995). Applicable to contaminated wastewaters, higher operating pressures allow 80% water recovery from solutions containing dissolved contaminant loadings of 5% and higher.

Carbon membranes are being developed to recover hydrogen from hydrocarbon gas streams present at low pressure (Rao and Sircar, 1995). In this process, hydrogen is recovered as retentate at a pressure similar to that of the feed. Though not commercial, the development of this process is potentially important because the demand for hydrogen is expected to increase as the need to upgrade heavier crude oils increases.

Long range research objectives should include developing nanoscale models to predict megascale performance. With membranes this would involve simulations to predict flux and separation factors from properties and structure of the membrane material and the operating conditions. This will not be easy but should remain a primary goal for longer term research programs.

## 1.4. Separative Reactor Processes

In separative reactor processes, the catalyst and mass separating agent are contained in a single unit and in some cases, such as in membrane reactor processes, they can be one and the same. The primary advantage of the separative reactor is that it takes the place of two individual units, a reactor and a separator. Separative reactor processes offer the advantages of a lower capital investment, which can be 20–30% of that for the traditional two-unit process (Humphrey and Keller, 1997). Moreover, by being able to remove product (or byproduct) as it is created, separative reactors hold the promise of improving the economics on a variety of equilibrium-limited reactions.

### 1.4.1. Factors Favoring Separative Reactor Processes

The following process characteristics favor separative reactor processes:

- A reversible reaction is involved. By removing product as it is formed, the reaction is driven more in the forward direction, thereby achieving a higher conversion to products.
- Reaction and separation temperatures overlap. For example, in reactive distillation, reaction temperatures must overlap with distillation temperatures.
- Reaction is exothermic. This can be an advantage if the heat generated by the reaction can be used to drive a separation process for which thermal heat is required.

### 1.4.2. Reactive Distillation

Reactive distillation, also called catalytic distillation, is finding important new applications including the production of methyl tertiary butyl ether (MTBE), an octane enhancer which is made by the reaction of methanol and isobutylene. The process concept is illustrated in Figure 1.10. Commercial MTBE plants are in

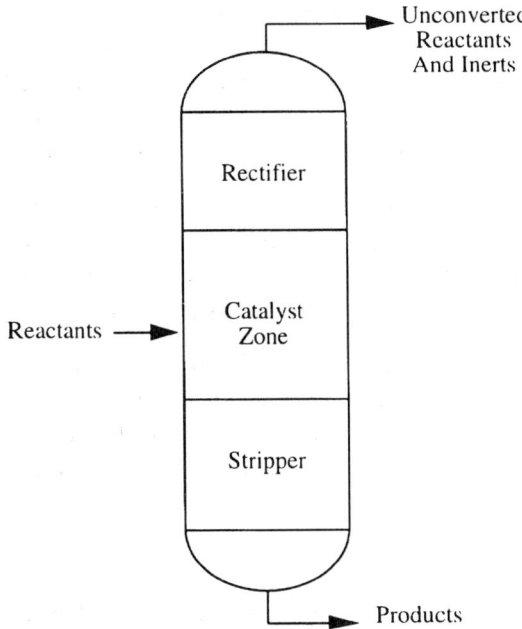

**Figure 1.10.** Reactive distillation (simplified).

operation in North America, Europe, the Middle East, and Asia (DeGarmo et al., 1992; Strauss, 1990). Another important commercial application is the production of methyl acetate, made from the reaction of methanol and acetic acid.

There are in excess of 70 operating reactive distillation units including 60 operating ether units. Although most of the ether units produce MTBE, there are two ethyl tertiary butyl ether and ten tertiary amyl ether units in operation. Additionally, there are eight commercial hydrogenation units which focus on the following applications (Gildert et al., 1997):

- In C4's, butadiene selective hydrogenation (3 units)
- In C5's, pentadiene selective hydrogenation (3 units)
- In C6's, hexadiene selective hydrogenation (1 unit)
- In aromatics, with benzene saturation (1 unit)

### 1.4.3 Absorption Reactors

*Bioventing:* The bioventing process is in the conceptual stage only and performance data and economics are needed to further evaluate the process. In bioventing, air would be injected below the zone of contaminated soil. As the air flows upward through the contaminated zone, organic contaminants would be stripped from the soil. As the air stream, which contains pollutants, flows through upper zones of soil, contaminants would be absorbed and biologically degraded. Nutrients would be sprayed on the soil surface to promote biological growth and biodegradation in the upper zones of soil (Humphrey, 1992).

***Biological Scrubbing:*** In biological scrubbing, a biologically active agent would be added to an absorber liquid which would react with gas phase pollutants. Separation and destruction would be accomplished in one process step. This process is not commercial and further developments are needed.

***Biostripping:*** The focus here is on development of stripping fluids containing reactive species which would convert contaminants to environmentally benign components. New stripping fluids would need to be developed which would remove pollutants from water streams and then react with them.

### 1.4.4. Adsorption Reactors

In this process the "adsorbent," which is normally used as the mass separating agent for the separation, also acts as the catalyst. This technology is still very much in its infancy though it has significant potential. The challenge for this technology is materials development involving catalysts/adsorbents. The matching of process conditions and regeneration cycles to achieve performance at the same temperature will also be a challenge.

### 1.4.5. Membrane Reactors

Using membranes to upset the equilibrium in a reversible reaction has been recognized since the late 1960s, but commercial applications of membrane reactors are largely limited to biological systems (Matson and Quinn, 1992). This is because enzyme immobilization techniques are highly developed and membranes are the natural environment for many enzymes. Membranes serve as a support for the catalytic enzymes and also provide a cost-effective means of separating the products. In this application, the membrane is a composite of a permselective layer for separation and a catalytic layer for reaction.

Conversion of the conventional membrane module to a membrane reactor requires altering the membrane surface to attain catalytic properties or by coating the surface with a catalytic material. Sometimes, it may be desirable to impregnate the membrane with a catalyst or pack the membrane canal with catalyst particles. Some reactions which may be amenable to catalytic membrane reactor processes are given in Table 1.2.

**Table 1.2.** Reactions Amenable to Catalytic Membrane Reactors

| Reaction Type | Specific Example | Conventional Reactor Conditions | Problems | Membrane Reactor Concept |
|---|---|---|---|---|
| Hydrogenation | $CO+2H_2 =$ $CH_3OH$ | 250°C, 50–100 Bar<br><br>Cu–Al–Zn oxide catalyst | Equilibria limited<br>Exothermic<br>Catalyst poisoning (S, Cl) | Product removal ($CH_3OH$) |
| Dehydrogenation | $\phi\, CH_2CH_3 =$ $\phi\, CHCH_2 + H_2$ | 500–600°C<br>Fe–Cr–K catalyst | Equilibria limited (60%)<br>Highly endothermic<br>Costly separation | Product removal ($H_2$) |
| Hydroformylation | $RCH = CH_2 + H_2 + CO =$ $RCH_2CH_2CHO$ | 100–250°C<br><br>200–450 Bar<br>Co, Ru, Rh homogeneous complex catalyst | Product separation<br><br>Catalyst recovery | Product removal<br><br>Catalyst retention |
| Hydrations | $CH_2 = CH_2 + H_2O =$ $CH_3CH_2OH$ | 300°C, 70 Bar<br><br>Acid catalysts ($H_3PO_4$) | Equilibria limited (<5%)<br>Large recycle | Product removal ($CH_3CH_2OH$) |

The concept of using a membrane module, both as a separator and as a reactor, is sometimes incompatible with reaction characteristics. For example, high-pressures favor membrane flux rates but can cause unfavorable equilibrium shifts in gaseous reactions. When the membrane layer is the catalyst, thin layers favor separation but could slow the reaction. The question of how heat must be removed or added during the reaction is a key consideration. In many applications, it will probably be necessary to incorporate intermediate heaters or coolers between individual membrane reactor modules.

*Dehydrogenation:* Dehydrogenation reactions are fast and reversible, and are probably good candidates for membrane reactors. The conceptual advantage of the membrane reactor concept in dehydrogenation has been demonstrated by Itoh who showed that an equilibrium conversion of 100% can be achieved using a membrane reactor for dehydrogenation of cyclohexane to benzene (Itoh, 1987). The common problems encountered in membrane-reactors for dehydrogenation are (1) the inability to produce an essentially pure hydrogen stream when dehydrogenating light paraffins such as ethane and propane, and (2) the inability of the membranes to operate at typical dehydrogenation temperatures (generally greater than 400°C). The first problem arises from the fact that, if hydrocarbons must be recovered from the permeate, then the savings in overall recovery and purification costs will suffer. To get around the second problem, as shown in Figure 1.11, it is possible to physically separate the dehydrogenation reaction from the membrane-separation function and therefore use heat exchange to reduce the gas temperature to the point that polymer membranes can be used. As the gas exits the membrane, it must be reheated through a second heat exchanger to achieve the desired dehydrogenation temperature. However, the investment cost of the heat exchangers, especially since they will be gas–gas

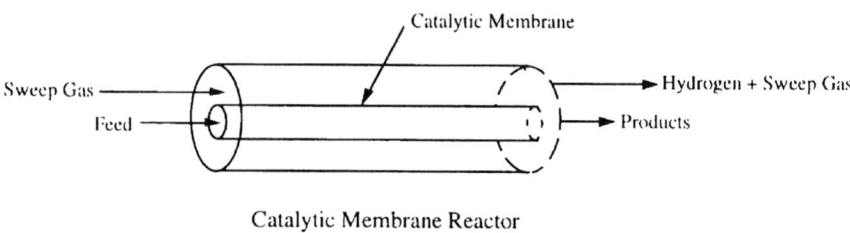

Catalytic Membrane Reactor

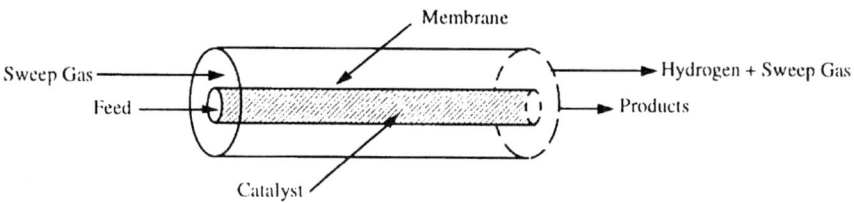

Inert Membrane Reactor

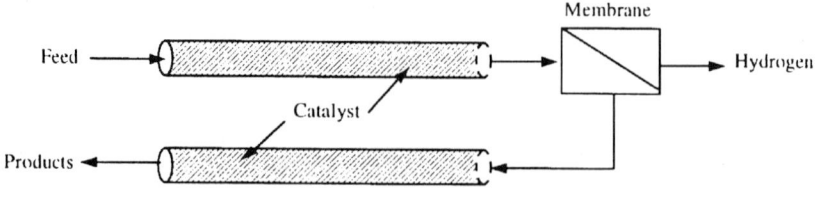

Reactor Plus Separate Membrane Unit

**Figure 1.11.** Reactor plus membrane combinations for dehydrogenation reactions.

exchangers and therefore of considerable area, will excessively add to the investment of the process (Humphrey and Keller, 1997). Recent work with high operating temperature ceramic-plus-polymer combinations gives indications of easing, but probably not eliminating this problem (Rezac et al., 1995).

*Hydrogen Sulfide:* With regard to pollution prevention, an interesting membrane reactor application is the dehydrogenation of hydrogen sulfide to form sulfur and hydrogen. The reaction is endothermic and the equilibrium is favored by high temperatures; a reaction temperature of 1500°F is favorable. For this reaction the equilibrium conversion is 55% at 850°F to 1000°F and 150 to 500 psig using $Al_2O_3$, $SiO_2$ catalyst. The current technology involves a tubular reactor located in a furnace hydrogen as the diluent. Conversions at lower temperatures such as 900°F are too low to be commercially significant. Thus, the commercial significance of the dehydrogenation reaction using traditional reactors is limited by equilibrium conversion and the high reaction temperature requirement (Kameyama et al., 1981). Hydrogen is valuable in terms of providing feedstock for hydrogenation or as a fuel, whereas sulfur is used to make sulfuric acid and sulfur-based drugs for the pharmaceutical industry. Membrane reactor technologies which would work for this application would be a significant contribution.

### 1.4.6. Future Directions in Separative Reactors

In its drive to minimize capital investments in future plants, industry is likely to support research and development programs which lead to commercialization of new separative reactor processes. Reactive distillation, which is at the forefront of this technology, will probably lead to new applications in the short term with new processes for production of esters, hydrogenated products, and perhaps dehydrogenated products. Though in the conceptual stage, newer concepts in reactive absorption and stripping should be pursued. Reactive adsorption and membrane processes, particularly reactive adsorption processes, are still largely untapped and new developments are needed to identify plausible process options.

## 1.5. References

Abrahim, G., Kane, J. T., Chester Engineers, Inc. (March 13, 1997) "Source Reduction of Volatile Organic Compounds (VOC) Emissions Using Separation Technologies," paper presented at the American Institute of Chemical Engineers Spring National Meeting, Houston, TX.

Cheryan, M. (July 22-27, 1995) "Application of Membranes in Corn Stillage Processing," paper presented at the Engineering Foundation Conference, "Separation Technology VI: Advances & Opportunities in Environmental Separations," Snowbird, UT.

DeGarmo, J. L. et al. (March, 1992) *Chemical Engineering Progress*.

Eisinger, R. S., Keller, G. E. II (November 7, 1989) "Electrosorption: A Case Study on Removal of Dilute Organics from Water," paper presented at the American Institute of Chemical Engineers Annual Meeting, San Francisco, CA.

Electric Power Research Institute (to be published 1997) EPRI TechApplication "Reverse Osmosis Reduces Costs and Protects Environment While Converting Wastewater to Boiler Feed Water."

Gildert, G. R., Rock, K., McGuirk, T. (March 10, 1997) "New Applications and Advantages of Catalytic Distillation," paper presented at the American Institute of Chemical Engineers Spring National Meeting, Houston, TX.

Humphrey, J. L., Keller, G. E. II (1997) *Separation Process Technology*, McGraw-Hill, New York, NY.

Humphrey, J. L., et al. (May, 1996) "Membranes Versus Competing Processes for Recovery of VOCs from Air," paper presented at the 8th Annual Meeting of the North American Membrane Society, Ottawa, Canada.

Humphrey, J. L. (August, 1992) *Separation Technologies in Pollution Prevention and Hazardous Waste Treatment*, Workshop Report prepared for the U.S. Environmental Protection Agency/RREL.

Humphrey, J. L. (February, 1989) *Utilization of Natural Gas in Large Scale Separation Processes*, Final Report prepared for Gas Research Institute.

Hussey, F., Gupta, A (March 13, 1997) "Removal of VOCs from Industrial Process Exhaust, Part II - Carbon and Zeolite Hybrid Systems," paper presented at the American Institute of Chemical Engineers Spring National Meeting, Houston, TX.

Itoh, N. (1987) *AIChE J.*, 33(9):1576.

Johnston, E. (June 23, 1995) Rochem Separation Systems, letter communication.

Kameyama, T., Dokiya, M., Fujishige, M., Yokokawa, H., Fukuda, K. (1981) *Ind. Eng. Chem. Fundam.*, 20:97.

Kochen, R. L., Navratil, J. D. (January 21, 1997) Removal of Radioactive Materials and Heavy Metals from Water Using Magnetic Resin, U.S. Patent No. 5,595,666.

Koros, W. J. (October, 1995) "Membranes: Learning a Lesson from Nature," *Chemical Engineering Progress*, pp 68-81.

Matson, S. L., Quinn, J. A. (1992) "Membrane Reactors," Chapter 43 of *Membrane Handbook*, Ho, W. S. Winston, Sirkar, K. K., Eds., Van Nostrand Reinhold, New York.

Miller, K. A., Stanford, P. T. (September, 1994) "Cleanup of Hazardous Waste Using an Advanced Reverse Osmosis System," presented at the American Chemical Society Emerging Technologies in Hazardous Waste Management VI Meeting, Atlanta, GA.

Mulder, Marcel (May, 1996) "Environmental Opportunities for Membrane Separations," plenary presentation, 8th Annual Meeting of the North American Membrane Society, Ottawa, Canada.

Niezgodski, Debra (March 26, 1997) Literature on The Polyad™ Process from Weatherly Inc., Chematur Engineering Group, Atlanta, GA.

Ninomiya, K. et al. (August, 1991) "Dehydration of Water-Alcohol Mixtures by Vapor Permeation Through Polyimide Membranes," presented at the American Institute of Chemical Engineers National Meeting, Pittsburgh, PA.

Rao, M. B., Sircar, S. (May, 1995) "Separation of Gases by Adsorption and Surface Flow Using Nanoporous Carbon Membranes," presented at the North American Membrane Society Meeting, Portland, OR.

Rezac, M. E., Koros, W. J., Miller, S. J. (1995) *Ind. Eng. Chem. Res.* 34, 862.

Strauss, Michael J. (July 12, 1990) *The Oil Daily*.

# 2

# Adsorption Technologies

**PRIMARY AUTHORS**

Kent Knaebel
Adsorption Research Inc.

Douglas Ruthven
University of Maine

Jimmy L. Humphrey
J. L. Humphrey & Associates

Robert Carr
University of Minnesota

**CONTRIBUTING AUTHORS**

Jeffrey R. Hufton
Air Products and Chemicals, Inc.

Alan L. Myers
University of Pennsylvania

John C. Crittenden
CenCITT/Michigan Technological University

John L. Bulloch
CenCITT/Michigan Technological University

**EDITORS**

John C. Crittenden
CenCITT/Michigan Technological University

John L. Bulloch
CenCITT/Michigan Technological University

## 2.1. Adsorption Technology Overview

### 2.1.1. Introduction

In the chemical and petroleum process industries, separations account for more than 40% of the total energy consumption (Humphrey et al., 1991). Accordingly, they constitute a major portion of processing costs. Improvements in separation technology are therefore key to the improvement of capital effectiveness as well as in efforts to reconfigure process technology to minimize noxious effluents and their environmental impact. Adsorption is inherently suited to applications in which the component to be removed or recovered is dilute and adsorbable, where there is little difference in volatility between the components to be separated, or where any of the components are permanent gases or nonvolatile liquids. Therefore, adsorption may be expected to play an increasing role in waste reduction for industries that employ gases, vapors, and liquids that must be recovered or purified.

With applications for product purifications and for removal (and recycle) of contaminants from air and water streams, adsorption processes play key roles in pollution prevention. Recovery and recycle of organics from plant effluents can reduce costs while eliminating sources of pollution. For example, polyethylene plants have compressor leak-off in streams that contain ethylene. Recovery credits of the ethylene can be significant while eliminating a source of pollution. This application is not currently feasible for a variety of reasons, including the lack of a suitable adsorbent. Another example is the elimination of high molecular weight organic contaminants that build up in ethylene and propylene recycle streams in polyethylene plants and polypropylene plants, respectively. To control concentrations of contaminants, these streams are "purged" to nearby olefin plants where the contaminants are removed by costly distillation processes. Use of selective adsorption processes would eliminate these purges.

In many situations, however, adsorption is feasible but, since the technology is considered esoteric, implementation is obstructed. Thus, adsorption as a unit operation is much less prevalent than it would be if it were better understood. Part of the explanation for this is that the principles of adsorptive separations and adsorbents have not been widely taught in universities. Rather, most learning occurs in industrial settings as a matter of necessity, or while doing postgraduate research at a few universities. There are, therefore, relatively few engineers who are familiar with the field and the relevant technology. Accordingly, in situations that are appropriate for adsorption, there may be no one involved who is sufficiently familiar with the subject to pursue this approach. This constrains applications and reduces market opportunities. Consequently, there are relatively few adsorbent manufacturers and process equipment manufacturers. Since they are the source of most adsorption technology and many technologists, a sort of vicious cycle exists that retards the growth of industrial applications. Hence, beyond the research and development needs that will open new opportunities for energy savings and environmental care, both education and increased implementation of adsorption in industrial applications are required to break this cycle. In view of that, and for the benefit of those readers who may not be familiar with the basic concepts of adsorption, a brief introduction is provided here.

## 2.1.2. Current Adsorption Processes

Adsorption separation processes can be categorized by applications, process equipment features, or adsorbent characteristics. Applications span the range of gas-phase and liquid-phase mixtures, from natural gas purification to air separation, and from municipal water purification to separation of liquid hydrocarbon mixtures. Some of the most important applications are emerging as solutions to environmental problems—for both air and water pollution clean-up or prevention.

Equipment for accomplishing these separations can be divided into two major groups: fixed-bed and fluidized-bed systems. The former is most prevalent and may be further subdivided into those that feature in situ regeneration versus those that require removal of the adsorbent for regeneration, reactivation, or disposal. Fluidized bed systems have been less prevalent until recently because of the lack of adsorbent particles that could withstand the impacts, abrasion, and grinding that occurs when the adsorbent is circulated between the uptake and regeneration units. An amalgam of these two types of systems employs a series of fixed beds that are connected so that they function as a moving bed (see Section 2.2.2.4).

Adsorption separation processes depend on the existence of a force field, which causes the molecules of some constituents to be attracted from the ambient fluid phase (gas or liquid) to the surface of the solid adsorbent. The strength of this attraction depends on the nature of both the solid surface and the molecules. The corresponding amount and rate of uptake (or release) depends on the driving force and equilibrium limit, while the rate also depends on the resistance. By a judicious choice of the solid and/or controlled modification of its structure or surface properties, it is possible to adjust the relative affinities or resistances for different molecular species. Accordingly, adsorptive separations exploit differences in either equilibrium or diffusion behavior.

The vast majority of adsorptive separations rely on equilibrium selectivity. Examples pertinent to waste reduction are the recovery of organic compounds from aqueous or gaseous streams. Specific instances are too numerous to cite, but a few general cases are recovery of solvents from industrial off-gases, secondary wastewater treatment to reduce chemical oxygen demand (COD), biological oxygen demand (BOD), and/or remove toxic or refractory organics from industrial wastewater.

Conversely, it is sometimes possible to separate gas mixtures based on differences in adsorption/desorption rates, and this is referred to as a kinetic separation. At present, separation of nitrogen from air is the only significant commercial application of this technique. Kinetic separations are generally practical only when the adsorption/desorption rates are controlled by sterically hindered diffusion in very small pores (micropores). In the limit where the molecules of one species are too large to enter the micropores, we have complete size exclusion or a molecular sieve separation. Kinetic separations may therefore be considered as partial molecular sieving in which the differences in molecular dimensions are large enough to give a substantial difference in sorption rates but not so extreme as to lead to complete exclusion of one species. Since equilibrium, kinetic, and molecular sieve separation processes operate in different ways, it is convenient and logical to use these three modes as a basis for classification of adsorption process technology.

Before describing specific processes and adsorbents in more detail, it is necessary to review some of the more important factors that govern adsorptive separations.

### 2.1.3. Adsorption Fundamentals

Adsorption is broadly characterized as physisorption or chemisorption. Physisorption depends on weak physical attraction of the solid phase for components in the fluid phase. There are several characteristics by which to distinguish physisorption:

- It is sensitive to temperature.
- Affinity for the adsorbent depends on the coulombic, van der Waals, or dipole interactions with the surface.
- There is nonspecificity among constituents in the fluid phase.
- Rate is limited by mass transfer because there is no activation energy barrier.
- Multiple layers of adsorbate may be found on the solid surface.
- The heat of adsorption is relatively small (i.e., $\Delta H_{ADS} < 2\Delta H_{VAP}$).

Conversely, chemisorption occurs by chemical bonding via electron transfer. Chemisorption also has some typical characteristics:

- It is constituent specific.
- The rate is slow due to the existence of an activation barrier (with associated chemical kinetics).

- A monolayer is formed on the solid surface.
- The heat of adsorption is relatively large, i.e., $\Delta H_{ADS} > 2\Delta H_{VAP}$.

Though certain specific instances of waste reduction exist that involve chemisorption, by far the majority involve physisorption. Thus, the remainder of this section is devoted mainly to that topic.

The forces of physisorption fall into two main categories: van der Waals (or dispersion) forces and electrostatic forces. Van der Waals forces are a basic property of all matter. The attractive van der Waals force field at a solid surface reflects the greater molecular density of the solid compared with the ambient fluid. The force constant is directly related to the polarizability. Thus, an approximate estimate of the relative strengths of the van der Waals attractions can be made based on a comparison of the size and polarizability of the sorbate molecules.

Electrostatic forces can be subdivided into polarization forces, field–dipole interactions, and field gradient–quadrupole interactions. These forces occur only if the surface is polar. Their strength depends on the size and polarity of the sorbate molecule, as well as on the strength of the electric field at the solid surface. Polarization forces arise when the electron shell of the sorbate is distorted by the surface field leading to a (favorable) interaction energy. The effect is generally relatively small, although it can be important for large, highly polarizable species. Field–dipole interactions occur when the sorbate has a permanent electric dipole (e.g., $H_2O$ or $NH_3$), which interacts directly with the electric field at the surface to give an additional (attractive) energy of interaction over and above the energies arising from van der Waals forces and polarization energy. Such effects can be very large and explain the strong preferential adsorption of small polar molecules on polar adsorbents such as zeolites. Field gradient–quadrupole interactions are more subtle but quite important. They can arise for electrically neutral molecules (such as $CO_2$ or $N_2$) which, nevertheless, have significant quadrupole moments. In the presence of a nonuniform electric field, the quadrupole moment will interact to yield an attractive energy that is proportional to the product of the field gradient and the quadrupole moment. This form of interaction does not depend on the actual strength of the electric field, only on its local gradient. Finally, hydrogen bonding is a specific form of electronic interaction that is important for hydroxylated surfaces, such as silica gel. Sorbates such as water or alcohols are strongly attracted to such surfaces, and this can be used as the basis for highly specific adsorptive separations.

An understanding of the forces of adsorption and the factors that promote or suppress the various kinds of interactions is critical to the selection and tailoring of adsorbents. Some examples illustrating the importance of the different kinds of surface interaction for selected systems are given in Table 2.1.

**Table 2.1.** Heats of Adsorption ($H_0$) at Zero Coverage for Three Adsorbents and Various Adsorbates

| Gas | Polarizability ($\times 10^{24}$) (cm³/molecule) | Dipole Moment (Debye) | Quadrupole Moment (Å²) | $H_0$ (kcal/mol) | | |
|---|---|---|---|---|---|---|
| | | | | Activated Carbon | 13X Zeolite | Silica Gel |
| He | 0.2 | 0 | 0 | 0.34 | — | — |
| Ne | 0.4 | 0 | 0 | 0.84 | — | — |
| Ar | 1.6 | 0 | 0 | 2.70 | — | — |
| Kr | 2.1 | 0 | 0 | 3.90 | — | — |
| $H_2O$ | 1.5 | 1.84 | — | 10.5 | 30 | 14.0 |
| $NH_3$ | 2.2 | 1.46 | — | 7.2 | 18.0 | — |
| $CO_2$ | 1.9 | 0 | 0.64 | — | 12.2 | — |
| $N_2$ | 1.4 | 0 | 0.31 | 3.08 | 6.50 | 2.50 |
| $O_2$ | 1.2 | 0 | 0.10 | 3.30 | 3.31 | — |

The equilibrium separation factor (discussed later) between nonpolar sorbates generally reflects mainly differences in polarizability. For such systems, tailoring generally involves simply the elimination of unwanted energetic heterogeneity and strong surface sites to minimize any electrostatic contributions. Separations between two polar/quadrupolar species or between a nonpolar species and a polar or quadrupolar species provide opportunities to tailor selectivity by adjusting the strength and heterogeneity of the surface field.

### 2.1.4. Adsorption Equilibrium Description

Adsorption equilibrium behavior can be correlated in a general way using thermodynamics, regardless of the detailed nature of the surface forces. Early contributions to this subject were made by Gibbs, Langmuir, and Dubinin (Ruthven, 1984). The amount of a component adsorbed at a specific temperature depends on its fluid-phase concentration or partial pressure, and is referred to as $q_i(C_i, T)$ or $q_i(P_i, T)$. Equilibrium data are commonly correlated as isotherms, which are simply plots of adsorbed phase concentration vs. fluid-phase concentration or partial pressure, at constant temperature. In general, fluid-phase concentration ($C_i$) and partial pressure ($P_i$) are interchangeable in the isotherm equations, depending on the application to gas or liquid phase systems. The data can also be

## 2. Adsorption Technologies

plotted holding constant thermodynamic variables other than temperature, giving rise to isosteres (plots of partial pressure vs. temperature for constant adsorbed phase loadings; see Section 2.1.4.1), and isobars (plots of adsorbed phase concentration versus temperature at constant partial pressure).

The shape of the equilibrium isotherm depends on the nature of the adsorbent and adsorbate. For microporous adsorbents, the isotherm is generally of the form sketched in Figure 2.1, varying from linear to highly favorable with increasing strength of adsorption. Since adsorption is exothermic, the slope and curvature of the isotherm decrease with increasing temperature. To construct the $X$–$Y$ diagram, the coordinates are made nondimensional using an appropriate (e.g., maximum) concentration, $C_0$, and the corresponding equilibrium loading, $q_0^*$.

The number of suggested isotherm equations is too numerous to discuss all of them here. However, they have been compiled in a number of texts included in the general reference section of this chapter. Only a few of the most important and prominently used isotherm equations will be introduced here.

The simplest case is when the adsorbent loading is linearly proportional to the fluid-phase concentration, as in *Henry's law*:

$$q_i(C_i, T) = K_i(T)C_i \qquad (2.1)$$

where $K_i(T)$ is called the Henry's law constant and is usually a fairly strong function of temperature as discussed in Section 2.1.4.1. This behavior is also observed in all systems at sufficiently low adsorbed phase concentration, as shown by the limits:

$$\lim_{q_i \to 0}\left(\frac{q_i}{C_i}\right) = K_i; \qquad \lim_{q_i \to 0}\left(\frac{q_i}{P_i}\right) = K_i' \qquad (2.2)$$

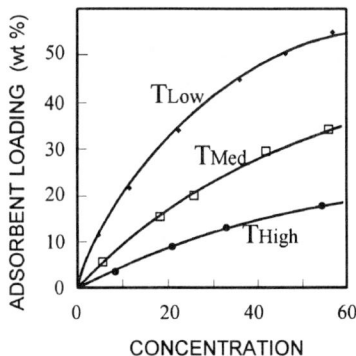 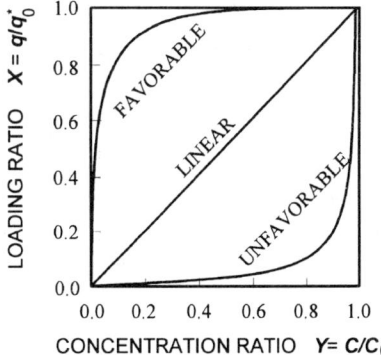

**Figure 2.1.** Schematic diagrams showing (a) the form of a typical (Type I) isotherm, characteristic of a microporous adsorbent; and (b) the representation as a dimensionless X–Y diagram.

The limiting slope of the isotherm ($K_i$ or $K_i'$) provides a direct measure of the affinity of the adsorbent for the sorbate since, at these low concentration levels the equilibrium is not significantly perturbed by sorbate–sorbate interaction effects.

When surface coverage is significant, the isotherms become curved. Furthermore, in some cases, the adsorbent surface appears to be homogeneous from the standpoint that the apparent force of attraction does not vary significantly as the extent of surface coverage changes. The ideal Langmuir isotherm model relies on this implicit assumption:

$$q_i(C_i, T) = \frac{q_{i,\max} K_i(T) C_i}{1 + K_i(T) C_i} \tag{2.3}$$

in which the monolayer saturation loading is $q_{i,\max}$. This isotherm equation reduces to Henry's law at the lower limit ($C_i \to 0$) and accounts for monolayer (or saturation) coverage of the adsorbent surface at high adsorbate concentrations. The assumptions from which the Langmuir expression was originally derived (array of identical surface sites; each site can hold only one adsorbed molecule; and no interaction between adsorbed molecules even when occupying neighboring sites) are more appropriate for chemisorption than for physical adsorption. Nevertheless, the Langmuir equation is widely used to represent physical adsorption isotherms. This is not, however, totally without justification since the same form of expression may be derived from the general Gibbs adsorption isotherm with assumptions that are more appropriate for physical adsorption. Although the Langmuir model provides an accurate quantitative representation for only very few systems, it provides a useful approximate semiquantitative representation for very many systems. Furthermore, it provides an excellent basis for predicting qualitative effects. For example, the model is easily extended to binary and multicomponent systems.

Other isotherm equations are commonly used that portray different features. One of the most popular in the environmental field is the Freundlich isotherm,

$$q_i(C_i, T) = A(T) C_i^{B(T)} \tag{2.4}$$

where $A$ and $B$ are empirically determined, but for virtually all situations, $B < 1$. This equation does not reduce to Henry's law. Rather, the nature of this isotherm depicts the inherent heterogeneity that characterizes many adsorbent–adsorbate combinations.

Another widely used equation that incorporates aspects of both the Langmuir and Freundlich models is called the Sips isotherm:

$$q_i(C_i, T) = \frac{q_{i,\max} B_i C_i^{d_i}}{1 + B_i C_i^{d_i}} A(T) C_i^{B(T)} \tag{2.5}$$

## 2. Adsorption Technologies

Like the Freundlich expression, this does not reduce to Henry's law in the low concentration limit. Having three parameters rather than two (for both the Langmuir and Freundlich isotherms), it can provide a more flexible fit of experimental data.

Most adsorption isotherms over the complete concentration range, from the Henry's law region to the saturated region, cannot be fitted within experimental error by equations containing only three parameters (Valenzuela and Myers, 1989). The Langmuir-virial equation is more flexible:

$$P_i = \left[\frac{q_i}{K_i}\right]\left[\frac{q_{i,\max}}{q_{i,\max} - q_i}\right]\exp(\alpha_1 q_i + \alpha_2 q_i^2 + \alpha_3 q_i^3 + \cdots) \quad (2.6)$$

$P_i$ is pressure and $q_i$ is loading (mol/kg). The constants are the Henry's constant ($K_i$), the saturation capacity ($q_{i,\max}$), and the virial coefficients ($\alpha_i$). The number of virial coefficients depends upon the isotherm, but three constants (up to $\alpha_3$) are usually sufficient to fit the data within experimental error. At low loading, Eq. (2.6) reduces to Henry's law, Eq. (2.1). Without the exponential factor, Eq. (2.6) reduces to the Langmuir Eq. (2.3); without the $q_{i,\max}/(q_{i,\max} - q_i)$ factor, Eq. (2.6) is the adsorption virial equation. The Langmuir-virial equation has the disadvantage that the loading is given implicitly, but inversion is a simple numerical procedure. Figure 2.2 shows a fit of Eq. (2.6) to experimental data (Siperstein and Myers, 1998) for adsorption of propane on NaX (Si/Al = 1.23) at 20°C; the average error is 1.3%. For comparison, the average error is 18% for the two-parameter Langmuir equation (2.3) and 13% for the three-parameter Toth equation.

Once isotherm data are available, a useful parameter that conveys adsorption separation capability for most applications is called the equilibrium selectivity. It is based on differences in the affinities of the solid for the different species, and is analogous to the relative volatility used in vapor–liquid equilibria. The basic definition of this parameter is obtained from pure component isotherms, normally for the more strongly adsorbed constituent (A) in a predominant but less strongly adsorbed carrier or solvent (B):

$$\alpha_{AB} = \frac{q_A}{q_B}\frac{C_B}{C_A} \rightarrow \frac{K_A}{K_B} \quad (2.7)$$

It may be shown that, for mixtures that obey the Langmuir model (and having identical values of $q_{i,\max}$), the separation factor, $\alpha_{AB}$, is independent of loading and corresponds to the ratio of the Henry's law constants. Generally, acceptable values of this parameter may range from 2 for a weakly adsorbed constituent up to 104 or higher for a very strongly adsorbed constituent. An efficient adsorption separation process is generally not possible when the selectivity is less than

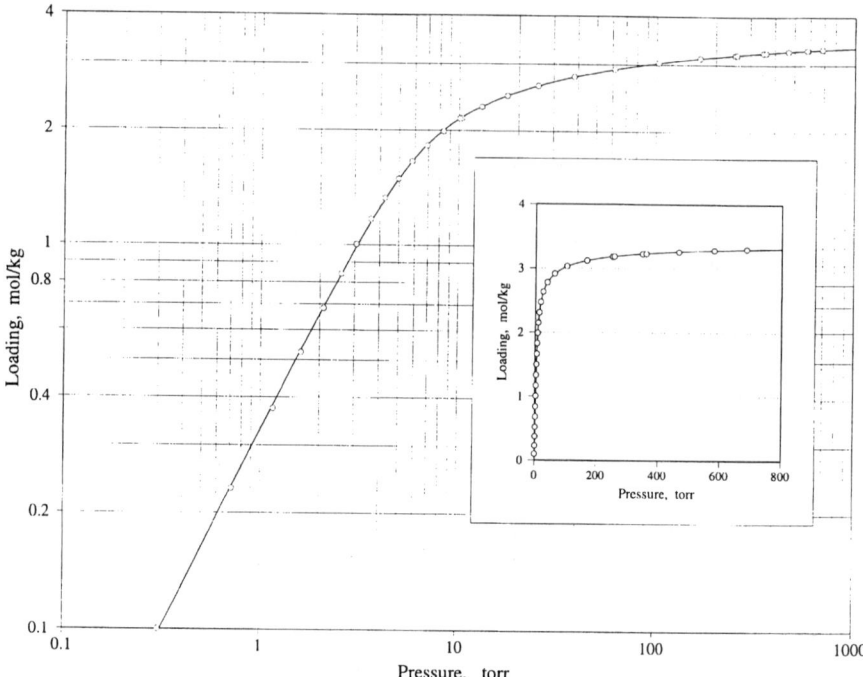

**Figure 2.2.** Adsorption of $C_3H_8$ on NaX at 20°C. Circles are experimental points and solid line is Eq. (2.6). The inset is a linear plot of the same data.

approximately 2. The problem with applications in the latter category is that regeneration becomes difficult and therefore expensive.

One of the complexities of this field is mixture equilibrium behavior. Frequently, such data can be represented approximately by a multicomponent version of the Langmuir isotherm, as shown in Eq. (2.8):

$$q_i(C_i, T) = \frac{q_{i,\max} K_i(T) C_i}{1 + \sum K_j(T) C_j} \qquad (2.8)$$

It should be noted that thermodynamic consistency requires that $q_{i,\max}$ be the same for all components, although this restriction is sometimes ignored in order to improve the fit of experimental data within a given concentration range. Clearly a small concentration of a strongly adsorbed species can greatly inhibit the adsorption of the other species (for example, water on a zeolite). Similarly, in liquid phase adsorption systems, even if the solvent is only weakly adsorbed, since it is present in large excess, it may greatly reduce the available capacity for the sorbate or sorbates of interest.

Many adsorbent–adsorbate combinations exhibit synergistic interactions. A simple equation that can accommodate such effects is the isotherm equation of Yon and Turnock (1971):

$$q_i(C_i, T) = \frac{q_{i,\max}(B_i/\eta_i)C_i^{d_i}}{1 + \sum(B_j/\eta_j)C_j^{d_j}} \tag{2.9}$$

This equation resembles a multicomponent version of the Sips isotherm, but with empirical affinity coefficients, $\eta_i$ and $\eta_j$. Another approach is the ideal adsorbed solution theory (IAST) of Myers and Prausnitz (1965), which is the adsorption analog of Raoult's law (which applies to vapor–liquid equilibria). As for Raoult's law, the IAS theory presumes ideality of both phases; hence, activity coefficients are unity. On the other hand, negative deviations from the IAS model frequently occur, implying that the activity coefficients are less than unity. However, this type of nonideal behavior can arise from heterogeneity of the adsorbent surface.

### 2.1.4.1 Heat of Adsorption

It should be noted that for physical adsorption, there is a relatively strong temperature dependence on the equilibrium, as evidenced by the temperature dependence in isotherm equations. The simplest way to illustrate this dependence is by examining the temperature dependence of the Henry's law equilibrium relationship.

The Henry's law constant is simply an equilibrium constant (or partition coefficient) and the temperature dependence is, therefore, given by the familiar van't Hoff relation:

$$\frac{\partial \ln K_i}{\partial T} = \frac{\Delta U_0}{RT^2}; \quad \frac{\partial \ln K'_i}{\partial T} = \frac{\Delta H_0}{RT^2} \tag{2.10}$$

in which $\Delta U_0$ and $\Delta H_0$, respectively, are the limiting internal energy and enthalpy of adsorption at zero coverage. Integration of these expressions assuming $\Delta U_0$ (or $\Delta H_0$) is independent of temperature yields:

$$K_i = K_{i\infty} \exp\left(-\frac{\Delta U_0}{RT}\right); \quad K'_i = K'_{i\infty} \exp\left(-\frac{\Delta H_0}{RT}\right) \tag{2.11}$$

showing that a plot of $\ln K_i$ (or $\ln K'_i$) vs $1/T$ should be linear with slope $-\Delta U_0/R$ (or $-\Delta H_0/R$). Such plots provide a simple and straightforward way of determining the limiting energy (or enthalpy) of adsorption. Moreover, it is this temperature dependence that gives rise to the behavior exploited by temperature swing adsorption, which is discussed in Section 2.2.2.3.

It can be shown from general thermodynamic arguments that physical adsorption from the gas phase is always an exothermic process ($\Delta H < 0$). The argument is less cogent for adsorption from the liquid phase, and indeed, examples of endothermic adsorption from the liquid phase are not especially uncommon.

Equilibrium between the adsorbed and fluid phases is governed by the Clausius equation:

$$\left.\frac{\partial \ln p}{\partial T}\right|_q = \frac{\Delta H}{RT^2} \qquad (2.12)$$

in which $\Delta H$ is the isosteric heat of adsorption. Integration yields the Clausius–Clapeyron equation:

$$\left.\ln\!\left(\frac{p_2}{p_1}\right)\right|_q = \frac{\Delta H}{R}\left[\frac{1}{T_1} - \frac{1}{T_2}\right] \qquad (2.13)$$

whence it is evident that the isosteres should be essentially linear with slope $-\Delta H/R$, when plotted in the form $\ln p$ vs. $1/T$ (for constant loading). Such plots therefore provide a convenient and reliable way of interpolating or extrapolating experimental equilibrium data as well as a straightforward way of measuring the loading dependence of the heat of adsorption.

The assumption that the heat of adsorption is independent of loading is implicit in the postulates of the Langmuir model. For most real systems, however, the heat of adsorption decreases with increasing loading which is due to the heterogeneity of the surface sites and pore structure.

## 2.1.5. Kinetics of Adsorption

The kinetics of physical adsorption are, in practice, always controlled by mass transfer rates since the intrinsic rate of equilibration at the surface is extremely rapid. However, this is not necessarily true for chemisorption since chemisorption is an activated process and the rates span many orders of magnitude from very fast to very slow, depending on the temperature and the magnitude of the activation energy.

Adsorbent particles typically have a biporous structure, as sketched in Figure 2.3. In an equilibrium selective process, it is desirable to minimize mass transfer resistance since any such resistance leads to inefficient utilization of the adsorbent as well as to loss of selectivity. In the common situation where major resistance to mass transfer is associated with macropore diffusion (or a combination of macropore and external diffusional resistance), the resistance may be decreased by decreasing the particle size. A practical limit is imposed, however, by the pressure drop that increases rapidly with decreasing particle size. When

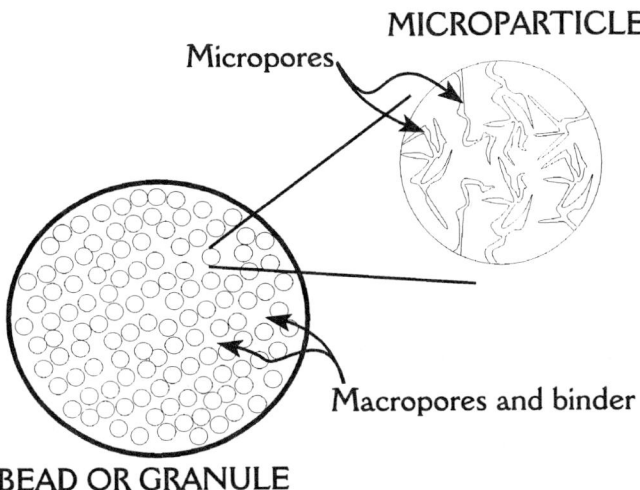

**Figure 2.3.** Conceptual view of a generic biporous particle. Zeolites have microcrystalline structures rather than irregular micropores.

intracrystalline or micropore diffusion is rate controlling, changing the particle size has no effect. Reducing the size of the microporous crystals (or char particles), however, reduces the micropore resistance. Hence, the selection of the correct crystal size is very important.

Differences in mass transfer rates may be utilized to obtain a desired separation using a kinetically selective adsorption process. In a kinetically selective process it is desirable to maximize the relative importance of the micropore resistance since kinetic selectivity depends on differences in micropore resistance. This may be achieved, in principle, by increasing the size of the microparticles. Any macropore or external film resistance reduces the effective kinetic selectivity. The limit is set, however, by the need to avoid sorption/desorption rates that are impractically small.

For a diffusion-controlled process, the relative rates of adsorption of two species depends not only on the ratio of their intracrystalline diffusivities but also on the ratio of the adsorption equilibrium constants (Kärger and Ruthven, 1992):

$$\alpha_K = \frac{\text{Rate of Adsorption of A}}{\text{Rate of Adsorption of B}} = \sqrt{\frac{D_A}{D_B} \cdot \frac{K_A}{K_B}} \qquad (2.14)$$

Clearly, an adverse equilibrium selectivity can reduce or even eliminate the effect of a substantial difference in micropore diffusivities.

The negative impact of an adverse equilibrium is actually less severe in the situation in which the sorption rate is controlled by surface resistance (at the surface of the microparticle) for under these conditions:

$$\alpha_K = \frac{D_A}{D_B} \cdot \frac{K_A}{K_B} \qquad (2.15)$$

This situation prevails in microporous adsorbents in which the pore mouth has been partially closed by procedures such as silanation or carbon deposition.

### 2.1.6. Molecular Simulation of Adsorption

The foundations of Monte Carlo simulation were laid in the famous paper by Metropolis et al. (1953). Early calculations were restricted to simple molecular models such as hard spheres, but during the 1960s and 1970s extended calculations were performed for molecules interacting with more accurate potentials such as that of Lennard-Jones. Early calculations required supercomputers but thermodynamic properties of complex molecules and mixtures can now be performed on 400 MHz personal computers available at a cost of a few thousand dollars. The increase in computer power has been accompanied by a dramatic increase in computer memory. Today's standard of 128 megabytes of RAM is sufficient to pretabulate the gas–solid potential energy and thus eliminate the most time-consuming step in the simulation: the summation of pairwise interactions of a molecule over all the atoms of the solid.

Molecular simulation is a powerful tool for understanding the behavior of adsorbed solutions. Given a set of intermolecular potentials for gas-gas and gas-solid interactions, the calculation is computer-intensive but straightforward. Most simulations are performed using the Monte Carlo technique, which is to sample equilibrium configurations with a probability proportional to the Boltzmann factor. A typical calculation of one point is the average of several million different configurations. Such calculations are costly in computer time but still relatively inexpensive compared to the cost of laboratory experiments.

The most challenging part of molecular simulation is not the computer calculation, but the selection of an accurate set of intermolecular potentials, or forcefield. Interactions between adsorbate molecules and adsorbents such as zeolites consist of two principal types: long-range electrostatic interactions between adsorbate molecules and ions in the zeolite and short-range van der Waals interactions between adsorbate molecules and the zeolite framework. Theoretically, it should be possible to estimate the forcefield from *ab initio* calculations. In practice, *ab initio* calculations are useful but usually must be adjusted in order to agree with experiment. For example, the charge distribution in an adsorbate molecule may be estimated from Hartree–Fock calculations and then scaled to reproduce the experimental dipole moment of the molecule. Similarly, the dispersion interaction ($r^{-6}$ coefficient) between an atom of an adsorbate-molecule and an atom in the framework of the adsorbent may be calculated from atomic polarizabilities, but then adjusted so that the Henry constant of the adsorbate agrees with the experimental Henry constant of the adsorption isotherm. In fairness, it must be said that molecular models of adsorption always contain certain approximations, such as the use of the Lennard-Jones 12–6 potential for atom–atom dispersion interactions and the assumption of pairwise additivity of intermolecular forces. Gas-solid intermolecular forces such as ion-induced dipole energy are frequently neglected even though they may constitute 5–10% of the total gas–solid interaction energy. On the one hand, even the most sophisticated simulations cannot be expected to agree exactly with experiment. On the other hand, a simulation technique capable of estimating adsorption isotherms and heats of adsorption within 20% error would be extremely useful.

The current state-of-the-art in computer simulations of adsorption is to construct a forcefield for a representative sample of the adsorbent containing several hundred atoms or, in the case of a zeolite, a few unit cells. The gas–solid energy of interaction between an adsorbate molecule and the adsorbent is obtained by summation over all pairs of atoms in the molecule and the solid. In the case of zeolites, the atoms (Si,Al,O) carry partial charges and the electrostatic energy is obtained by summing pairwise interactions using Coulomb's law. Since electrostatic energies are long-range, special procedures (Ewald sum) are needed to correct for the finite size of the solid sample.

In addition to its predictive power, molecular simulation can provide detailed information about:

1. Locations of strong and weak adsorption sites;
2. Resolution of total energy of adsorption into gas–gas and gas–solid interaction energies;
3. Increase in heat of adsorption with decreasing pore size;
4. Variation of selectivity with temperature, composition, and loading;
5. Division of gas–solid energy of adsorption into van der Waals and electrostatic terms;
6. Segregation of adsorbed mixtures inside micropores;
7. Orientation of adsorbed molecules from host–guest pair distribution functions.

Another powerful feature of molecular simulation that has hardly been explored is the prediction of preferential adsorption for the design of separation processes. Using either *ab initio* calculations or potentials derived from single-gas adsorption isotherms, mixture adsorption (selectivity, loading, individual heats of adsorption) can be predicted without making any additional assumptions about intermolecular forces.

Several software companies offer packages that perform molecular simulation, but interpretation of the reliability of the results requires extensive knowledge of molecular modeling and molecular simulation. Therefore many companies have hired an expert who can perform molecular simulations using proprietary programs and offer advice about the reliability of commercial software.

The most important need for research is clearly the development of accurate forcefields. The goal is a handbook of atom–atom potentials and charge distributions for molecules and zeolites that has been verified by comparison with experiment for selected thermodynamic and transport properties.

### 2.1.7. Introduction of Different Adsorbents and Their Usage

Adsorbents are separating agents that exploit differences between species based on equilibrium isotherms (most commonly) or between diffusion resistances within micropores (sometimes achieving a "sieving" effect). Most commercially successful adsorbents have large internal surface areas (e.g., 200 to 2000 $m^2/g$) and significant macroporosity. These translate into large adsorption capacities and fast diffusion rates. Some general classes of adsorbents are described below, and some of their properties are listed in Table 2.2.

Adsorptive separations can be categorized according to the adsorbents that are employed because they are inherently tied to the constituents they remove. In fact, as outlined earlier adsorption capacity, selectivity, and kinetics are all specific to the combination. Broad groups of adsorbents include: activated aluminas, activated carbons, polymeric adsorbents, silica gels, zeolites, and many others. All of these include subcategories comprising several individual types. Brief descriptions follow in corresponding sections.

#### 2.1.7.1 Carbon Adsorbents

Activated carbon is arguably the most ubiquitous type of adsorbent. It is used for both gas-phase separations and liquid-phase separations, including water-borne contaminants. It can be made from wood, coal, petroleum coke, peat, coconut shells, used tires, or other materials by pyrolysis or carbonization at 400 to 500°C, followed by activation with air, steam, flue gas, etc. at 800 to 900°C. Typical products have internal surface areas of 1000 to 1500 $m^2/g$. The physical properties (pore size distribution, capacity, chemical reactivity, etc.) of the activated carbon are highly dependent on the initial starting material.

Activated carbon may be used in a variety of different forms. It may be used directly as the particles that are formed after activation. Alternatively, a bead or

## 2. Adsorption Technologies

**Table 2.2.** Typical Properties of a Few Common Adsorbents

| Adsorbent Type | | Surface Area, m²/g | Average Pore Size, Å | H₂O Capacity Wt. % @25 C, 25% R.H. |
|---|---|---|---|---|
| Activated Alumina | | 200 to 400 | 10 to 140 | 10 |
| Activated Carbon from: | | | | |
| Coal | | 800 to 1500 | 20 to 200 | 2 to 5 |
| Coconut Shells | | 1000 to 1500 | 10 to 100 | 2 to 5 |
| Petroleum Coke | | 1500 to 3000 | 5 to 50 | 2 to 5 |
| Wood | | 800 to 1200 | 20 to 200 | 2 to 5 |
| Polymers | | 250 to 1200 | 10 to 150 | 2 to 5 |
| Cation Exchange Resins | | 200 to 800 | 50 to 200 | 5 to 15 |
| Silica Gel | | 400 to 800 | 10 to 400 | 7 to 15 |
| Zeolite | Si/Al | | | |
| Type A | 1 | N.A. | 3 to 5 | 20 |
| Type X | 1 to 1.5 | N.A. | 8 | 20 |
| Type Y | 1.5 to 3.0 | N.A. | 8 | 20 to 10 |
| USY | >3.0 | N.A. | 8 | <10 |
| U ZSM-5 | >30 | N.A. | 6 | <2 |

granule containing several activated carbon particles may be created using a binder as shown in Figure 2.3. Activated carbon fibers have also increased in use during recent years. Activated carbon fibers have an advantage of very low mass transfer resistances. Impregnated activated carbons are also produced for removing specific contaminants by reacting with them (e.g., formaldehyde, $H_2S$, $SO_2$, Hg).

Carbon molecular sieves have been developed since the mid-1970s and are used exclusively in gas separations (e.g., nitrogen from air). They are made by pyrolysis (at 700 to 1200°C) of PVDC, Saran, coal, sugar, or coconut shell; or by depositing a pyrolized polymer in the pore mouths of "ordinary" activated carbon.

### 2.1.7.2. Zeolites

The zeolites comprise a large and important class of about 60 different microporous structures, many of which occur in nature. In a true zeolite, the framework has the composition of an aluminosilicate:

$$M_{x/n}[(AlO_2)_x (SiO_2)_y] \cdot wH_2O \tag{2.16}$$

where M is the cation of valence n; $y/x$ is the Si-to-Al (atomic) ratio; $(x + y)$ is the total number of tetrahedra per unit cell; and, w is the number of water molecules per unit cell required to completely fill the intracrystalline pores. This water may be removed by heating and/or evacuation to yield a high capacity microporous adsorbent. Examples include NaA (4A) {Type A}, for which M = Na, $x = 12 = y$, and $w = 27$; NaX (13X) {Type X} for which M = Na, $x = 86$, $y = 106$, and $w = 276$. The cation, M, is easily changed by direct ion exchange. The silica to alumina ratio, $y/x$, controls the affinity for water: low ratios (e.g., 1:1) yield highly hydrophilic materials, such as NaA or NaX, which are commonly used as desiccants. For Si/Al ratios greater than about 5, the zeolite becomes increasingly hydrophobic, and the water isotherm assumes an unfavorable form. The high silica zeolites, such as silicalite or dealuminized USY are, therefore, useful as adsorbents under high humidity conditions or in an aqueous environment.

The range and variety of zeolite structures are greatly increased by framework substitution. Analogous structures may be prepared with boron in place of aluminum, and to a limited extent, with germanium in place of silicon. Furthermore, $AlO_2 \cdot SiO_2$ is isoelectronic with $AlPO_4$, so many zeolite structures can be formed as aluminophosphates or silicon–aluminum phosphates. These frameworks are electrically neutral, and therefore contain no exchangeable cations. Such materials are less polar than zeolites and show quite different adsorptive properties.

The aluminum-rich zeolites are usually synthesized from sodium silicate and sodium aluminate by a hydrogel process. Synthesis of the high silica materials is less straightforward and often requires the addition of a template to promote crystallization of the desired phase. Further reduction of the aluminum content can be achieved by direct extraction from the crystallized material. As a result, the high silica zeolites are generally substantially more expensive (i.e., 3 to 4 times) than the common aluminum-rich varieties such as NaA and NaX.

Another name for zeolites is molecular sieves; certain homologous series of zeolites contain uniform pores for separating molecules of slightly different size or shape using the sieving effect. For example, the series TON (ZSM-22), MTW (ZSM-12), and UTD-1 high-silica zeolites (Meier and Olson, 1992) have straight one-dimensional pores composed of 10-, 12-, and 14-membered rings, respectively. The straight pores in this homologous series of zeolites have elliptical cross-sections: $5.5 \times 4.4$, $6.2 \times 5.5$, and $10.0 \times 7.5$ Angstroms, respectively. As expected, heats of adsorption decrease with increasing pore size: for methane, the isosteric heats at the limit of zero coverage (Savitz et al., 1998) are {27.2, 20.9, 14.2} kJ/mol for {TON, MTW, UTD-1}.

### 2.1.7.3. Polymeric Adsorbents and Ion Exchange Resins

Polymeric adsorbents are among the most rapidly developing category of adsorbents, due to advances in synthesis techniques that have led to improvements in properties and diversity in product types. Their structure is generally similar to that of an inert ion exchange resin. That is, they are commonly made from polystyrene–divinyl benzene or acrylic beads, having 4 to 20% crosslinking, which have been specially treated to achieve high porosity and high surface area, yet retaining high strength. Because they lack functional groups they are generally hydrophobic; ion exchange resins are hydrophilic due to their acidic or basic functional groups. Applications include decolorization of sucrose and high fructose corn syrup and VOC recovery from air or other gases. Some varieties, however, are pyrolyzed to increase surface area and porosity. These tend to be more expensive than other polymeric adsorbents and are used for specialty purifications.

Conversely, ion exchange resins have evolved into mature products, which exist in weak or strong acidic or basic forms. They are used for anionic or cationic separations, such as water softening and boiler feed water clean-up. They are made from phenolic or polystyrene–divinyl benzene beads that have been functionalized with acidic and/or basic groups that are chemically bound to the polymeric backbone. Their affinity for water depends on the nature of the polymer and the functional groups present. They appear mostly as beads, typically having diameters of 0.1 to 4 mm. Ion exchange resins have evolved from the so-called gel-type that was popular before 1970 to include macroporous resins that were introduced then. The former type was made typically from polymethylmethacrylate, while the latter type is made of polystyrene with typically 8 to 12% divinylbenzene for crosslinking. They are specially treated to achieve high porosity and high surface area, yet to retain high strength.

### 2.1.7.4 Activated Alumina

Activated alumina is very hydrophilic, and is primarily used as a desiccant, both for gases and liquids. Most commercial grade activated alumina is made from $Al(OH)_3$ by rapid heating to 400 to 800°C. This generally forms an amorphous - alumina with an internal surface area of about 300 $m^2/g$. Other crystalline forms of alumina exist and are used as adsorbents, but they are less common. The structures of all aluminas depend on the residual water content and the temperature of formation. Aluminas are commonly formed into beads (up to 1 cm dia.) or less frequently into extrudates or granules. Very fine particles are frequently used to pack gas chromatography columns.

Aluminas have traditionally been known as adsorbents for removing impurities and as media for gas chromatography. Perhaps the most prevalent impurity has been moisture in gas streams, e.g., air and natural gas, as well as heavier hydrocarbon streams and even paraffinic, aromatic, and halogenated hydrocarbon liquids. They are, however, also used for treating sulfur-bearing petrochemical streams, acid gases, municipal wastes, and purifying polymers and pharmaceuticals (including proteins, steroids, and vitamins).

According to Goodboy and Fleming (1984), there are five stable phases of alumina, as well as other metastable transition forms. The stable forms generally have low porosity and surface area, so they are not very useful as adsorbents. Rather, transition forms are the most useful types of aluminas. Most types are formed from the Bayer process by precipitation from a sodium aluminate solution. As a result, they possess surface hydroxyls that impart acidity and affect adsorption properties. Transition forms have undergone thermal decomposition of intracrystalline hydroxyls, leaving primarily aluminum and oxygen in the lattice. For example, $\gamma$-alumina has a defect spinel structure in which the oxygen ions are in a well ordered cubic close pack arrangement, but the aluminum ions are less ordered. Similarly, $\eta$-alumina has a spinel structure but exhibits a one-dimensional disorder that results in a greater concentration of surface acid sites than in $\gamma$-alumina.

### 2.1.7.5. Silica Gel

This is another common drying agent, being very hydrophilic, but it is also used for hydrocarbon separations. It is made from sodium silicate ($Na_2SiO_3$) mixed with aqueous HCl to get silica ($SiO_2$) and NaCl, in the form of a hydrosol, which is dried and activated at 250°C. Typical products have internal surface areas of 400 to 800 m$^2$/g. Most of the products are in the form of beads, up to 3 mm in diameter, but fine particles are used to pack GC columns. Beads are known to shatter upon contact with liquid water. To counteract that, silica gel is sometimes alloyed with alumina to increase its tolerance for liquid water. Aside from gas- and liquid-phase drying applications, silica gel is also used for recovery of heavier hydrocarbons from natural gas.

### 2.1.7.6 Clays

Bentonite is the largest class of adsorbent clays. It has volcanic origins and is comprised primarily of montmorillonite, $Al_2O_3 \cdot 4\ SiO_2 \cdot H_2O$. Due to unbalanced surface charges, spaces between the platelets of clay are occupied by exchangeable cations, such as $Mg^+$, $Na^+$, $K^+$, etc. As a result, they exhibit cation exchange capacities of 70 to 100 meq/g, and the platelets swell significantly in water. Clays have been impregnated with a variety of substances to modify their properties. The starting materials are usually bentonites, which are altered, for example, to take up organic vapors from air. For that application, Harper and Purnell (1990) suggested exchanging the metal cations for organic cations, that is, tetramethyl- or tetraethyl-ammonium ions. Likewise, to remove dilute, free oil from water,

Alther (1996) suggested impregnating clay with dimethyl (di-hydrogenated) tallow ammonium chloride with chains of $C_{12}$ to $C_{22}$. The nitrogen-end of the quaternary amine replaces the sodium cation. The long organic chains are oriented vertically from the clay platelet surface, and they make the clay strongly hydrophobic and organophilic. As a result, these materials swell significantly upon exposure to hydrocarbon liquids, and may take up 60 to 80% by weight of organic compounds. The most frequent commercial applications are removal of oil from: storm-water runoff, air compressor condensate or contaminated groundwater, and primary wastewater treatment for pesticides, PCB, phenolic contaminants, etc.

Pillared clays have recently been promoted for gas-phase adsorptive separations (Yang and Baksh, 1991). None is commercial yet, however. They were first synthesized in the 1970s (Brindley and Semples, 1977). On a microscopic scale, they are two dimensional materials—planar silicate layers that are separated by large, inorganic hydroxycations. Upon heating, the hydroxycations decompose forming stable metal oxide clusters that are the "pillars" separating the silicate layers. Pore sizes in pillared clays appear to be limited by interpillar spacing rather than interlayer spacing, which has been measured at about 15 to 20Å. Thus, a challenge has been to reduce the density of pillars and to open up the structure.

## 2.1.8. Selection of an Adsorbent

Although adsorbents are widely known by their generic properties, as outlined in the previous section, finding the precise adsorbent exhibiting the best combination of properties for a new application is subtle. Even the priorities involved in selecting an adsorbent depend on the type of application. The following characteristics are usually among the most desired, but their relative importance varies:

- high capacity and selectivity (equilibrium and/or kinetic) or separation factor
- sufficiently fast intraparticle diffusion
- isotherm shape: linear for bulk separations; rectangular (full plateau) for contaminant removal without regeneration (explained in more detail later)
- high crush strength, chemical and physical stability, attrition resistance
- regenerability, e.g., via temperature or pressure swing, at low cost
- minimal pretreatment or protection required to avoid early replacement
- low cost per unit volume or mass.

Secondary characteristics include surface area (which is related to adsorption capacity), average pore size and pore size distribution (which affect capacity and kinetics), and particle shape. For equilibrium-based separations, it is important that adsorption kinetics be adequate, which implies that the pore size distribution be adequate to minimize internal diffusional resistances. That simply means that the dimensions of both micropores and macropores should be large enough to achieve equilibrium quickly. For kinetic or molecular sieve adsorbents, however, the issue of pore size is key. This is because the kinetic selectivity, and hence the viability of the process, depends closely on the correct choice of micropore dimensions.

For equilibrium-based separation processes, the choice of adsorbent, the selection of the process cycle, and the choice of process conditions can all be understood by considering the shape of the equilibrium isotherm. Some of the basic isotherm equations were discussed in Section 2.1.4. In that context, a highly favorable isotherm (represented by $\alpha \to \infty$, which corresponds to $K_i \to \infty$ for the Langmuir isotherm or $B_i \to 0$ for the Freundlich isotherm) is useful for removal of trace compounds since, with this form of isotherm, the adsorption step is very efficient and gives nearly complete utilization of the full capacity of the adsorbent bed, even at very low concentrations in the fluid phase. Chemisorption isotherms are often of this form. The disadvantage is that adsorption is practically irreversible; and desorption can only be achieved by raising the temperature or by displacement with a competitively adsorbed eluent. Reduction of the pressure or partial pressure does not achieve significant desorption, so a system with this form of isotherm could not be used effectively in a pressure swing process. Its use would be limited to nonregenerative or thermal swing systems. Conversely, for most cyclic adsorption systems, an isotherm that does not deviate too greatly from linearity is best (represented by $\alpha \geq 1$, which corresponds to $K_i \to 0$ for the Langmuir isotherm or $B_i \to 1$ for the Freundlich isotherm). This is because, for such a system, adsorption and desorption processes are symmetric, leading to an efficient overall process.

## 2.2. Adsorption Process Design—Engineering, Economic, Environmental, and Energy Considerations

### 2.2.1. Mechanisms of Adsorptive Separations

The fundamental mechanisms of adsorptive separations depend on the properties that were discussed in Section 2.1.3. This section covers the exploitation of those properties to effect separations by various techniques. In the past, many if not most applications of adsorption were based on a single use, that is, by discarding the spent adsorbent or possibly regenerating it off-site. The most prominent of those is municipal water purification using activated carbon. Since the early 1970s, however, enormous strides have been made in the area of cyclic adsorption processes. The most prominent of those are thermal swing and pressure swing systems. The former can be used for either gas- or liquid-phase applications, but the latter is restricted to gas-phase applications—generally involving highly volatile gases or vapors.

## 2.2.2. Process Configuration

A wide variety of process configurations are possible. The most important distinction relates to the choice between continuous operation (with moving adsorbent or simulated moving adsorbent) and cyclic operation (semicontinuous, with parallel adsorbers being simultaneously exhausted and regenerated). A third choice, in which the adsorbent is used only once and then removed and discarded or regenerated off-site, is sometimes found in pollution control applications involving trace component removal. Such processes, however, do not utilize the adsorbent very efficiently. Other distinctions relate to the nature of the vessel, the mode of regeneration, the fluid phase, and the adsorbent type.

### 2.2.2.1. Nonregenerative Adsorption Processes

In most adsorption separations, the process cycle is determined largely by the way in which the adsorbent is regenerated. Nonregenerative systems operate intermittently, that is, in a semibatch-wise mode. That is, once a specific amount of feed has been processed, the adsorbent is exhausted and must be replaced. At that point the system is shut down. Applications include municipal water treatment, odor removal, recovery of biological products and pharmaceuticals, and decolorization of sugar or other sweeteners. Production rates range from less than 10 g/day for pharmaceuticals to more than 10 million gallons/day for water treatment. Early fractions are nearly pure solvent or carrier. Later fractions contain increasing amounts of the least strongly adsorbed solute, then each successively more strongly adsorbed component, until ultimately the product composition approaches that of the feed. The fate of adsorbents used in such processes is discussed in Section 2.2.5. Irreversible adsorbents are often used in these processes.

A rough rule-of-thumb is that irreversible adsorbents may be a candidate if the amount of contaminant to be removed does not exceed 50 lb/day, because it is not economically favorable to operate a regeneration and recovery system for such a small quantities. These "adsorbents" are a special case in that they are not regenerated in place, but are replaced by the supplier and regenerated by special processes at the suppliers location. Applications for irreversible adsorbents include removal of the following contaminants (ICI Katalco, 1993):

- Sulfur compounds: $H_2S$ (although Calgon Carbon does make a water regenerable catalytic carbon), COS, $SO_2$, organic sulfur compounds
- Halogen compounds: HF, HCl, $Cl_2$, organochlorides

- Organometallics: $AsH_3$, $As(CH_3)_3$
- Mercury and its compounds, metal carbonyls
- Nitrogen compounds: $NO_x$, HCN, $NH_3$, organonitrogen compounds
- Unsaturated hydrocarbons: olefins, di-olefins, acetylenes
- Oxygenates: $O_2$, $H_2O$, methanol, carbonyls, organic acids
- Miscellaneous: $H_2$, CO, $CO_2$

*2.2.2.1.1. Rejuvenation and recycling of adsorbents*

Adsorbents that are used in nonregenerative systems or those that become deactivated or poisoned in a regenerative system can pose a significant problem. Frequently, the applications contain enough toxic or noxious materials so that spent adsorbent is categorized as hazardous waste. Hence, the disposal charges for incineration, landfilling, or other destinations can be enormous. To help customers who face those issues, many activated carbon manufacturers offer to reactivate the spent carbon that renders it a nonhazardous reactivated carbon. Many manufacturers offer both virgin and reactivated carbon. This reflects the tendency of firms to prefer virgin material, but for environmental or economic reasons, they return the spent material. That leaves others who are strictly economy-minded to purchase slightly inferior, reactivated carbon with substantial cost savings. Activated carbon is reactivated using steam and heating in a kiln under a nitrogen purge to drive off and chemically destroy strongly adsorbed contaminants, especially organic compounds. There is often a significant loss of adsorbent during the replacement of adsorbent in a bed (e.g., 10 to 20%).

Another rejuvenation method that can be used for inorganic oxide adsorbents, especially zeolites, is to burn off any high molecular weight or low volatility compounds by periodically administering hot gas containing oxygen. Of course, the vessel and ancillary equipment must be able to withstand the resulting high temperatures as well as the possibility of corrosive decomposition products. This procedure reactivates the adsorbent without removing it and can appreciably extend its useful life. This procedure is impractical for activated carbon since the reaction of the adsorbed materials could easily be extended to uncontrolled oxidation of the carbon itself, resulting in what is referred to as a bed fire.

### 2.2.2.2 Regenerative Processes

For regenerative systems, there are three common choices: thermal swing adsorption (TSA), pressure swing adsorption (PSA), and displacement purge adsorption (DPA). The factors governing this choice are summarized in Table 2.3. Examples of each, in terms of application, cycle type, regenerant, contactor type, and adsorbent are summarized in Table 2.3. Although the list is not comprehensive, it shows the diversity of options from which one may infer the major factors governing the choices.

## 2. Adsorption Technologies

**Table 2.3.** Representative Examples Processes Employing Different Regeneration Methods

| Separation | Cycle Type | Regenerant | Contactor | Adsorbent |
|---|---|---|---|---|
| Natural Gas Drying | TSA[1] | Hot Dry $CH_4$ | Fixed Bed | 4A Zeolite |
| Cracked Gas Drying | TSA | Hot $N_2$ | Fixed Bed | 3A Zeolite |
| Clean-up SLA[2] | TSA | Hot Air or $N_2$ or Steam | Fixed Bed | Activated Carbon or Zeolite |
| Clean-up SLA (Blizzard®) | TSA | Combustion Gas | Fluidized Bed | Polymeric Adsorbent |
| Gas Drying | PSA[3] | Low P / Dry Gas | Fixed Bed | Silica Gel, Zeolite or Alumina |
| $H_2$ purification | PSA | Low P / $H_2$ | Fixed Bed | 5A Zeolite |
| $O_2$ from air | PSA | Low P / $O_2$ | Fixed Bed | 5A, 10X, 13X, or LiX Zeolite |
| $N_2$ from air | PSA | Low P / $N_2$ | Fixed Bed | Carbon Mol. Sieve |
| p-Xylene from Mixed Xylenes (Parex®) | Displacement | $NH_3$ | SMB[4] | Zeolite |
| Fructose from Corn Syrup (Sarex®) | Displacement | Water | SMB | $Ca^{++}$ Resin or CaX Zeolite |
| Water Clean-Up | Displacement | Isopropanol | Fixed Bed | Amberlite XAD-4 |
| Linear paraffins from branched isomers | Displacement | $NH_3$ | Fixed Bed | 5A Zeolite |

[1] Temperature Swing Adsorption
[2] Solvent Laden Air
[3] Pressure Swing Adsorption
[4] Simulated Moving Bed

### 2.2.2.3 Fixed Bed Systems

**A. Inert–Purge Cycle.** This is the simplest, two step adsorption cycle, usually of a few minutes or hours duration. A schematic diagram is shown in Figure 2.4. The adsorption step is often accompanied by heat evolution as the heat of adsorption is released. That is followed by regeneration with an inert fluid. Regeneration reduces the contaminant concentration in the interstices and adsorbent. It also tends to cool the fluid and adsorbent as desorption proceeds by consuming the

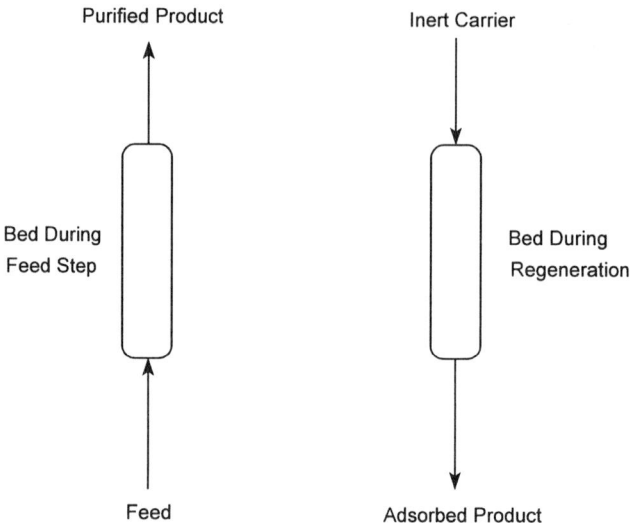

**Figure 2.4.** Simple two-column inert–purge adsorption system.

heat of adsorption, though heat retained by the adsorbent that was released during uptake may be recovered, which would mitigate that effect. The overall effect is to dilute, instead of enrich, the adsorbate. The last aspect implies that, although a specific amount of feed fluid may be purified, the quantity of regenerant fluid is greater than that of the purified feed, and its concentration is lower than that of the original feed.

**B. Displacement-Purge Cycle.** This is another two step process consisting of adsorption followed by regeneration. A sketch of this cycle is shown in Figure 2.5. In this case a carrier fluid ("C") that contains a weakly adsorbed component ("A") is the feed. The adsorbent bed is initially loaded with a strongly adsorbed component ("B"). During the adsorption step, "B" is gradually displaced by "A." This leads to a diffuse or proportional front. Conversely, during regeneration, "B" displaces "A," which leads to a sharp or constant pattern front. The net result is that the cycle enriches the desired adsorbate "A." The main problem with this type of cycle is that a separate separation system (e.g., distillation) is required to split the regenerant from the desired component.

**C. Temperature Swing Adsorption.** Temperature swing adsorption (TSA) cycles consist of two or three steps, of several minutes to several days duration. A simple two-step system is shown in Figure 2.6. First, adsorption occurs at low

## 2. Adsorption Technologies

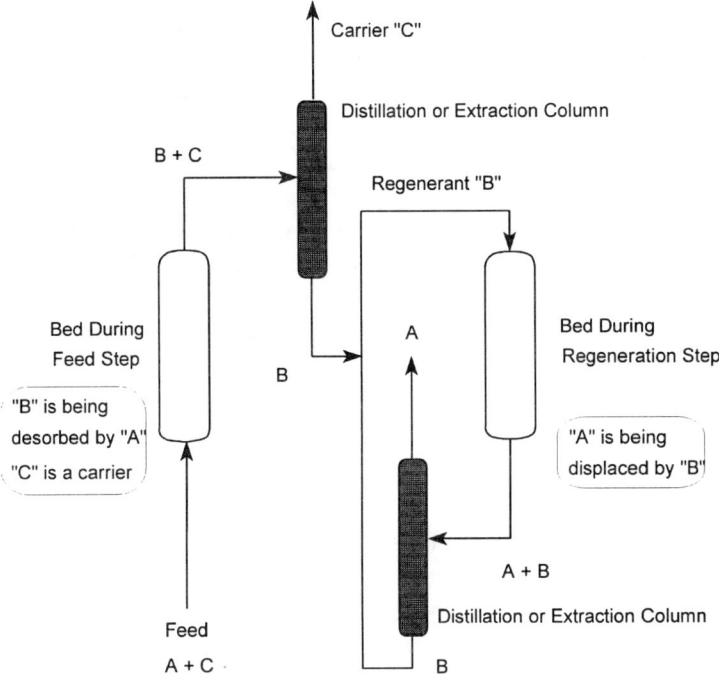

**Figure 2.5.** Two-column displacement–purge adsorption system.

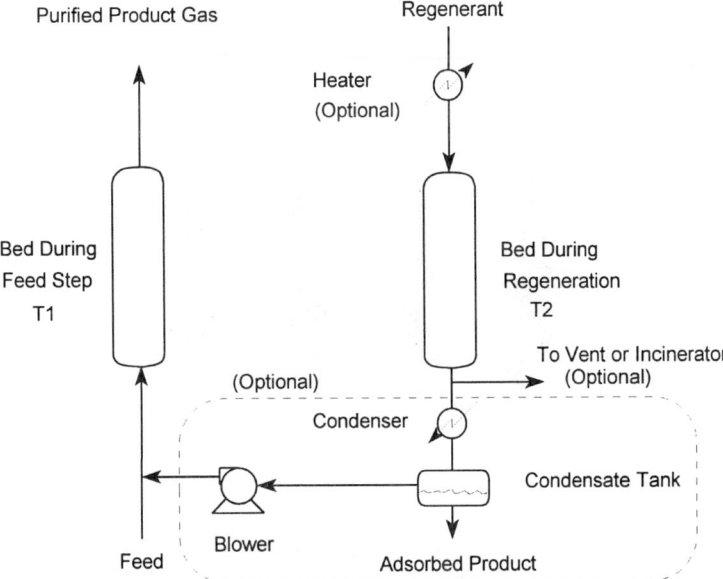

**Figure 2.6.** Two-column temperature swing adsorption system.

temperature. Second, regeneration is performed at high temperature. Optionally, one may use heated feed, hot solvent (or carrier), a separate hot inert fluid, electrical current (in conjunction with a small amount of purge), microwaves (also with purge), or steam as a regenerant. All of these reduce the adsorbate concentration by conveyance rather than displacement, so the regeneration profile is diffuse, rather than sharp as in the previous case. Conversely, TSA cycles enrich the adsorbate because the temperature is high enough that the amount of fluid required for desorption is less than that during uptake. Third, cooling is an optional step and may be performed with solvent (or carrier) or a separate inert fluid. Commonly, the shift in adsorbent loading may be as much as 20 wt.% (or more) for concentrated feeds. In general, the energy requirement for regeneration is the major factor in the overall cost, and the energy requirement may be so large that it may overshadow the adsorbent cost and the losses due to pressure drop.

**D. *Pressure Swing Adsorption.*** Pressure swing adsorption (PSA) cycles usually consist of four or more steps, sometimes as many as twelve, of a few seconds to several minutes duration. A simple four-step system is shown in Figure 2.7. The general idea is to separate the "heavy" (more adsorbable) component from the "light" (less adsorbable component) by cycling the pressure, influent composi-

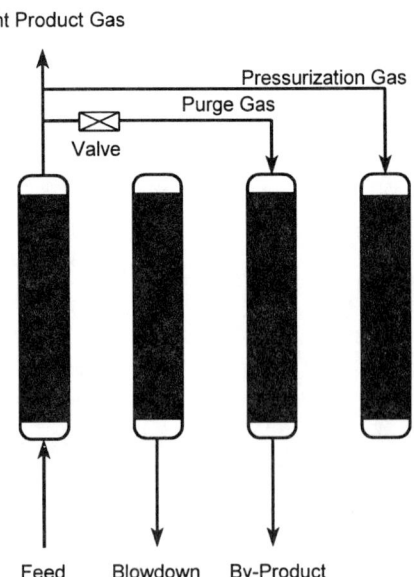

**Figure 2.7.** Simple four-step pressure swing adsorption system.

## 2. Adsorption Technologies

tions, and flow directions in a synchronous manner. Perhaps ten unique steps exist that have different variations of relative pressure, pressure shifts, influent composition, initial composition, and flow direction. Permutations of those steps comprise cycles, of which perhaps 50 unique combinations exist. It is not clear whether any truly unique cycles remain to be invented. The most common steps are: pressurization (with feed or purified light product), feed (at high pressure), blowdown, and purge (with purified product at low pressure). Feed involves a "wave" of the heavy (more adsorbable) component moving toward the product-end at high pressure. Flow stops when breakthrough is imminent and usually yields the "light" product. Regeneration always occurs by depressurization and may be supplemented by purging so the bed is ready for pressurization and usually yields "heavy" byproduct. Rinse (at high pressure) selectively adsorbs the "heavy" component and allows the residual feed to be recovered; the heavy component can be recovered by depressurizing. A simple five step PSA system including a rinse step is shown in Figure 2.8. For large-scale applications, it is common to employ "equalization" steps in which a bed at high pressure exhausts into a bed at lower pressure. That conserves the energy (though at a lower level) and the material, whether feed or product.

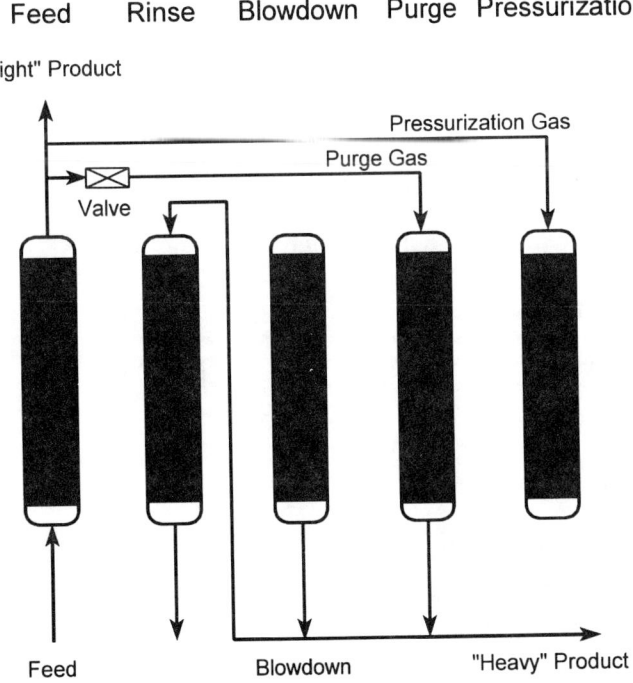

**Figure 2.8.** Simple five-step pressure swing adsorption system

Most cycles exploit equilibrium selectivity differences between species. Some exploit differences in intraparticle diffusion rates. These are becoming more prevalent with advances in adsorbent modification techniques. Efficient PSA cycles result from: good adsorbent, full exploitation of the input power, and careful control. The input power is exploited fully when gas product or byproduct is used to accomplish tasks as "reversibly" as possible. Shifts in adsorbent loading may be 1 to 10 wt. %, for concentrated feeds. Finally, the energy requirements of pressurizing the feed, evacuating (during regeneration), and optionally repressurizing the heavy product (for rinsing) strongly affect overall cost.

Of interest is the use of the pressure swing cycle for removal of organics from air. Based on this process concept, Dow Chemical Company has installed 35 commercial Sorbathene® units for removal of hydrocarbons, chloroflourocarbons (CFCs), chlorinated solvents, aromatics, and monomers from air. More recently, this process has been commercialized for removal of gasoline vapors from air (Pezolt et al., 1996). For example, it has recently been installed in England to recover gasoline vapors from an air stream flowing at a rate of 3000 scfm. Of the 35 Sorbathene® units installed to date, 30 of them have been installed in Dow plants (Robbins, 1997). A partial listing of organics that may be removed from vent streams by PSA, and dates of installation of some of the earlier units are given in Table 2.4.

With the pressure swing process, solvents such as CFCs can be recovered and recycled to the process. Only electric energy and a cooling fluid are required, no steam or other form of energy is needed. However, because vacuum pumps are expensive at large volumes, PSA units have an upper economic capacity of about 2000 cfm (Robbins, 1993 and Hall, 1993). On the low end of the capacity limit, high performance valves become expensive at small sizes and thus, begin to limit the practical low value of capacity, units less than 40–50 cfm may not be

**Table 2.4.** Partial Listing of Commercial Sorbathene Units

| Organic Contaminant | Unit Start-Up | Organic Contaminant | Unit Start-Up |
|---|---|---|---|
| Trichloroethylene | Jul-88 | Carbon Tet/MeCl$_2$ | Jul-91 |
| Styrene (2) | Aug-88 | Perchloroethylene | Jul-91 |
| Acetone (2) | Feb-90 | Carbon Tet | Oct-91 |
| Epichlorohydrin | Dec-90 | MeCl$_2$ | Oct-91 |
| Freon R-11 | Mar-91 | Acetone | Nov-91 |
| Carbon Tet | Apr-91 | Benzene | Dec-91 |
| Chlorobenzene | Jun-91 | Carbon Tet/MeCl$_2$ | Dec-91 |

## 2. Adsorption Technologies

cost-effective. Feed streams must be kept free of solids and contaminants that would foul the beds.

### 2.2.2.4. Moving-Bed Adsorbers

The optimal contacting scheme is a countercurrent system in which adsorbent and fluid move through the adsorber in countercurrent plug flow. Several such systems have been developed but they are more expensive than a simple cyclic batch system and so generally prove economic only for bulk separations.

The obvious way of circulating a solid adsorbent is in fluidized form. This necessarily involves considerable back-mixing, so a "staged" fluidized bed is generally used in which each stage is essentially well mixed but with virtually no back-mixing between stages. An example shown in Figure 2.9 is the Blizzard™ system, offered by On-Demand Environmental Systems, Inc. A similar system, called Polyad™, is offered by Weatherly, Inc. (Atlanta, Georgia). There are five units in commercial operation in the U.S. and about 10 units are in operation in Europe. These designs employ a tower that contains trays which allow stagewise contacting between the VOC-laden gas and the adsorbent. The spherical shape of polymer bead adsorbents allows the design of the tower to mimic a distillation operation, with polymer beads overflowing each tray onto subsequent trays while the gas stream passes countercurrently through the adsorbent on the trays. The trays are similar to sieve trays used in distillation. Fresh adsorbent enters the

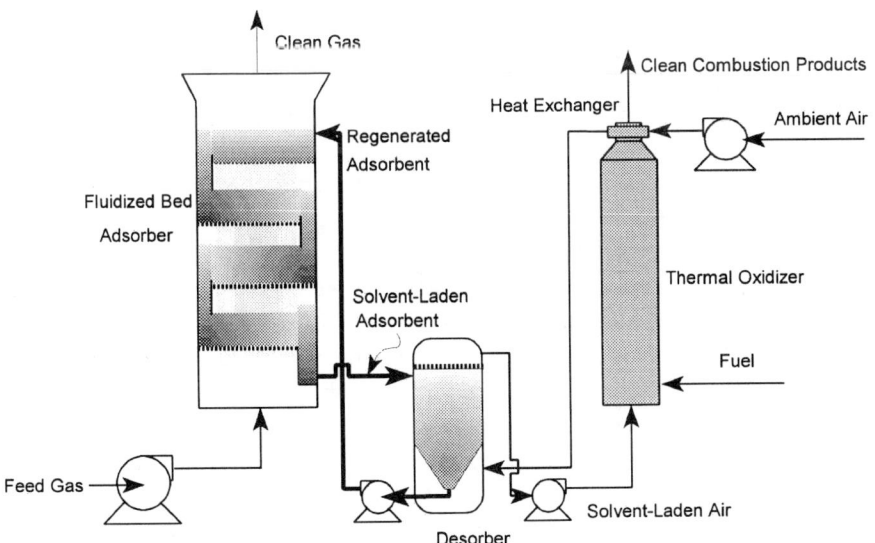

**Figure 2.9.** "Blizzard" Adsorption system for clean-up of VOC-laden gas, by On-Demand Environmental Systems, Inc.

uppermost adsorber tray and overflows to subsequent trays. The VOC-laden gas stream first contacts the adsorbent in the lowest tray of the adsorption tower. The tower includes an entrainment zone above the uppermost tray to minimize loss of adsorbent in the off-gas. The adsorbent in the lowest tray also overflows, and is conveyed to the top of a separate unit that is the desorber tower. The desorber operates with heated purge gas, and the regenerated polymer beads are returned to the adsorber tower. Indirect heat may be applied to assist the purge during desorption. The goal of this process is to obtain a reject stream that will have a concentration of 10 to 50 times that of the original adsorber inlet.

Another approach is to use a moving adsorbent bed, one embodiment being the rotary wheel adsorbers commonly used for VOC removal, up to about 1000 ppm from large volumes (5000–30000 scfm) of low pressure air streams. These wheels can reduce original volumes by 90–95% and thereby concentrate organics by a factor of 10–20 times. The concentrate from the wheel is generally fed to a thermal oxidizer or other type of destruction unit. Figure 2.10, which is oversimplified, shows that as the wheel rotates, the adsorbent is alternately contacted by the feed gas and by a regeneration gas at an elevated temperature. Rotary wheels rotate slowly—about two revolutions per hour. In a typical application, the volume of regeneration gas is about 10% of the feed and thus feed components are concentrated by a factor of 10. About 75-90% of the wheel is used for adsorption while the remainder is regenerated. The desorption section is isolated from the rest of the rotor by precision seals. Not only do wheels offer a degree of process simplicity through the use of only one unit, but also the cycle times are considerably shorter than for traditional fixed-bed TSA processes. Instead of cycle times of several hours or more for traditional processes, wheels can have regeneration times of only 5–10 minutes. Wheels are made with diameters in excess of 13 feet and are available with carbon, silicalite, alumina, and silica gel. In some cases, two adsorbents, such as carbon and silicalite, are used on the same wheel. The carbon adsorbs hydrocarbon solvents such as xylene, while silicalite adsorbs low molecular weight oxygenates such as methanol (Hussey and Gupta, 1997).

A complication associated with adsorbent wheels is that there can be leakage of process gas into the regeneration section and of regeneration gas into the adsorption section. This cross contamination can limit practical removals that are lower than that which can be achieved with fixed-bed processes. In addition, the mass-transfer rate per unit length of bed traversed is not as high as that in a fixed-bed of adsorbent. These factors combine to limit wheel-based removals to 95–97%.

Yet another approach is the "simulated counter-current" system, exemplified by UOP's Sorbex process, in which the adsorbent is contained in a fixed bed, divided into multiple elements, and effective counter-current contact is achieved

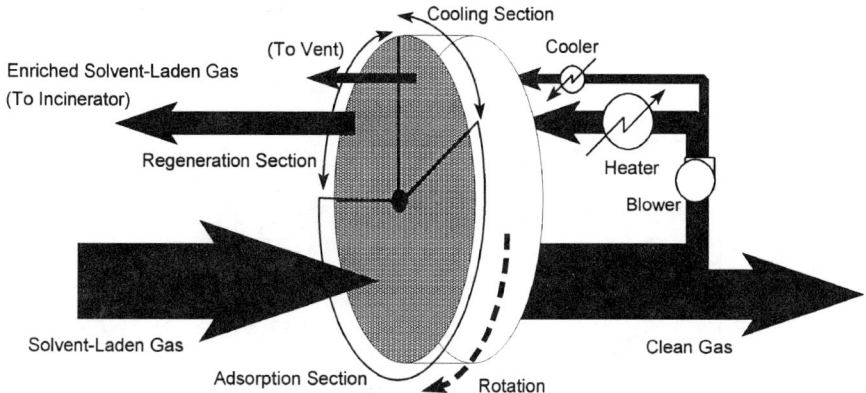

**Figure 2.10.** Adsorbent coated monolith (honeycomb) rotor for cleanup of solvent-laden gas (or for dehumidification).

by advancing the inlet and outlet lines by one stage in the direction of flow at fixed time intervals. A schematic of this type of system is shown in Figure 2.11. Other companies such as US Filter/IWT have also commercialized SMB processes, and their applications are similar in nature to UOP's. Such systems have proved highly effective for carrying out "difficult" bulk liquid separations, i.e., separations where the selectivity is limited but the capital cost is high so that such processes are generally justified economically only for high "value added" separations rather than waste reduction. The best known example is the separation of xylene isomers and ethyl benzene to produce pure para xylene (the Parex process) but this approach is being increasingly applied in the pharmaceutical field, especially for resolution of chiral isomers. Bulk separations that are possibilities for SMB processes include:

- *Separation of close-boiling isomers.* Such separations can be made on the basis of differences in molecular geometry (the MOLEX process, for example, can separate straight-chain paraffins from others that need to have no more than a methyl branch) or polarity (the PAREX process for example).
- *Separation of materials that differ substantially in polarity.* This can be seen in the OLEX process, which takes advantage of the extra polarity added by one double bond to discriminate between long-chain paraffins and olefins.
- *Separation of very-high-boiling materials.* The SAREX process separates essentially nonvolatile sugars, while the CITREX process separates high-boiling, fermentation-derived citric acid from other high-boiling components of the broth.

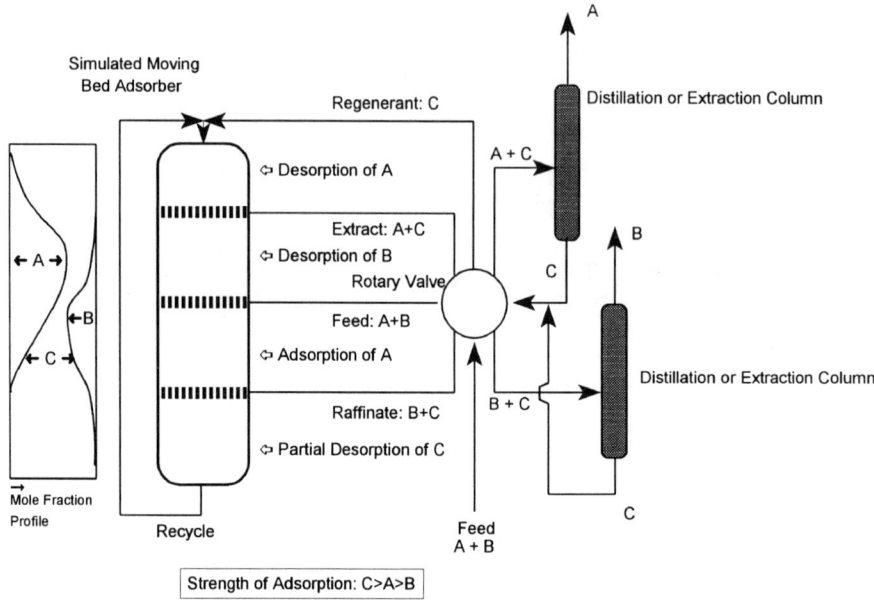

**Figure 2.11.** Simulated moving-bed adsorption system. After UOP Sorbex Concept.

- *Separation of relatively expensive materials.* It is not likely that SMB processes will be commonly used to separate, for example, azeotropic mixtures of water and low-cost organics, or waste streams from which various species must be removed. The investment and operating-cost implications are just too great for such separations to bear. On the other hand, as the value of a product increases, SMB technology becomes a more realistic option. Separations of products in the flavors and fragrances area, the personal-care area, the pharmaceutical and biochemical areas, the electronic-chemicals area and a number of others should be investigated for resolution by SMB technology if the separations are inherently difficult by other means.

A more specific listing of the various types and number of commercial applications of the Sorbex process are given in Table 2.5 while other separations demonstrated by UOP at the pilot and laboratory scales are included in Table 2.6.

### 2.2.2.5. Pressure Drop / Cost Issues—Packed Bed Problems

The standard adsorbent contactor is a randomly packed bed in which the adsorbent is in the form of pellets, granules, or beads of a convenient size. The size and shape are chosen to provide a suitable compromise between pressure drop and

**Table 2.5.** Commercial Separations with UOP Sorbex Technology

| Process Name | Separation | No. of Units |
|---|---|---|
| PAREX | $p$-Xylene/C8 Aromatics | 53 |
| MOLEX | $n$-Paraffins/$i$-paraffins, $C_{10}$–$C_{14}$ | 24 |
|  | $C_4$–$C_6$ | 9 |
| OLEX | Olefins/paraffins | 6 |
| CYMEX | $p$- or $m$-Cymene | 1 |
| CRESEX | $p$- or $m$-Cresol | 1 |
| SAREX | Fructose/glucose | 5 |
| CITREX | Citric acid/fermentation broth | 1 |
|  |  | 100 |

mass transfer resistance. Packed beds are cheap and versatile, but pressure drop tends to be high leading to high operating costs, and adsorbent inventory tends to be large with high associated capital costs.

Pressure drop is rarely a dominant economic problem, except for the largest systems. Generally, those are conducted in pressure vessels for which the cost is proportional to volume, within some narrow constraints on length-to-diameter ratio ($L/D$). In order to minimize pressure drop, the cross-sectional area must be increased, which leads to a small $L/D$. On the other hand, the required pressure rating and vessel diameter strongly affect the vessel wall thickness, which, along with the volume, dictates the vessel cost and, it is generally better to have a large $L/D$. Unfortunately, few general approaches exist for choosing such design parameters. In most large-scale systems, a fairly tedious optimization is required. In small systems or high value-added systems, pressure drop is relatively less important than performance in terms of separation efficiency.

Another approach to avoid high pressure drop is to use alternate flow geometries to the conventional axial flow, fixed bed. Magnetically stabilized fluidized bed systems are a special case of this. They rely on either imbuing a magnetic particle with adsorptive characteristics, for example, adding an adsorbent coating to a magnetic particle, or vice versa.

*2.2.2.5.1 Parallel Passage / Monolith Contactors*
It is easy to show theoretically that a parallel passage contactor coated with a thin layer of adsorbent can yield a substantially higher efficiency in terms of pressure drop per theoretical stage compared with a conventional packed bed. The device might consist of a monolith or parallel sheets, evenly spaced with small gaps

**Table 2.6.** Separations Demonstrated with UOP Sorbex Process (Adapted from UOP Literature)

**Hydrocarbon Separations**
  $m$-Xylene/$C_8$ aromatics
  Ethylbenzene/$C_8$ aromatics
  $o$-Xylene/$C_8$ aromatics
  $i$-$C_6$ olefins/n-$C_6$ aromatics
  1, 3-Butadiene/$C_4$ hydrocarbons
  Indene/alkyl aromatics
  b-Pinene/a-pinene
  3, 5-Diethyltoluene/diethyltoluene isomers
  2, 6-Diethyltoluene/diethyltoluene isomers
  Isoprene/C5 naphtha and gas oil cracking fractions
  1, 3-Butadiene/butadiene isomers
  $p$-Ethyltoluene/ethyltoluene isomers
  $o$-Diethylbenzene/diethylbenzene isomers

**Industrial Chemical Separations**
  $p$-Chloronitrobenzene/other chloronitrobenzene isomers
  $o$-Chloronitrobenzene/other chloronitrobenzene isomers
  2, 6-Toluene diisocyanate/toluene diisocyanate isomers
  2, 4-Toluene diisocyanate/toluene diisocyanate isomers
  4, 4-Dichlorodiphenylsulfone/dichlorodiphenyl sulfone isomers
  Hydroxyparaffinic dicarboxylic acids/olefinic dicarboxylic acids
  Dihydroxybenzene isomers
  Coumarone/indene
  $o$-Nitrotoluene/$p$-nitrotoluene
  $p$-Toluidine/toluidine
  Picoline isomers
  Nitrobenzaldehyde isomers
  2, 5-Dichlorotoluene/dichlorotoluene isomers
  $m$-Dichlorobenzene/dichlorobenzene isomers
  Toluenediamine isomers
  Methyl $p$-hydroxybenzoate/methyl $o$-hydroxybenzoate
  Thiophene, pyridine, phenol/naphtha

**Fatty Chemical Separations**
  Unsaturated fatty acid methyl esters/saturated fatty acid methyl esters
  Saturated fatty acids/unsaturated fatty acids
  Fatty acids/unsaponifiables
  Stearic acid/palmitic acid

**Table 2.6.** continued

| |
|---|
| **Fatty Chemical Separations** (continued) |
|   Oleic acid/linoleic acid |
|   Fatty acids/rosin acids |
|   Triglycerides by degree of unsaturation |
|   Diglycerides/Triglycerides |
|   Monoglycerides/triglycerides |
|   Monoglycerides/diglycerides |
|   Trans olefins/cis olefins |
|   Fatty acid esters/rosin acid esters |
| **Biochemical Separations** |
|   Lactic acid/fermentation broth |
|   Phenylalanine/fermentation broth |
|   Ethanol/water |
| **Carbohydrate Separations** |
|   Glucose/mannose |
|   Glucose/polysaccharides |
|   Sucrose/molasses |
|   Arabinose/pemtose pluse hexose |
|   Maltose/glucose |
|   Psicose/saccharides |
|   Monosaccharides/disaccharides |

between the sheets. Such contactors are obviously more capital intensive than the traditional packed bed, though they can offer substantially reduced adsorbent inventories. Nevertheless, where flow rates are high and value added is small, the energy cost associated with pressure drop becomes important, leading to a substantial advantage for parallel passage or monolith contactors. This situation occurs commonly with dilute effluent streams such as in VOC removal and in modern VOC adsorbers a parallel passage contactor is often arranged in a rotary system to allow continuous thermal regeneration (see Figure 2.10).

### 2.2.3. Equilibrium and Mass Transfer Parameters for Adsorber Design

Regardless of the process choice, even the most elementary adsorber design procedure uses crude estimates for equilibrium capacity and mass transfer performance to estimate adsorber size. That is the "length of unused bed" or "length of the mass transfer zone" approach. Those methods ascribe relative amounts of adsorbent to complete exhaustion and to partial utilization (due to slow kinetics), respectively. The reasoning is that, at imminent breakthrough, a fraction of the bed (toward the feed end) has been fully loaded and the remainder (at the product end) is undergoing a gradual transition when the flow must be stopped to preserve product quality. For applications in which the feed composition and flow rate are constant, and the adsorption capacity does not deteriorate due to "poisons," this approach is fairly reliable. In other cases, however, fairly significant efforts are necessary to collect equilibrium and mass transfer data to permit accurate design.

For state-of-the-art industrial designs of systems to separate binary mixtures, there are two extremes of data required. First, for conceptual designs, it is sufficient to use shortcut methods that have a few empirical kinetic parameters or analytical methods that may have no empirical kinetic parameters. Both, however, would require equilibrium data, usually in the form of an isotherm equation for the adsorbable component(s). Somewhat more sophisticated design or simulation methods would typically incorporate an intraparticle diffusion and a film diffusion, and these require empirical rate coefficients, though sometimes they are lumped together as a single factor. Still more complex are methods that account for multiple adsorbable constituents and/or heat effects. The final degree of complexity is reached when dealing with unconventional flow geometry, axial dispersion, thermal disturbances, and the like. Again, empirical parameters are generally required to account for each of these effects, and the only way to assess them accurately is via tests with very close analogs or full-scale equipment.

One of the most overlooked needs of adsorption is the development of relevant mathematical models that have been developed for suitable boundary conditions and a data base that can be used to determine reliable transport coefficients. Mathematical models for adsorbers, in which composition, velocity, temperature, and pressure all vary to a significant extent, are typically solved using classic boundary conditions, such as those of Danckwerts or Wehner-Wilhelm. Unfortunately, their validity is questionable because they were derived for steady conditions, which are never fully legitimate for cyclic adsorption applications. In the same vein, reliable methods need to be developed to determine the film diffusion coefficients, axial dispersion coefficients, friction factors, and the like under these transient conditions. For example, Wakao and Kaguei (1982) have discussed the effect of intraparticle diffusion on axial dispersion under transient conditions with nonlinear equilibrium, but additional research is needed. The problem is that virtually all published correlations were obtained from experiments conducted under steady conditions, while adsorption by definition is practically never at steady-state. To remedy this, research should be conducted to explore the validity of established correlations for adsorption under cyclic operations.

## 2. Adsorption Technologies

## 2.2.4. Economic Viability of Adsorptive Separations

### 2.2.4.1. Capital Cost and Operating Cost vs. Other Separation Technologies

It is impossible to generalize about the costs of adsorptive separations, whether concerning alternate adsorbents, cycle options, or simply operating conditions. Likewise, it is practically impossible to make meaningful or reliable statements about the magnitudes relative to other technologies.

On the other hand, some generalizations can be made about the three significant cost categories: vessels, adsorbent, and energy. One of the common rules-of-thumb is that capital costs for chemical process equipment follow an exponential dependence on capacity, that is,

$$C_2 = C_1 \left(\frac{S_2}{S_1}\right)^n \tag{2.17}$$

where $C_i$ is the cost associated with the system having capacity $S_i$, and $n$ is an empirical index that is traditionally 0.6 for most chemical operations, but may be 0.7 to 0.8 for adsorptive systems. This is because the index for adsorbents is generally about 0.9 for purchases of 100 to 100,000 lb, while that for vessels, pipes, etc. is about 0.6. Frequently, the adsorbent cost is roughly equivalent to that of the other hardware. The main reason that generalization is senseless is that the vessel dimensions and the amount of adsorbent required are strongly dependent on the nature of the adsorption cycle, the operating conditions, and the adsorbent and adsorbate. All of those parameters and variables are widely different between different applications. In contrast, for membrane systems, the index, $n$, is closer to 1.0, indicating practically no economy of scale. This is because the cost of a membrane process is dominated by the cost of the membrane itself, which scales linearly with the membrane area. For adsorption processes, as for other process applications, the costs of pumps or compressors, piping, instrumentation, safety features, maintenance, engineering, installation, as well as those for ancillary equipment such as heat exchangers (e.g., for TSA systems), filters, etc., are by no means negligible. They are, however, not particularly high or low for adsorptive systems. Hence, they are ignored here.

Another major component of the total cost is the cost of the energy to drive the process. This depends on the nature of the cycle, for example, PSA, TSA, displacement, or nonregenerative. In the last instance, that is, for systems in which the exhausted adsorbent is removed and replaced, the direct energy cost is negligible and the major operating cost is associated with replacement of adsorbent. Cyclic processes, however, require energy input in some form to continue operation. For a TSA system, the energy requirement depends on the manner in which the temperature is driven (e.g., directly—with steam or a hot fluid, or indirectly—via a heat transfer medium or electric heater in the vessel wall or inserted as coils). The cost for TSA is proportional to the magnitude and duration of the temperature swing, which depends on the contaminant level and the nature of the isotherms. Generally, estimates are based on the amount of energy consumed per unit of contaminant removed (though sometimes it is more reliable to use a ratio based on the mass of adsorbent present). Typical values are listed in Table 2.7. Those values were compiled from a wide variety of studies on steam-driven TSA.

Conversely, for PSA the energy requirement depends on the availability of feed at an acceptably high pressure and on the necessity of pressures different from atmospheric, or what ever the ultimate low pressure destination requires. Otherwise, when compression or vacuum is required, the cost depends on the ratio of the absolute pressures.

### 2.2.5. Environmental Benefits and Challenges

Adsorption has potential for increased use as a separation technology to facilitate in-process pollution prevention. Whether this is achieved by separating and recy-

**Table 2.7.** Steaming Efficiencies

| Solvent | Steam Consumption (mass steam/mass solvent) |
|---|---|
| MEK | 20 |
| THF / Toluene | 5.2–6.5 |
| THF/ Toluene/MEK | 6.0 |
| Hexane | 0.8–11 |
| Toluene | 8.2–15 |
| Toluene/IPAc | 0.8–1.0 |
| 1,1,1 Trichloroethane | 3.4 |
| n-Butanol/MIBK | 4.0–5.0 |
| Unspecified | 1.9 |
| General Use | 1.0–10 |
| | (mass steam/mass carbon) |
| Vinyl Chloride | 0.00163 |
| General Use | 0.25–0.35 |

cling reactants and feedstocks, eliminating solvents from gas cleanup, selectively removing impurities from process streams, or other applications, adsorption could benefit the environment in a number of ways. However, selection of the adsorbent and process configuration (e.g., choice of a regenerative or nonregenerative process, displacement adsorption, temperature or pressure swing adsorption) has environmental considerations which need to be considered. The adsorbent and process selection determines how much energy and the size of the physical plant required to recover or destroy organics that are adsorbed.

Many industries use a substantial part of their total energy requirements for oxidation of organic emissions and may face the problem of low concentration, high volume VOC gas streams. For example, the 3M Company (Minneapolis, MN) reportedly uses 12% of its total energy expenditure for catalytic oxidation of gaseous streams containing VOCs (Maves, 1996). 3M is committed to eliminating VOCs from production but does not anticipate that this will occur until 2020. Until then, cyclic adsorption processes could potentially be used to concentrate these streams prior to oxidation (reducing energy requirements) or for possible economic recovery of the organics. Such designs are commercially available.

The choice of a regenerative or nonregenerative adsorption process is an important environmental decision. For a nonregenerative system, the loaded adsorbent may be classified as a hazardous waste for handling, transportation, disposal in a landfill, reactivation or incineration. In this case, no in-process pollution prevention has occurred since the contaminants have simply been transferred to a different phase and must be destroyed by reactivating the adsorbent. For regenerative processes, the method of regeneration must be carefully analyzed. If steam regeneration with solvent recovery is to be performed, an organic concentrated aqueous condensate will be collected that must be further treated in some manner or disposed. Innovative process schemes have included a "superloading" step where this concentrated aqueous stream is separated and treated using spent adsorbent prior to its regeneration. If solvent regeneration is used, the regenerant and desorbed organic must be separated in another process. For temperature or pressure swing adsorption, one big concern is the presence of noncondensible gases that contain VOCs. Thus, the environmental considerations of the regeneration method and process configuration must be analyzed on a case-by-case basis.

Adsorbent selection is closely tied to the decision whether to use a regenerative process or reactivation. If reactivation is used, an adsorbent with a very high adsorption capacity must be used. Moreover, onsite reactivation is not economically feasible unless 500,000 pounds or more of activated carbon is reactivated each year (Sontheimer et al., 1988) and this consideration may entail returning the adsorbent to the manufacturer for reactivation. In this case, the recovery of adsorbed compounds is not possible and this may expose the industry to hazardous waste shipping and handling by a second party. Presence of fouling agents in the feed stream can also cause premature bed replacement and adsorbent regeneration, reactivation, or disposal. Careful choice of the adsorbent media may significantly reduce fouling effects and increase bed life and reduce adsorbent disposal.

Another important environmental consideration in adsorbent choice is its reactivity with the feed stream. Reactivity of the adsorbent with the feed stream can potentially cause bed fires (e.g., in the case of VOC recovery in air streams), extreme corrosion problems, and failure of systems. This can especially be a problem when recovering chlorinated solvents with activated carbon made from natural materials (often containing transition metals) that typically exhibit some catalytic activity at the temperatures of thermal regeneration. A good choice of adsorbent and regeneration technique can minimize these problems. For example, carbon fiber adsorbents or polymeric adsorbents can be manufactured without transition metal impurities, thus virtually eliminating the catalytic oxidation and mineral acid production during thermal regeneration.

The choice of a fixed-bed or fluidized/moving bed system has environmental considerations. A fixed-bed system, since it typically requires at least two adsorbers, requires more adsorbent material than a fluidized or moving bed system. This larger amount of materials uses a larger amount of raw materials to produce and also creates more waste for disposal should the beds become exhausted or fouled by contaminants.

The implementation of adsorption processes require careful analysis of the environmental benefits and considerations on a case by case basis. The environmental implications of different adsorbates, adsorbents, bed configurations, and process operation combinations must be analyzed to simultaneously optimize the environmental benefits, safety and the process economics.

## 2.3. Progress toward Implementation: Incentives and Impediments, Research Needs

The preceding discussion indicates that there is great potential for in-process pollution prevention through current and future applications of adsorption. However, several technology impediments must be overcome for adsorption to fulfill its potential for in process pollution prevention. The research needs may broadly be categorized in three areas: (1) adsorbent material development, (2) adsorption process improvements, and (3) advances in engineering design information.

### 2.3.1. Adsorbent Material Development

Although several different types of adsorbents were discussed in Section 2.1.4, there are many advances to be made in adsorbent design and materials. The example given in the introduction provides an excellent illustration of this. Recovery of ethylene from compressor leak off streams in polyethylene plants is currently not possible due to the lack of a suitable adsorbent. The following identified research needs in adsorbent material development will help lead to pollution prevention through cleaner separations.

***Improved Adsorbents to Upgrade Low BTU Natural Gas.*** There are extensive deposits of natural gas that contain methane ($CH_4$) at varying concentration levels, diluted by nitrogen and/or $CO_2$. Present processes for $CO_2$ removal depend on either thermal swing adsorption or absorption in amine solution. Neither of these processes is very efficient. The possibility of developing an improved PSA process with high recovery of methane at pipeline purity (96%) is an attractive but elusive goal. Selective adsorbents for $CO_2/CH_4$ are available but they generally require low pressure (even vacuum) desorption, which does not appear economically attractive for large-scale operation. For nitrogen-rich gas, the problem is to develop an adsorbent with high selectivity for nitrogen relative to $CH_4$. (Most adsorbents adsorb $CH_4$ in preference to nitrogen but for an efficient process it is preferable to adsorb the nitrogen rather than the $CH_4$.)

***Nanofiber Adsorption Technology for High-Density Fuel Storage and Novel in-Process Separations.*** A challenge facing alternative fuel vehicles is the capacity of fuel storage devices that are small enough to be feasible for automotive use and of sufficient capacity to provide long driving distances between refueling. Some recent evidence suggests it may be possible to utilize graphite nanofibers to overcome this challenge for systems involving fuel cell technology and hydrogen fuel. Such technology may also provide means to build in-process adsorbers to reduce purge requirements, improve feed stream quality and remove contaminants leading to "cleaner" chemical processing. In addition, the technology should be investigated as a means to reduce volume and pressure requirements for transportation of commodity gases.

***Selective Adsorbents for Separation and Purification of Dilute Streams.*** Dilute separations are an important problem for the chemical processing industry as they are commonly encountered as waste contaminants, valuable products, or recyclable feedstocks. Stricter environmental regulations for gaseous and aqueous emissions are driving the need for improved separations of waste by-products, desired products, and recyclable feedstocks. Improved adsorbents with increased selectivity and shorter transport paths increase the opportunity for recovery and reuse of materials from dilute streams.

***Advancement of Processes for Tailoring Adsorbents.*** There are many options available for adsorbent materials as discussed in Section 2.1.4. It may be preferable to tailor the surface chemistry and pore structure of an available adsorbent to achieve a desired separation rather than finding a completely new adsorbent material. Techniques are needed to very accurately tailor the pore-size distribution of carbon adsorbents as well as surface chemistry and heterogeneity to meet a user-specified requirement.

## 2.3.2. Adsorption Process Improvements

Once an adsorbent is identified that is capable of exhibiting the required selectivity for a given separation, an efficient process of contacting the fluid and adsorbent must be identified. Although there are many different options for adsorption process operation (fixed bed, moving bed, thermal swing, pressure swing, etc.), many opportunities exist for process improvement. The following identified research needs in adsorption processes will identify some of these opportunities for improvement.

***Parallel Passage or Monolith Contactors.*** In most adsorption processes the gas–solid contactor is a randomly packed adsorbent bed. Such contactors are relatively cheap and have reasonably good mass transfer characteristics. However, the pressure drop is relatively high. Where the "value added" in the process is high, the cost of the power needed to overcome the pressure drop is insignificant, but in air conditioning and pollution control processes, which involve large air flow rates and low concentrations of the absorbable species, this is not the case. Theoretical considerations suggest that, in terms of the pressure drop per theoretical stage, a parallel passage contactor (monolith, honeycomb, etc.) can provide improved performance relative to a conventional packed bed and such contactors are indeed used in some VOC removal and air conditioning processes. However, further improvements are needed to reduce the dimensions of the gas channels, to increase the adsorbent density, to develop a means for coating highly porous adsorbents on the channels and to improve the regularity of the structure as well as to reduce the cost of fabrication.

***Radial Flow Contactors.*** Another promising configuration involves radial flow geometry that was designed to minimize dead-volume, because dead-volume can be disastrous in situations exhibiting low selectivity. Radial flow offers advantages in pressure drop and possibly for maintaining sharp fronts that are not available in ordinary axial flow systems. Advanced contactors would also be useful for rapid cycle PSA processes that are at present limited by the performance of the packed bed.

2. Adsorption Technologies

***Rotary Wheel Adsorbers.*** Rotary wheel adsorbers have greatly increased the feasibility of using TSA. This is primarily due to their process simplicity and the requirement of only one unit instead of at least two fixed beds typically required for TSA. They have also greatly reduced the cycle time associated with TSA processes. However, as stated earlier, a complication associated with adsorbent wheels is that there can be leakage of process gas into the regeneration section and of regeneration gas into the adsorption section. This cross contamination can limit practical removals that are lower than that which can be achieved with fixed-bed processes. In addition, the mass-transfer rate per unit length of bed traversed is not as high as that in a fixed-bed of adsorbent. These factors combine to limit wheel-based removals to 95–97%. Improvement of both the mass transfer characteristics of the rotary wheel adsorption beds and the precision seals that allow contamination of the adsorption section by the regeneration gas would allow even greater use of this technology.

***Novel Continuous Flow Contactors***. One of the main drawbacks of adsorption has always been that it is typically operated as a semibatch process. In contrast, countercurrent adsorbent flow and feed would be the ideal situation. Several such systems have been developed. However, they typically turn out to be more expensive than a simple cyclic batch system and generally prove economically competitive only for bulk separations. Rotary wheel adsorbers, "staged" fluidized beds (i.e., Blizzard™ and Polyad™ systems), and "simulated countercurrent" (i.e., Sorbex™) systems have all made advances toward the optimal countercurrent contacting scheme. However, a truly countercurrent contactor that is economically competitive with semibatch processes would result in greater use of adsorptive separations. Research is needed to identify a countercurrent contacting scheme such as this.

### 2.3.3. Advances in Engineering Design Information

Due to the complexity of adsorption, a large amount of information is required to completely evaluate the potential of using adsorption for a given separation. The information required includes as a minimum the fluid-solid equilibrium. Increasing degrees of complexity are added by accounting for mass transfer, fluid dispersion, heat effects, competitive adsorption, unconventional flow geometries, and the like. In the literature, many models for adsorption have been proposed. However, many needs for research in modeling of adsorption still exist.

***Single-Component Adsorption Equilibrium***. For the most simple analysis of adsorption, the adsorption equilibrium must be known. The adsorption equilibrium is typically determined by measuring an isotherm for the desired adsorent–adsorbate combination. This requires a good deal of effort and cost for simply screening a technology. Nanoscale models are needed to predict equilibrium parameters based on the molecular structure of adsorbents and their interactions with a potential adsorbate. Recent advances in computational chemistry may help facilitate such calculations. This would help to more readily evaluate an adsorbent for use in a particular separation as well as to identify characteristics an adsorbent needs to facilitate a given separation.

*Molecular Simulation of Adsorption.* Molecular simulation of adsorption has undergone many advances recently, somewhat due to recent advances in computing power. Molecular simulation still holds a great deal of potential for understanding adsorption and predicting adsorption equilibria based on *ab initio* calculations. Prediction of preferential adsorption using molecular simulation for the design of separation processes has hardly been explored and could have large benefits to industry. Using either *ab initio* calculations or potentials derived from single-gas adsorption isotherms, mixture adsorption (selectivity, loading, individual heats of adsorption) can be predicted without making any additional assumptions about intermolecular forces. Further development in this area will enable more accurate design of adsorption processes and facilitate design of adsorption processes for specific applications. The most important need for research is clearly the development of accurate force fields. The optimum research would be to develop a handbook of atom–atom potentials and charge distributions for molecules and zeolites that has been verified by comparison with experiment for selected thermodynamic and transport properties.

*Multicomponent Adsorption Modeling.* It would be valuable for design engineers to have a means to estimate multicomponent adsorption equilibria, including pressure and temperature effects, quickly enough to incorporate in computer models of adsorption systems. Currently, there is a plethora of thermodynamically valid, but mathematically cumbersome methods. Many of those are best suited for assessing equilibrium data, or at best for process simulations of noncyclic systems involving a few relatively similar adsorbates, at constant temperature, constant total pressure, or both. Needless to say, those assumptions are not valid in many, if any, adsorption systems that are of practical value.

*Transient Adsorption Process Modeling.* There are no universally accepted software packages for pressure swing adsorption (PSA) and temperature swing adsorption (TSA) that have been put in the public domain. There are very few that are commercially available, and for those that are, the inner workings are proprietary, so the only way to learn about them is to purchase or license them.

A good deal of research in modeling adsorption has been done as is evident in the reference list of this document. Many different models have been presented and verified with laboratory tests. However, very few have been verified with full-scale adsorption processes that can differ drastically from laboratory-scale adsorption processes. The different models that have been presented are often unique to individual situations and require a high degree of expertise in their correct application.

Projects to develop simulation packages are needed that could handle PSA, TSA, displacement, or just conventional adsorption in a fixed bed, with multicomponent capability, energy balance, pressure drop, dispersion, etc. Due to the large number of process options in adsorption, this simulation package should include heuristics to guide users who are not adsorption experts in choosing the applicable model options. These collaborative projects should involve industry, academia, and government agencies to maximize the value of the research to industry. An important distinction in this simulation package development is keeping the cost relatively low while having an adequate level of complexity involved. This is essential since many smaller companies could use adsorption beneficially but have no expert in-house and do not have a large computer simulation budget.

As part of the effort, and to increase its scientific merits, it would be appropriate to test the validity of many assumptions that are now taken for granted in several proposed models. For example, many proposed models incorporate the classic boundary conditions suggested by Danckwerts or Wehner and Wilhelm. Those boundary conditions were developed assuming steady-state, isothermal, isobaric conditions, none of which are necessarily valid for cyclical adsorption processes. Likewise, correlations of pressure drop may not follow the Ergun equation, especially under conditions of PSA or TSA (especially if the adsorbent is subject to swelling upon uptake). Again, that correlation was produced for isothermal, steady-state conditions. The proposed models often are verified by describing fixed-bed data where several model parameters are altered simultaneously. This leaves open the possibility of concluding that the assumed boundary conditions are correct when in fact, they may not.

## 2.4. Selected Emerging and Proven Nonreactive Uses of Adsorption

It is useful to break down adsorption applications into purifications and bulk separations. Descriptions of some key applications follow that are complemented by Table 2.8, which provides a summary of applications for both gas and liquid mixtures (Humphrey and Keller, 1997). Following the introduction of many applications for adsorption, more detailed descriptions are provided for applications that hold particular source reduction opportunities for industrial pollution prevention.

**Table 2.8.** Examples of Adsorption Separations

| Separation | Adsorbent |
|---|---|
| **Gases/Bulk Separations** | |
| Normal Paraffins/isoparaffins, Aromatics | Zeolite |
| $N_2/O_2$ | Zeolite |
| $O_2/N_2$ | Carbon molecular sieve |
| CO, $CH_4$, $CO_2$, $N_2$, A, $NH_3/H_2$ | Zeolite, activated carbon |
| Hydrocarbons/vent streams | Activated carbon |
| $H_2O$/ethanol | Zeolite |
| **Gases/Purifications** | |
| $H_2O$/olefin-containing cracked gas, natural gas, synthesis gas, air, etc. | Silica, alumina, zeolite |
| $CO_2/C_2H_4$, natural gas, etc. | Zeolite |
| Hydrocarbons, halogenated materials, solvents/vent streams | Activated carbon, silicalite, others |
| Sulfur compounds/natural gas, hydrogen, LPG, etc. | Zeolite |
| $SO_2$/vent streams | Zeolite |
| Hg/chlor-alkali cell gas effluent | Zeolite |
| Indoor air pollutants—VOCs | Activated carbon, silicalite |
| Tank-vent emissions/air or nitrogen | Activated carbon, silicalite |
| Odors/air | Silicalite, etc. |
| **Liquids/Bulk Separations** | |
| Normal paraffins/isoparaffins, aromatics | Zeolite |
| $p$-Xylene/$o$-xylene, $m$-xylene | Zeolite |
| Detergent-range olefins/paraffins | Zeolite |
| $p$-Diethyl benzene/isomer mixture | Zeolite |
| Fructose/glucose | Zeolite |
| Chromatographic analytical separations | Wide range of inorganic, polymeric, and affinity agents |

| Liquids/Purifications | |
|---|---|
| $H_2O$/organics, oxygenated organics, halogenated organics, etc.—Dehydration | Silica, alumina, zeolite, corn grits |
| Organics, oxygenated organics, halogenated organics, etc./$H_2O$—water purification | Activated carbon, silicate |
| Odor and taste bodies/$H_2O$ | Activated carbon |
| Sulfur compounds/organics | Zeolite, others |
| Decolorizing petroleum fractions, syrups, vegetable oils, etc. | Activated carbon |
| Various fermentation products/fermentor effluent | Activated carbon, affinity agents |
| Drug detoxification in the body | Activated carbon |

*Notes:*
Adsorbates are listed first.
Bulk separations: adsorbate concentration is greater than about 10 wt% in the feed.
Purifications: adsorbate concentration is generally less than about 3–5 wt% in the feed.

## 2.4.1 Gases/Bulk Separations

***n-Paraffin Separations.*** Linear paraffins are separated from branched chain hydrocarbons with 5Å molecular sieves. These sieves have pores that allow only linear paraffins to pass and molecules of larger cross section are excluded. *n*-Paraffins are used for production of biodegradable detergents, specialty solvents, plasticizers, fatty acids, synthetic protein. The normal paraffins are used for enhancement of the octane value of low boiling gasoline fractions. The pressure swing cycle is effective for low molecular weight *n*-paraffins, but not for higher boiling compounds because their equilibrium loading is insensitive to pressure variations. For the desorption of higher molecular weight compounds, displacement purge is the preferred cycle.

***Unsaturated Hydrocarbon Separations.*** Molecular sieves preferentially adsorb the most unsaturated hydrocarbons; thus, selective separation is based on degree of saturation of the molecules, such as separation of acetylene from ethylene or ethylene recovery from saturated hydrocarbons. An example is 5Å molecular sieves used for the separation of *n*-butylenes from isobutylene. Separations here are based on molecular size as in the case of *n*-paraffin separation.

***Nitrogen Production.*** A highly successful application is the marriage of carbon molecular sieves with the PSA cycle to produce nitrogen from air. Units are used to provide blanket gas and for other plant needs and applications. Thousands of units to produce nitrogen by this process are in commercial operation.

*Oxygen Production.* With the increasing demand for oxygen enriched air for pollution control and prevention, this application is growing. Molecular sieves (generally 5A sieves) are used to produce high purity oxygen from air or oxygen-rich air employing a pressure swing adsorption process. PSA units are carried on small ships which disperse the oxygen produced to contaminated waters to enhance oxidization of contaminants. Pressure swing has an economic advantage over the conventional cryogenic air separation process for on-site oxygen production at 25 tons/day and lower and is competitive up to 100 tons/day and higher.

### 2.4.2. Gases/Purifications

*Natural Gas Processing.* Adsorption processes are designed to purify or alter natural gas composition or to extract valuable by-products. They also encompass applications such as dehydration and removals of carbon dioxide and sulfur. Molecular sieve dehydration, carbon dioxide removal, and sweetening may be applied to any size gas stream and performance of sieves is not affected by the degree of saturation of feed. This application has significant pollution prevention implications because the use of adsorption can displace the use of solvent-based processes, which can themselves be sources of pollution.

*Natural Gas Drying.* The use of type 3A and 4A sieves in natural gas drying has expanded steadily in recent years. Molecular sieves are used for dehydration and cryogenic production of LNG and for ethane recovery from natural gas. Use of adsorption for this application and for the "sweetening" process and other gas drying processes that follow, can in some cases take the place of solvent-based processes where the solvents themselves become polluted and become sources of pollution.

*Natural Gas Sweetening.* The broad term "sweetening" has a variety of meanings. In gas treating plants hydrogen sulfide is removed from gas, which also contains appreciable amounts of carbon dioxide. Removal of mercaptans using 13X sieves, may also be involved in sweetening. When the gas contains lesser amounts of acid components, molecular sieves adsorb both water and the acid components, thus eliminating the need for dehydration, a final processing requirement in most liquid sweetening plants.

*Cracked Gas Drying.* Cracked gas can mean any complex gas mixture containing olefins produced for ethylene manufacture from a variety of feedstocks. Mixed-olefin streams are dried by a variety of desiccants including glycols, silica gels, aluminas, and 3A and 4A sieves.

*Air Drying.* There are many applications in which the presence of moisture in air is detrimental, causing metals to corrode or ice plugs to form in low temperature equipment. Air drying is accomplished using molecular sieves, silica gels, and activated alumina. For normal drying, silica gels and activated alumina are adequate; however, 3A and 4A molecular sieves are used for extreme drying, particularly when low inlet water concentrations are involved or when air streams are at a high temperature.

In cryogenic applications, water and carbon dioxide have to be removed prior to air liquefaction to avoid heat exchanger freeze-up. The 13X molecular sieves possess high adsorption capacities and a fast rate of carbon dioxide adsorption; they are ideally suited for simultaneous removal of moisture and carbon dioxide.

## 2.4.3. Adsorption for BTU Adjustment in Natural Gas

As "high quality" natural gas reserves are depleted (i.e., those that meet interstate pipeline standards, chiefly 95% methane), the nation will be forced to exploit those that are substandard. Of the current, domestic natural gas reserves, roughly 12% are estimated to exceed the limit for nitrogen, and an additional 13% exceed the limit for carbon dioxide. In many reservoirs, those contaminants may constitute 5 to 25% or more of the gas. Removing those constituents to meet pipeline standards generally requires an economical means for rejecting nitrogen and/or carbon dioxide. Until recently, that was only possible using cryogenic distillation for nitrogen or amine-absorption for carbon dioxide. Recent advances in adsorbents and cycles, however, have made it feasible to separate them via PSA because natural gas is normally available at relatively high pressure, which implies that it contains the energy required to perform the separation.

### 2.4.3.1. Technology Description

Natural gas upgrading to remove nitrogen via PSA has only been commercialized in this decade. Since nitrogen is the "light" or less strongly adsorbed component, and methane is the "heavy" or more strongly adsorbed component on virtually all adsorbents, the required cycle involves recompressing the methane to pipeline pressure. In fact, the cycle can employ an "equalization" or "rinse" step to recover methane, as described in Section 2.2.2.3.D. Systems that treat approximately 5 to 10 million std. cu. ft per day of natural gas containing 10% nitrogen are currently in operation by Nitrotec Energy Corp. The product contains less than 2% nitrogen and is recovered at about 97% of the feed methane. Their cycle employs a rinse step.

Research on cycles to split carbon dioxide from methane by PSA is currently underway. In contrast to nitrogen, carbon dioxide is usually more strongly adsorbed than is methane. Likewise, it is also known that carbon dioxide diffuses more quickly than methane, leading to the possibility of a kinetic separation. So in this case, equilibrium and kinetic selectivities are complementary. Thus, by either purely equilibrium-based PSA or equilibrium-kinetic-based PSA, it is possible to produce methane as the pure "light" product at high pressure. This provides a substantial economic incentive for this application.

### 2.4.3.2. Engineering, Economic, Environmental, and Energy Considerations

The principal engineering, economic, environmental, and energy concerns vary from case to case. For example, these revolve around the nature of the source: location, quantity, composition, value of the product, and process specifics. That is, if a source that is heavily contaminated by carbon dioxide or nitrogen is near a pipeline or potential consumer, there may be sufficient economic incentive to purify it. On the other hand, for a source that is not near a pipeline or potential consumer, given the market for natural gas, it would probably be impractical to purify it, given the added expense. Of course, the larger and purer the source, the greater the distance that may be economically covered with a new pipeline. Since costs increase as the percentage of contamination increases, there is an associated constraint to the economic viability of highly contaminated sources.

One of the troublesome issues of natural gas purification is dealing with sources that contain noxious components, such a hydrogen sulfide, mercaptans, etc. Currently, these highly contaminated sources are avoided when possible due to the added expense of controlling them in an environmentally acceptable manner. They may be removed by a "guard bed," though sizing those requires an extra degree of caution, not to mention trouble to ensure that they perform as designed. When those constituents are chemisorbed or catalytically decomposed, the adsorbent must be periodically replaced or the byproducts must be accommodated. Otherwise, if the noxious constituents are simply removed, they may be destroyed in a flare or released into the environment, if that is permissible.

For all that, as new adsorbents that have improved selectivity and kinetics become available, the inherent costs of purifying natural gas will decrease. Coupled with improved PSA cycles, plus additional technology to remove other, more strongly adsorbed or noxious components such as hydrocarbon liquids and sulfur-bearing compounds, the extent of purification that is economically feasible will increase.

Finally, it is recognized that gas hydrate reservoirs dwarf those of ordinary natural gas (Collett, 1997). In fact, the estimated ratio of hydrate sources to those of ordinary sources is nearly 35:1. These sources contain methane in a cage-like structure of water molecules, and are typically located beneath the oceans or permafrost, so extracting them is more difficult than land-based wells. Once extracted to the surface, adsorption may be used to separate the methane and water mixture.

## 2.4.3.3 Contacts, References, and Suggested Vendors

Nitrotec Energy Corp., 16360 Park Ten Place Drive, Houston, TX 281-398-3391

UOP Process Plants, Algonquin, Des Plaines, IL 847-391-3475

## 2.4.4. $H_2$ Separation from Low-Grade Refinery Gases

Hydrogen purification from refinery off-gases by PSA has been practiced for nearly thirty years. The early patents (Erdmann, 1941; Skarstrom, 1960; Skarstrom and Heilman, 1963; Hoke et al., 1964; Stark, 1966) laid much of the foundation for cycles that are still in operation. Perhaps the most enduring technology was suggested by Wagner (1969), since it is still in use today for small (1–100 SCFM) to moderate (100–5000 SCFM) sized units. Perhaps the most important recent developments have been in reaction engineering. Reforming units have been optimized for medium to large (>5000 SCFM) sized systems, while partial oxidation units have become very successful for small-scale units.

### 2.4.4.1 Technology Description

Following along the lines suggested by Wagner (1969), the basic PSA–hydrogen recovery cycle consists of the following steps:

1. *Feed*—The feed mixture is admitted under pressure to column 1. The contaminants are adsorbed and purified hydrogen is produced and withdrawn.
2. *Blowdown*—Concurrent blowdown from column 1 enters the product end of column 3. Then, column 1 undergoes countercurrent blowdown, the effluent being part of the byproduct.
3. *Purge*—Column 1 is purged (countercurrently to the feed) at atmospheric pressure, and the effluent is discharged as byproduct.
4. *Pressurization*—Part of the contents of column 3 are admitted to column 1, equalizing their pressures. Subsequently, part of the high-pressure product from column 2 is admitted to complete the pressurization.

### 2.4.4.2 Engineering, Economic, Environmental, and Energy Considerations

It is not possible to generalize much about the engineering, economic, environmental, and energy considerations of PSA-hydrogen recovery cycles. These depend heavily on the feed flow rate and composition, the supply pressure, and the desired product purity. A rule-of- thumb is that for large streams, it is generally more feasible to operate with several beds in parallel (e.g., as many as 12). On the contrary, for small streams, it would be typical to use only four parallel beds, though only two or three could be preferred to minimize capital cost and complexity.

By capturing hydrogen, it is basically diverting it from consumption as a fuel. There are no particular noxious effluents that are avoided. In terms of energy, the fuel value is less than the value as a chemical feedstock, though recovery for that purpose typically has virtually no associated energy cost.

### 2.4.4.3 Contacts, References, and Suggested Vendors

Air Products & Chemicals, 7201 Hamilton Blvd., Allentown, PA 800-851-8021.
BOC Process Plants, 460 Mountain Ave., New Providence, NJ 908-464-8100.
Questor Industries, 3650 Wesbrook Mall, Vancouver, BC Canada.
Phoenix Gas Systems, 3925 Vernon St., Long Beach, CA 90815 562-597-2442.
Praxair Inc., 200 Colorado Ave., Buffalo, NY 716-893-6706.

## 2.5. General Listing of Adsorbent and Adsorption Process Suppliers

The number of companies supplying adsorbents and adsorption processes is large and far too numerous to list all of them here. The companies listed below are among those with substantial businesses in the sale of adsorbents and adsorption processes. Much more extensive lists of companies can be found, for example, in (1) the *Thomas Register* under Adsorbents, Adsorbers, Adsorbers: Carbon, Dryers: Adsorptive, Filters: Activated Carbon, Filters: Adsorption, and Filters: Air; and (2) the *Chemical Week Annual Buyers' Guide* under Adsorbents, and Adsorption Systems. Although a few international companies are listed in Table 2.9, the coverage is more complete for U.S. companies.

**Table 2.9.** Suppliers of Adsorbents and Adsorption Processes

| Company | Address | Products |
|---|---|---|
| Advanced Separation Technologies, Inc. | 5315 Great Oak Drive Lakeland, FL 33801-3180 | ISEP continuous adsorption process |
| Air Products and Chemicals, Inc. | 7201 Hamilton Blvd. Allentown, PA 18195 | Carbon molecular sieves, PSA and VSA processes |
| Alcoa Industrial Chemicals Div. | P.O. Box 300 Bauxite, AR 72011 | Activated $Al_2O_3$, other $Al_2O_3$-based adsorbents |
| American Norit Co. | 420 Agmac Ave. Jacksonville, Fl 32205 | Activated carbon |

## 2. Adsorption Technologies

| Company | Address | Products |
|---|---|---|
| Artisan Industries | 73 Pond St.<br>Waltham, MA 02254 | Activated carbon-based processes |
| Atochem, Inc. | 266 Harristown Rd.<br>Glen Rock, NJ 07452 | Activated carbon |
| Barnebey & Sutcliffe Corp. | 835 N. Cassady Ave.<br>Columbus, OH 43216 | Activated carbon, activated carbon regeneration, activated carbon-based processes |
| Bergbau Forschung GmbH | Franz-Fischer-Weg<br>4300 Essen 13 (Kray)<br>Germany | Carbon molecular sieves (CMS) and CMS-based processes |
| Calgon Carbon Corp. | P. O. Box 717<br>Pittsburgh, PA 15230 | Activated carbon and activated carbon-based processes |
| Chematur Engineering AB | Box 430, S-691<br>27 Karlskoga<br>Sweden | Moving-bed adsorption processes |
| C. M. Kemp Mfg. Co. | 490 Baltimore-Annapolis Blvd.<br>Glen Burnie, MD 21061 | CMS, CMS-based processes, activated carbon and other adsorbent-based processes |
| Culligan International | 1 Culligan Parkway<br>Northbrook, IL 60082 | Activated carbon and activated carbon-based processes |
| Durr Industries, Inc. | 14492 Sheldon Road<br>Plymouth, MI 48170 | Wheel-based processes |
| Envirogen, Inc. | Princeton Res. Ctr.<br>4100 Quakerbridge Rd.<br>Lawrenceville, NJ 08648 | Biosorption processes |
| ICI Americas, Inc. | P.O. Box 15391<br>Wilmington, DE 19850 | Activated carbon |
| ICI Katalco | Two Transom Plaza Dr.<br>Oakbrook Terrace, IL 60181 | Irreversible adsorbents |
| LaRoche Chem. Co. | P.O. Box 1031<br>Baton Rouge, LA 70821 | Activated $Al_2O_3$ |
| Munters Zeol | Kalksteenscagen 1<br>S-22 378 Lund, Sweden | Adsorbent-wheel-based processes |
| Norton Co. | 60 E. 42nd St.<br>New York, NY 10017 | Activated $Al_2O_3$ |
| Progress Water Technologies, Inc. | P.O. Box 33042<br>St. Petersburg, FL 33733 | ISEP continuous adsorption processes |

| Company | Address | Products |
|---|---|---|
| Seibu Giken Co., Ltd. | 1043-5 Wada Sasaguri-Machi Kasuya-Gun, Fukuoka Japan/T811-24 | Adsorbent wheels and wheel-based processes |
| Seitetsu Kagaku Co., Ltd. | Sumitomo Bldg, No. 2 4-7-28, Kitahama Chuo-ku, Osaka 541 Japan | PSA processes |
| Tigg Corp. | P.O. Box 11661 Pittsburgh, PA 15228 | Activated carbon and activated carbon-based processes |
| UOP | 25 E. Algonquin Rd. Des Plaines, IL 60017 | ZMS, silicalite, and ZMS- and silicalite-based processes |
| U.S. Filter/IWT | 4669 Shepherd Trail Box 560 Rockford, IL 61105 | Various ZMS- and activated carbon-based processes |
| Westates Carbon | 2130 Leo Ave. Los Angeles, CA 90040 | Carbon, impregnated carbons, aluminas, solvent recovery, and regeneration |
| Westvaco Corp. | 299 Park Ave. New York, NY 10171 | Activated carbon |
| W. R. Grace & Co. (Davidson) | 5500 Chemical Blvd. Baltimore, MD 21226 | Activated $Al_2O_3$, $SiO_2$, ZMS |

## 2.6. References for Nonreactive Adsorption

Alther, G. R. (1996) Organoclays boost activated carbon efficiency. The National Environ. J., 6 (1), 38–44.

Brindley, G. M., Semples, R. E. (1977) Preparation and Properties of Some Hydroxy-Aluminum Beidellites. *Clay Minerals,* 12, 229.

Collett, T. S. (1997) Quoted in "ACS Meeting Briefs," *Chem. and Eng. News,* 75(18) 60.

Erdmann, K. (1941) Process for the Removal of Carbon Monoxide from Mixtures thereof with Hydrogen, U.S. Patent No. 2,254,799.

Goodboy, K. P., and Fleming, H. L. (1984) Trends in adsorption with aluminas. *Chem. Eng. Prog.,* 80(11), 63–68.

Hall, T. L. (July 21 and August 24, 1993) Personal communication to J. L. Humphrey, AWD Technologies, Inc., Houston, TX.

Harper, M., Purnell, C. J. (1990) Alkylammonium montmorillonites as adsorbents for organic vapors from air. Environ. Sci. Technol., 24(1), 55–60.

Hoke, R. C., Marsh, W. D., Bernstein, J., and Pramuk, F. S. (1964) Hydrogen Purification Process, U.S. Patent No. 3,141,748.

Humphrey, J. L., Keller, G. E. II (1997) *Separation Process Technology,* McGraw-Hill, New York, NY.

Humphrey, J.L., Seibert, F. A., and Goodpastor, C. V. (1991) Separation Technologies—Marketing Factors. U.S. DOE Report DOE/ID/12920-2.

ICI Katalco (1993) PURASPEC Purification Process, promotional literature.

Kärger, J., and Ruthven, D. M. (1992) *Diffusion in Zeolites.* Wiley, New York.

Maves, F. (1996) Verbal communication from 3M.

Meier, W. M., and D. H. Olson (1992) *Atlas of Zeolite Structure Types,* Butterworth-Heinemann, London.

Metropolis, N., A. W. Rosenbluth, M. N. Rosenbluth, A. H. Teller, and E. Teller (1953) *J.Chem.Phys.* **21**, 1087–1092.

Myers, A .L., Prausnitz, J. M. (1965) Thermodynamics of mixed-gas adsorption. AIChE J., 9(1), 121–133.

Pezolt, D. J., Collick, S. J., Johnson, H. A., and Robbins, L. A. (November 10, 1996) "Pressure Swing Adsorption for VOC Recovery at Gasoline Loading Terminals," paper presented at the American Institute of Chemical Engineers Annual Meeting, Chicago, IL.

Robbins, L. A. (July 14, 1993) Personal communication to J. L. Humphrey, Dow Chemical Co., Midland, MI.

Robbins, L. A. (March 26, 1997) Personal communication to J. L. Humphrey, Dow Chemical Co., Midland, MI.

Savitz, S., Siperstein, F. Gorte, R. J., and Myers, A. L. (1998) Calorimetric Study of Adsorption of Alkanes in High-Silica Zeolites, *J.Phys.Chem.,* in press

Siperstein, F. and Myers, A.L. (1998) "Effect of Temperature on Adsorption Equilibria of Gas Mixtures," *Fundamentals of Adsorption VI,* Giens, France, May.

Skarstrom, C. W. (1960) Method and Apparatus for Fractionating Gaseous Mixtures by Adsorption, U.S. Patent No. 2,944,627.

Skarstrom, C. W. and Heilman, W. O. (1963) Technique with the Fractionation of Separation of Components in a Gaseous Feed Stream, U.S. Patent No. 3,086,339.

Stark, T. M. (1966) Gas Separation by Adsorption Process, U.S. Patent No. 3,252,268.

Sontheimer, H., J.C. Crittenden & S. Summers (1988) *Activated Carbon for Water Treatment.* Second Ed., DVGW-Forschungssttelle, FRG.

Valenzuela, D.P. and A.L. Myers (1989) *Adsorption Equilibrium Data Handbook,* Prentice-Hall, Englewood Cliffs, NJ.

Wagner, J. L. (1969) Selective Adsorption Process, U.S. Patent No. 3,430,418.

Wakao, N. and Kaguei, S. (1982) *Heat and Mass Transfer in Packed Beds,* Ch. 2, Gordon and Breach Science Publishers, New York

Yang, R.T., Baksh, M.S.A. (1991) Pillared clays as a new class of sorbents for separation. AIChE J., 37(5), 679-684.

Yon and Turnock (1971) Multicomponent adsorption equilibria on molecular sieves. AIChE Symp. Ser., 67(117), 75-83.

## 2.7. Adsorptive Chemical Reactors

### 2.7.1. Introduction

Recently, there has been considerable interest in chemical processing where chemical reaction and separation are carried out simultaneously in combined reactor separators. Examples of separative reactors are reactive distillation, membrane reactors and adsorptive reactors, in which distillation, mass transfer through a porous membrane or selective adsorption on a solid cause separation to occur simultaneously during reaction. When chemical reaction and separation occur in concert, the requirements for downstream processing may be eliminated or greatly reduced, resulting in less complex and less costly chemical plants. A significant advantage of integrated reaction and separation operations is the possible reduction in capital expenditure.

The combined separation of reactants and products provides other advantages. Unfavorable chemical equilibria can be shifted to enhance reactant conversions and desired product yields, reducing recycle streams. Unwanted reaction products can be reduced or eliminated because contacting of reactants with products can be minimized. Furthermore, when dealing with reaction networks, it may be possible to optimize the selectivity for a desired, intermediate product. There would not only be economic benefits from realization of these goals, but environmental benefits from reduced waste streams, lower $CO_2$ emissions and energy requirements. With the increasing emphasis upon environmental issues, integrated reaction separation processes will be of increasing importance. Membrane reactors are the subject of other chapters in this book, while adsorptive reactors are discussed in this chapter.

In adsorptive reactors chemical reactions are carried out in the presence of a solid adsorbent capable of selective adsorption of the components of the reaction mixture. Differences in adsorption selectivity cause separation to occur during the course of the reaction. The reaction may be catalyzed by the adsorbent or by a catalyst that is added to the adsorbent, or the reaction may occur homogeneously in the fluid phase.

There are several ways in which adsorptive reactors have been configured. It has been known for more than three decades that if a chemical reaction occurs in a chromatographic column following pulsed introduction of reactant(s), reaction and separation occur simultaneously in a process called reaction chromatography. Recent developments have extended this concept to continuous flow processing, overcoming the essentially batch nature of reaction chromatography, and the low throughput, which is the principle objection to practical applications. Continuous flow reaction chromatography has been demonstrated in rotating cylindrical annulus, countercurrent moving bed and simulated countercurrent moving bed configurations. Reactive separations have also been carried out in pressure swing adsorbers and in trickle bed reactors.

If an equilibrium limited chemical reaction is carried out in an adsorptive reactor, the separation can shift the equilibrium, increasing the conversion of reactant(s). For example, in a reaction of the type $A \Leftrightarrow B + C$, separation of B from C suppresses the reverse reaction and the conversion of A can exceed the equilibrium limit that would be the maximum obtainable in a nonseparative reactor. Similar conclusions are reached for other reversible reaction types. Under favorable circumstances conversions closely approaching unity can be obtained, and the recycle that would otherwise be necessary can be avoided or greatly reduced. This reduction of process complexity has clear economic implications, both in lower capital costs and in lower energy requirements. Adsorptive separations are less energy intensive than many other types of separations, such as distillation.

Adsorptive reactors also provide more flexibility in temperature optimization than do nonseparative reactors. For reversible endothermic reactions the increase of equilibrium conversion with increasing temperature means that nonseparative reactors should be run at high temperature to avoid or minimize recycle. If equilibrium is circumvented by means of an adsorptive reactor, however, high conversion can be obtained at lower temperature, giving further savings on process energy requirements, and most likely reducing the rate of catalyst deactivation. There will be an optimum temperature depending upon the reaction rate and the separation achieved. In nonseparative reactors reversible exothermic reactions should be run at low temperature for high conversion, but attaining adequate reaction rates may dictate higher temperature operation, resulting in lower conversion. Adsorptive reactors permit optimization of the temperature without loss of conversion and without need for recycle.

A central issue in adsorptive reactors is adsorbent selection. The adsorbent should be stable, should maintain enough selectivity to accomplish the separation at the reaction temperature, and should preferably have a well-behaved adsorption isotherm. In particular, the distribution of adsorption energies should be narrow and relatively uniform. If high energy surface sites are present it will be difficult to regenerate the adsorbent, and reactor performance will be adversely affected. The desired product will contain impurities and conversion will suffer. There is extensive literature on adsorbents, coming from years of development of commercial adsorptive separations and from analytical gas and liquid chromatography. Although the vast majority of adsorption data has been collected at temperatures lower than those typically employed in adsorptive reactors, these data can still be used as a first-pass screen for potential adsorbents. Conventional solid, porous adsorbents are available in numerous materials ranging from inorganics such as alumina and zeolites to porous polymer beads. Various pore sizes are available for some materials. In addition, there are many liquid stationary phases available for coating on inert solids. These have been utilized mostly in analytical applications, and do not seem to have been much investigated for commercial-scale separations. Some zeolites that are shape or size selective have been developed, as well as shape selective zeolite catalysts. With such a wealth of adsorbents, there is considerable scope and flexibility in adsorbent selection for adsorptive reactors.

### 2.7.2. Reaction Chromatography

Since the beginnings of gas chromatography in the 1950s there have been reports of chemical reactions associated with analytical applications of chromatography. Early work on pre- or post-analytical column chemical reactions facilitated analytical measurements (Beroza and Coad, 1967). This gave way to the development of catalytic microreactors with chromatographic analysis of the reaction product mixtures (Kokes et al., 1955), and to integration of the catalytic reactor with the chromatographic analysis (Basset & Habgood, 1960). This is termed reaction chromatography, whereby reaction and separation occur in concert. In its original conception, a pulse of reactant(s) is injected into a column packed with a chromatographic adsorbent and possibly a solid catalyst. Chemical reaction occurs as reactants and products are swept through the column by an inert mobile phase. Much of the research on reaction chromatography has been concerned with the determination of reaction kinetics and reaction rate coefficients, as well as numerous other physicochemical measurements (Paryjczak, 1986).

It has been pointed out that use of chromatographic columns as reactors for preparative purposes has been limited in spite of potential usefulness for laboratory-scale work (Coca et al., 1993). The pulsed input mode of column reaction chromatography is an inherently batch process that is unattractive for larger-scale processes. However, if relative motion between the feed and the bed can be arranged, a continuous flow process for simultaneous reaction and separation can be realized. This holds potential for preparative work or commercial-scale operations.

The continuous flow reaction chromatographs reported in the literature are derived from developments in continuous flow separations by chromatographic means. These have generally been achieved by arranging for relative motion between the feed stream and the packed bed. First among these, the rotating cylindrical annulus was given conceptual birth in 1949 (Martin, 1949), theoretical analysis in 1962 (Giddings, 1962), and experimental demonstrations in later years. These developments have been recently reviewed (Bridges and Barker,

## 2. Adsorption Technologies

1993; Goto and Takahashi, 1993). Another arrangement, the countercurrent moving bed, was commercialized for light hydrocarbon separations in the Hypersorption process of the 1940s (Berg, 1946), but was soon supplanted by the more economical low temperature distillation. Subsequent research on countercurrent moving bed separations appears to have been confined to theoretical and laboratory-scale experimental work. A summary of this work is given by Fish et al. (1989). On the other hand, the simulated countercurrent moving bed has been extensively researched (Ganetsos and Barker, 1993), and has been developed on a commercial scale by UOP (Broughton, 1984). These three configurations have all been shown to be effective for reaction chromatography. The rotating cylindrical annulus and the countercurrent moving bed have recently been reviewed (Carr, 1993) and are only briefly described below. The simulated countercurrent moving bed chromatographic reactor, developed more recently and of greater interest for practical processes, is discussed in more detail here.

### 2.7.3. The Rotating Cylindrical Annulus Chromatographic Reactor

In the rotating cylindrical annulus the solid adsorbent or the adsorbent and catalyst mixture occupies the annular space between two concentric cylinders. The mobile phase enters the bed uniformly distributed about the entrance to the annular region, while the feed stream, which is stationary and not rotating with the bed, is confined to a small sector. Rotation of the bed about its axis causes adsorbents to describe helical paths resulting from axial motion imparted by the mobile phase and circumferential motion imparted by adsorption on the solids. This arrangement is illustrated in Figure 2.12. Selective adsorption causes separa-

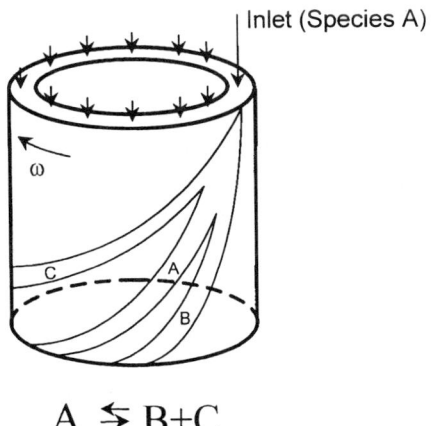

**Figure 2.12.** Rotating cylindrical annulus chromatographic reactor.

tion, since adsorbates that are most tenaciously held by the solid elute at larger angular rotation than those that are more weakly adsorbed. Product streams can be continuously collected at fixed locations about the annular exit. Alternatively, it is possible to hold the bed stationary and continuously rotate the feed and product locations. The latter arrangement is advantageous in larger scale operations, since rotating the entire assembly would not only be cumbersome, but would also require significant motive power to overcome the friction of sliding seals between the annulus and the headers. This is a lesser problem with the rotating feed, since the sliding seals would only encompass the drive shaft.

Investigation of the acid catalyzed hydrolysis of aqueous methyl formate in a rotating feed port cylindrical annulus demonstrated the efficacy of this configuration as a reaction chromatograph (Cho et al., 1980). This homogeneous reaction was carried out in the presence of activated charcoal, and good separation of the reaction products, methanol and formic acid, was achieved. At the feed conditions, 25°C and 3.0 $M$ methyl formate, the equilibrium conversion in a nonseparative reactor would have been 75%. Experiments revealed 100% conversion in the cylindrical annulus chromatographic reactor. The reactant, methyl formate, was not experimentally detected in the reactor effluent, probably because it was below the approximately 0.1% detection limit of the analytical method, and the apparent conversion was 100%. Furthermore, a mathematical model of the reactor predicted 99% conversion when kinetic and adsorption data from independent experiments were used in numerical simulations. To our knowledge, this was the first experimental demonstration of simultaneous separation and reaction in the rotating annulus configuration.

A gas–solid reaction, the dehydrogenation of cyclohexane to benzene over $Pt/Al_2O_3$ catalyst, has also been investigated in a cylindrical annulus (Wardwell et al., 1982). In this work, the bed was rotated under a stationary feedport. However, the reactor was incapable of complete conversion because dispersion prevented complete separation of reactant from products. Nevertheless, conversions greater than equilibrium were obtained. The dispersion may be resolved into components in the axial and circumferential directions. The circumferential component could be limited if the packed annular region were replaced with a circular array of packed columns, the walls of which would present a barrier to transport. This might provide improved performance for gas phase reactions.

### 2.7.4. The Countercurrent Moving Bed Chromatographic Reactor

An alternative continuous contacting configuration for reaction chromatography is the countercurrent moving bed chromatographic reactor (CMCR), where adsorbent and catalyst particles flow against a counterpropagating mobile phase.

## 2. Adsorption Technologies

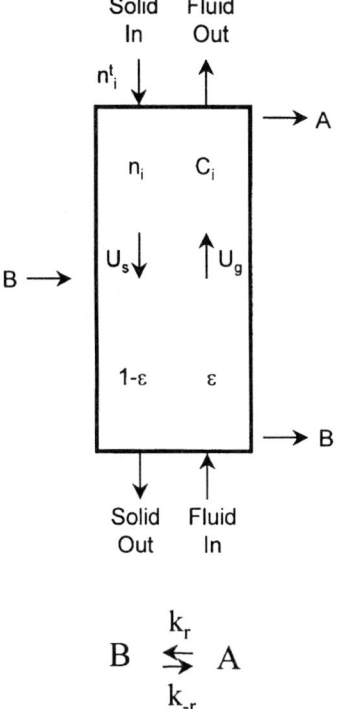

**Figure 2.13.** Countercurrent moving bed chromatrographic reactor.

A convenient arrangement, illustrated in Figure 2.13 is a vertical tube with controlled, gravity driven solids flow into which the reactant may be introduced at any point along the tube. As with the rotating annulus, the separation that occurs simultaneously with reaction can give a reaction product stream that is relatively uncontaminated with reactant(s) and other products, and permits chemical equilibrium to be shifted to favor a desired product. The solids and mobile phase flow rates may be adjusted so that the most strongly adsorbed species are carried down with the solids and those with less affinity for the adsorbent are carried up with the mobile phase. This can be seen with Eqs. (2.18), (2.19), and (2.20). The velocity of a concentration front ($V_f$) for a species adsorbing according to the Langmuir isotherm in a fixed bed is given by

$$V_f = \frac{U_g}{1 + \frac{(1-\varepsilon)NK}{\varepsilon(1+CK)}} \tag{2.18}$$

and in a countercurrent moving bed by

$$V_f = \frac{U_g(1-\sigma)}{1+\frac{(1-\varepsilon)NK}{\varepsilon(1+CK)}} \qquad (2.19)$$

$$\sigma = \frac{U_s}{U_g}\frac{1-\varepsilon}{\varepsilon}\frac{NK}{1+CK} \qquad (2.20)$$

Equation (2.20) shows how the parameter sigma depends upon the solid and mobile phase flow rates, $U_s$ and $U_g$, respectively. The value of sigma may be greater than or less than 1, but must be a positive value. According to Eq. (2.18) this determines whether $V_f$ is positive or negative, hence whether an adsorbate travels up or down the column.

In Eqs. (2.18), (2.19), and (2.20), $\varepsilon$ is the bed porosity, $N$ is the total concentration of surface sites, $K$ is the adsorption equilibrium constant, and $C$ is the mobile phase concentration of adsorbate. The CMCR was introduced by Viswanathan and Aris (1974) with the development of a mathematical model for an irreversible, first order chemical reaction occurring on the surface of a catalyst. It was predicted that this reactor is capable of driving the reaction to completion in a finite length of bed. Subsequent theoretical and experimental work has shown that reversible reactions can be taken to conversions that are substantially in excess of chemical equilibrium, with model predictions of nearly unit conversion under optimal conditions. High purity reaction product streams that are necessarily diluted with mobile phase are obtained. Theoretical considerations show that A $\leftrightarrow$ B reactions can be successfully carried out in a CMCR due to the development of steep concentration fronts that force the separation of A and B even when the reverse reaction rate is not slow compared to the forward rate (Cho et al., 1982). In contrast, the rotating annulus would not be effective in this situation. The theoretical predictions have been corroborated by experimental results on the catalytic hydrogenation of mesitylene (MES) to 1,3,5 trimethylcyclohexane (TMC). The product stream emerging from the CMCR only contained about 1% MES, and the conversion was in the vicinity of 90%, significantly greater than the approximately 50% that would have been obtained at equilibrium (Fish and Carr, 1989).

The flow of solids in the CMCR presents challenges for large-scale processing. Removal of fines formed by attrition, solids recycle, and nonuniform solids flow in large diameter columns all present difficulties that the design must take into account. An alternative configuration that circumvents these factors is the simulated countercurrent moving bed, which is discussed next.

## 2.7.5. The Simulated Countercurrent Moving Bed Chromatographic Reactor

To preserve the advantages of countercurrency while avoiding the problems associated with solids movement, simulated countercurrency can be implemented. This reactor is more readily scaleable than the CMCR, making it more attractive for commercial-scale operations. A simulated countercurrent moving bed reactor (SCMCR) may be configured as a fixed bed with several inlets and outlets located at intervals along the axial direction. A mobile phase flows through the entire column. The reactant stream is routed to the first inlet at the end where the mobile phase enters. After a period of time the reactant flow is shifted to the next inlet in line. Outlet positions are simultaneously shifted. The inlet and outlet flow switching is continued at regular intervals at a rate that is slower than the speed of a fluid element in the bed. The shift of inlet position in the direction of mobile phase flow looks, to an observer moving with the inlet position, like countercurrent movement of solids against the mobile phase, simulating countercurrent flow of fluid and solid phases without movement of the solid. Thus, the true countercurrent arrangement where the bed flows past the inlet is replaced in simulated countercurrency by moving the inlet past the bed. It should be noted that the apparent solids motion occurs in discrete steps. Figure 2.14 is a schematic showing subdivision of the packed bed into sections between adjacent inlets and outlets. Only one inlet and one outlet position are illustrated in the figure. The indicated direction of inlet and outlet movement is meant to suggest the other inlets and outlets along the column. This arrangement is conceptually similar to the UOP Sorbex processes, which have been commercially available for separa-

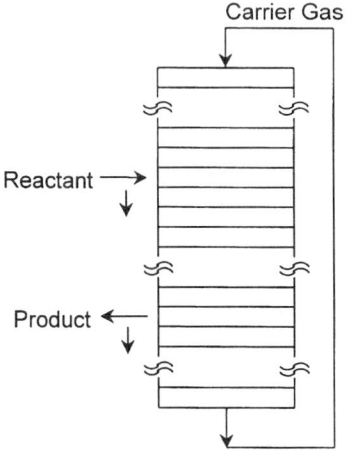

**Figure 2.14.** Simulated countercurrent moving bed chromatrographic reactor.

tions since the 1960s (Broughton, 1984). It is only recently that simulated countercurrent moving beds have been suggested for reactive separations (Ray et al., 1990).

The velocity of adsorbates in the simulated countercurrent moving bed can be described by Eq. (2.18) if $U_s$ is replaced with the pseudosolids velocity, $U_{ps} = L/t_s$, where $L$ is the distance between inlets and $t_s$ is the feed switching period. By analogy with the countercurrent moving bed, $t_s$ and the mobile phase flow rate can be manipulated so that some species in the reacting mixture will be carried ahead of the feed point, while others will lag behind it. Reaction products and unreacted reactants can be removed at certain outlets that must be advanced in position simultaneously with the inlet. When the feed point reaches the end of the bed, it is returned to the beginning and another cycle is started. As with the rotating annulus and the countercurrent moving bed, reaction and separation occur simultaneously, production of a high purity product (diluted by mobile phase) is possible, and chemical equilibria can be shifted to favor products.

There is an important distinction between the simulated countercurrent moving bed and the other two continuous chromatographic reactors discussed above. The SCMCR operates in a transient mode, rather than the steady state characteristic of the rotating annulus and the countercurrent moving bed, even though the feed flow rate is steady. This occurs because each advancement of feed position introduces a concentration disturbance that will decay toward the steady state, but may not attain it before the next switch. The simulated countercurrent moving bed will exhibit periodic concentration transients that propagate along the bed, and concentrations of species emerging from the outlets will have waveforms of regular period. The reactor operates in a mode that may be called a periodic steady state. This is discussed in more detail in Section 2.4.5.3.

### 2.7.5.1. Equilibrium Stage Model

A mathematical model of the SCMCR, with rate coefficients and adsorption equilibrium constants for the $Pt/Al_2O_3$ catalyzed hydrogenation of 1,3,5-trimethylbenzene vapor to 1,3,5-trimethylcyclohexane, has confirmed the performance expected of the reactor (Ray et al., 1994). This reaction was selected since with excess $H_2$ present it is an example of the reaction type A ↔ B, which was the subject of earlier theoretical investigations of the CMCR (Cho et al., 1982; Petroulas et al., 1985). It was also used in the experimental investigation of the CMCR (Fish and Carr, 1989). The SCMCR was modeled as a series of equilibrium stages with one stage associated with each of the 20 inlets and outlets.

## 2. Adsorption Technologies

The material balance equations for the fluid and solid phases of the isothermal model with first-order reversible kinetics are given by the following equations:

$$\varepsilon \frac{\partial C_i}{\partial t} + \varepsilon u j \frac{\partial C_i}{\partial x} + (1-\varepsilon)k_{ai}(N - n_A - n_B)C_i - (1-\varepsilon)k_{di}n_i = 0 \quad (2.21)$$

$$\frac{\partial n_i}{\partial t} - k_{ai}(N - n_A - n_B)C_i + k_{di}n_i + \alpha_i(k_f n_A - k_b n_B) = 0 \quad (2.22)$$

where $C$ is gas phase concentration, $N$ is total surface site concentration, $n$ is surface concentration, $k_a$ and $k_d$ are adsorption and desorption rate coefficients, $k_f$ and $k_b$ are forward and back reaction rate coefficients, $i$ is chemical species, and the $\alpha$ is a stoichiometric coefficient. Gas phase concentrations are related to surface concentrations by the Langmuir isotherm. The width of the stages provides a numerical simulation of axial dispersion. Feed is assumed to be uniformly mixed into the feed stage, and the model equations are integrated by a fourth-order Runge–Kutta method. The feed and product take off locations were advanced at $t/t_s = 1, 2, 3, \ldots$, where $t_s$ is the switching interval.

Model calculations of 190°C concentration profiles for a 400-cm reactor with 20 stages, $t_s = 5$ s, and 15% removal of the total gas phase flow 9 stages ahead of the feed stage are shown in Figures 2.15 to 2.17. The profiles presented are those existing an instant before the feed is advanced to the next stage. In Figure 2.15 the MES concentration profiles after 6, 12, and 60 switches of feed position (feed stages 7, 13, and 1) illustrate how the reactant moves through the reactor. A 15% make up feed is added to compensate for product removal. Thus the MES shows no decay due to reaction as $t/t_s$ increases. In contrast, Figure 2.16 shows TMC profiles increasing due to reaction over the same period. The initial transient due to start up has reached a periodic steady state by $t/t_s = 60$. Figure 2.17 displays the axial reactant and product distribution at $t/t_s = 60$, and predicts that TMC can be withdrawn in a stream that is only slightly contaminated with MES if the product is taken somewhere between four and nine stages ahead of the feed, which is on stage 1 of this figure. It was found that stage 9 gives the highest purity, 98.5%, calculated without accounting for the amount of mobile phase present. The corresponding MES conversion is 0.98. Trial calculations show that the conversion and purity depend upon the values of $\sigma$ for MES, and TMC, $t_s$, $U_{ps}$, $L$, fraction of flow withdrawn and inlet concentration. A range of operating conditions predicted conversions greater than 0.995 and purities greater than 99%. In comparison, the equilibrium conversion that would be obtained in a nonseparative reactor at 190°C is 0.62. Thus the theoretical analysis shows that for an equilibrium limited reaction the chemical reaction and adsorp-

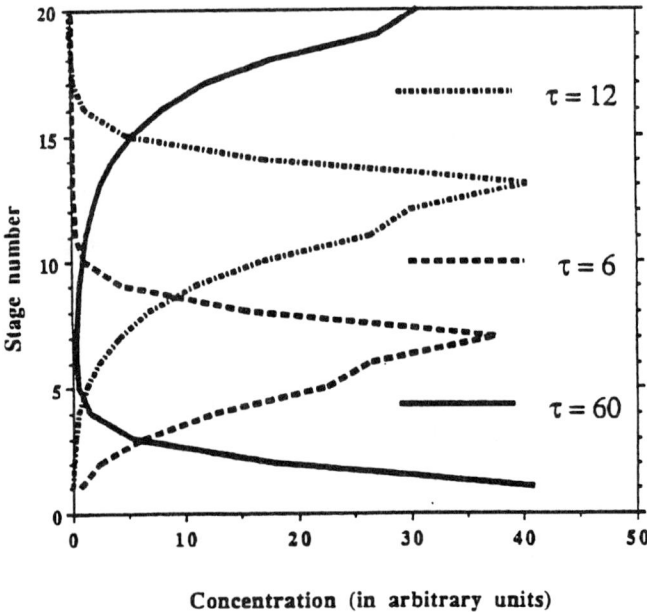

**Figure 2.15.** Model calculation of mesitylene concentrations in the SCMCR.

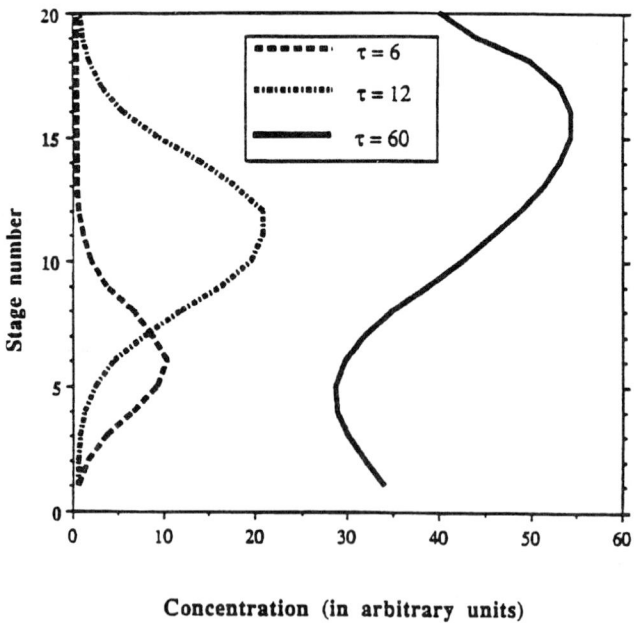

**Figure 2.16.** Model calculation of trimethylcyclohexane concentrations in the SCMCR.

## 2. Adsorption Technologies

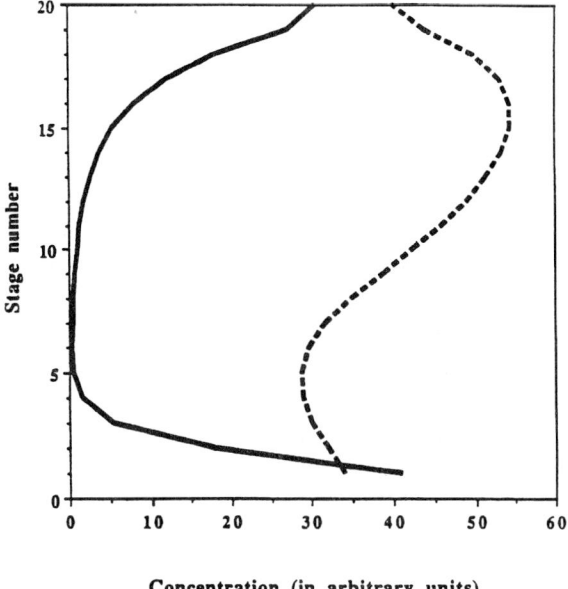

**Figure 2.17.** Model calculation of mesitylene (solid line) and trimethylcyclohexane concentration (dashed line) in the SCMCR.

tion on the solid surface interact to break the local chemical equilibrium. The SCMCR model predicts that it is possible to obtain improved conversions (closely approaching unity for some conditions), and a high purity product.

### 2.7.5.2. Multiple Column Configuration

An experimental investigation of the hydrogenation of MES was undertaken to test model predictions, to determine the effect of variables on performance, and to compare the SCMCR with the CMCR (Ray and Carr, 1995a). For laboratory experimentation a reactor consisting of packed columns in series is more convenient than the single fixed bed, multiple inlet/outlet configuration of the above model calculations. In the multiple column arrangement there is provision for feeding reactant or removing product between each column. Thus, each column corresponds to a section between adjacent inlets of the single fixed bed. An advantage here is the flexibility afforded since the number of columns (sections) can be readily increased or decreased as the demands of the experiment dictate. It is important to understand that the reactant enters a regenerated feed column as a square wave and exits with a waveform that depends upon the adsorption characteristics and fluid-phase dispersion. The concentration fronts are normally steep, and proper description of the reactor falls into the realm of adsorber dynamics.

A leading design consideration is the appropriate number of packed columns. The minimum number has been shown to be three columns for binary separations when the ratio of breakthrough times is 2, but the optimum number of columns is four (Fish et al., 1993). For the MES + 3H$_2$ = TMC reaction, preliminary model calculations indicated that five columns would be satisfactory. The SCMCR configuration for the experiments is shown in Figure 2.18. It was computer controlled to handle switching of reactant, product and N$_2$ carrier gas flows via solenoid valves, and rapid GC sampling and data analysis of the effluent streams.

The reactor is divided into two parts. In the first part, consisting of columns 1 and 2, the N$_2$ carrier gas and the reactants, a mixture of MES in excess H$_2$, are introduced. Since the product, TMC, is less strongly adsorbed than the reactant, it elutes from the feed column first. The TMC is collected until just before MES breaks through, at which time the feed and port A are switched to the next column in line. All streams, including both carriers and port B are simultaneously advanced to the next column at TMC breakthrough. The make up carrier gas introduced at the top of column 3 in Figure 2.18 sweeps out unreacted MES, preparing column 3 to receive feed. An elevated make up carrier flow rate can be used to remove as much MES as possible, minimizing contamination of the collected product. It was found that the Al$_2$O$_3$ surface has some high energy sites for MES adsorption, and as a consequence it was difficult to remove the last traces of MES from column 3.

Figure 2.19 shows experimental results at 473°K. The data points are compositions of the effluent from port A and port B as a function of time. The major component at each port clearly shows a repetitive waveform that is characteristic of a periodic steady state. The leading edges of the TMC undulations in Figure 2.19a correspond to breakthrough from successive columns every 4 minutes, the

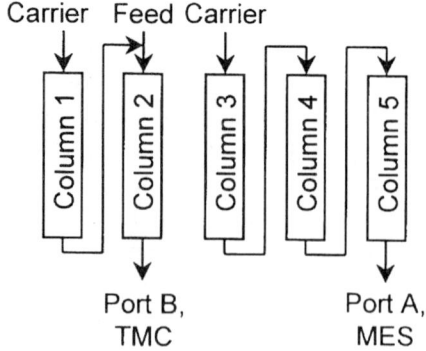

**Figure 2.18**. Multiple column configuration of the SCMCR for the hydrogenation of mesitylene.

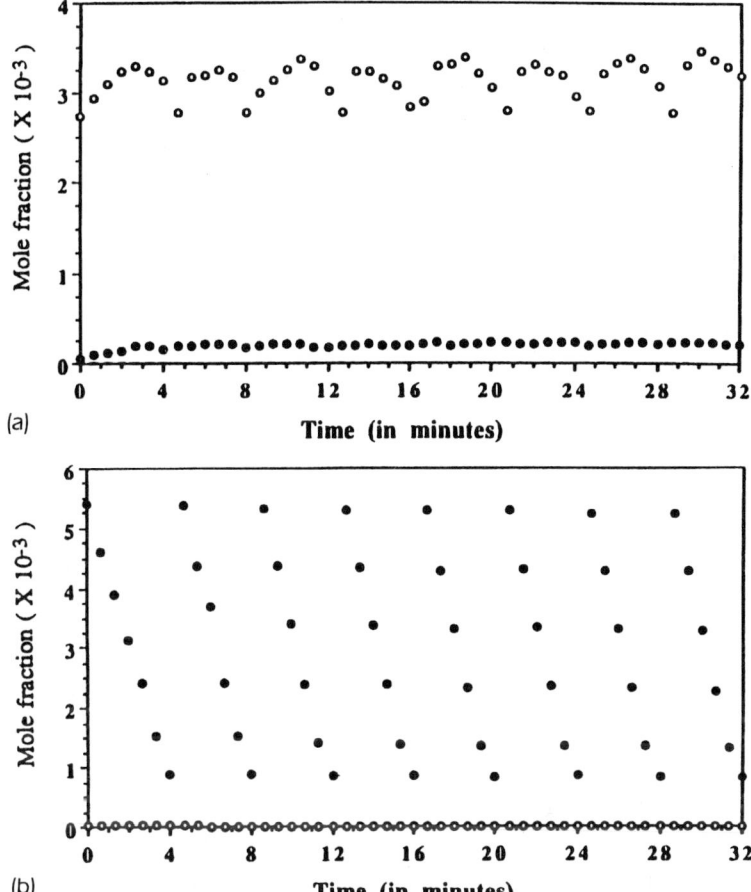

**Figure 2.19** Trimethylcyclohexane (open circles) and mesitylene (filled circles) data from the multiple column SCMCR for mesitylene hydrogenation. Top, Port B. Bottom, Port A.

switching time interval in this experiment. Note the background MES that is present due to incomplete purging. The MES waveforms in Figure 2.19b correspond to the desorption tail as this species is purged from the reactor by $N_2$. Only traces of TMC appear in the purge stream. These data illustrate the relative difficulty of desorbing MES compared with TMC. This is discussed further in the next section. The conversion, $X$, and purity, $Y$, were calculated by averaging MES and TMC over each 4-minute interval—$X_{MES} = n_{TMC}/(n_{TMC} + n_{MES}')$ where the data from both exits were used, and $Y_{TMC} = n_{TMC}/(n_{TMC} + n_{MES})$, where data from only the feed column effluent were used. The values obtained from this experiment were $X = 0.83$ and $Y = 0.96$. The conversion is smaller than model predictions for both the single column, multiple inlet/outlet SCMCR discussed above and the multiple column SCMCR discussed below. It is evident that this is due to loss of MES from the desorption tail in the purge stream. This might be remedied by modifying the regeneration procedure, employing a different adsorbent, or some combination of these. The importance of adsorbent selection and regeneration to achievement of good SCMCR performance cannot be overemphasized.

Table 2.10 gives a comparison of experimental performance and model predictions for MES hydrogenation in the CMCR and the SCMCR. The models predict nearly unit conversion and a nearly MES free product stream from a reaction that would have given an equilibrium conversion of only about 50% at the reactor operating conditions. The experiments corroborate the model predictions, although the observed performance is not as good as the predictions. The discrepancies between experimental and theoretical conversions and purities can be attributed to the behavior of adsorbed MES. This is discussed further in the next section.

### 2.7.5.3. Esterification of Acetic Acid

The liquid phase reaction of acetic acid with ethyl alcohol on a cross linked sulfonic acid ion exchange resin in a SCMCR has recently been reported (Mazzoth et al., 1996). This contrasts with the gas–solid MES hydrogenation and demonstrates that condensed phase reaction systems can be carried out to good advantage in simulated moving beds. The resin acts both as a catalyst and as a selective adsorbent. Ethyl acetate was obtained in high purity, and conversions

**Table 2.10.** Comparison of Reactor Performance: Mesitylene Hydrogenation

| Reactor | $X_{MES}$ | $Y_{TMC}$ |
|---|---|---|
| CMCR[a] | 0.88 | 0.95 |
| CMCR[b] | 0.97 | 1.0 |
| SCMCR[c] | 0.83 | 0.96 |
| SCMCR[d] | 0.97 | 0.98 |
| SCMCR[e] | 0.98 | 0.99 |

[a]Experiment, Fish and Carr, 1989.
[b]Model prediction, Fish and Carr, 1989.
[c]Experiment, Ray and Carr, 1995a.
[d]Model prediction, Ray et al., 1994.
[e]Model prediction, Ray and Carr, 1995b.

that are greater than those possible in the absence of separation were achieved for this equilibrium limited reaction. Experimental studies of multicomponent adsorption, reaction kinetics and resin swelling were done, and the results were described with appropriate models. Experiments in a fixed-bed chromatographic reactor packed with the resin were conducted to characterize its behavior. A mathematical model of the reactor, incorporating both kinetic and thermodynamic descriptions of the reacting system, has been developed. The model has predictive capability, and is able to describe experimental behavior with reasonable accuracy.

### 2.7.5.4. Reactor Dynamics

Each column (section) of a SCMCR is a fixed bed through which concentration waves propagate at regular intervals. The description of this transient situation is a dynamic problem that must be dealt with in realistic mathematical models. The dynamics of adsorption and desorption in packed beds has been extensively treated in the literature. Ruthven (1984), Rhee et al. (1986), and Yang (1987) give excellent reviews of this subject. The material balance equation for each chemical species, $i$, in axially dispersed plug flow with concentration independent dispersion coefficients is

$$\frac{\partial C_i}{\partial t} + \frac{\partial (vC_i)}{\partial z} - D_i \frac{\partial^2 (C_i)}{\partial z^2} + \frac{1-\varepsilon}{\varepsilon} \frac{\partial q_i}{\partial t} = 0 \qquad (2.23)$$

in which, $C$ is fluid-phase concentration; $q$ is adsorbed phase concentration; $t$ is time; $z$ is axial distance, $\varepsilon$ is the bed porosity, and $D$ is the dispersion coefficient. If the SCMCR is of the co-packed catalyst and adsorbent variety, the reaction rate expression must be included. Examples are Eqs. (2.21) and (2.22). Otherwise, Eq. (2.23) and an expression for the rate of adsorption are sufficient to describe concentration waves in isothermal adsorbers.

Each section of a SCMCR is subjected to periodic concentration changes. The concentration profile entering a freshly prepared feed section depends upon the exit profile from the previous section, and the shape of the makeup feed profile. Figure 2.20 is a sketch of a profile where rapidly opening and closing the valve to the feed section produces steep rising and falling edges that can be idealized as step functions. The initial and boundary conditions then are

$$t < 0, \quad C_i = C_{t=0,i} \qquad 0 \leq z \leq L \qquad (2.24)$$

$$z = 0, \quad C_i = C_{in,i} \qquad t > 0 \qquad (2.25)$$

$$z = L, \quad \frac{\partial C_i}{\partial z} = 0 \qquad t > 0 \qquad (2.26)$$

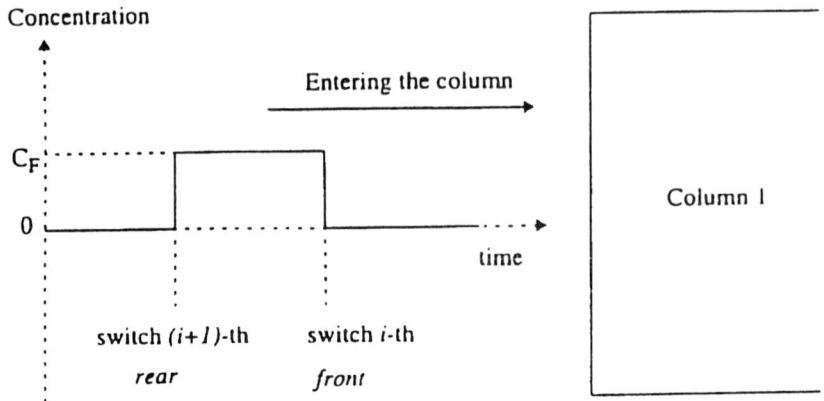

**Figure 2.20.** Idealized entering concentration waveform for the multiple column SCMCR.

in which $C_{t=0,i}$ is the concentration of component $i$ at $t = 0$; and $C_{in,i}$ is the concentration at the inlet, which varies with time. As the square wave traverses the column, dispersion and adsorption modify the waveform.

In equilibrium adsorption models the effect of adsorption on the concentration profiles can be understood from adsorption isotherms. The speed of a concentration front, neglecting dispersion, is given by

$$V_f = \frac{U_g}{1 + \frac{(1-\varepsilon)}{\varepsilon} \frac{dq_i}{dC_i}} \qquad (2.27)$$

where $(dq_i/dC_i)$ is the slope of the adsorption isotherm. For a linear isotherm, $q_i = K_i C_i$, $V_f$ is independent of concentration and the only influence on the shape of the front is dispersion. With nonlinear isotherms the propagation speed is concentration dependent. For example, for the Langmuir isotherm

$$q_i = \frac{K_i C_i}{1 + \sum K_i C_i} \qquad (2.28)$$

$d^2 q_i/dC_i^2 < 0$, so $dq_i/dC_i$ decreases with increasing $C_i$ and $V_f$ increases with increasing concentration. If a broad concentration front is introduced into the adsorbent bed, it will self sharpen along the column. Isotherms with this behavior are referred to as favorable isotherms. In the SCMCR we must also pay attention to the trailing edge of the wave. It is evident that the concentration effect of the Langmuir and other "favorable" isotherms that causes the front to sharpen will broaden the tail. The other case of interest involves isotherms with curvature

$d^2q_i/dC_i^2 > 0$, making $dq_i/dC_i$ an increasing function of increasing concentration, and the leading edge of the input square wave of Figure 2.20 would spread along the column, while a dispersed trailing edge would sharpen.

The experimentally observed output waveforms from MES hydrogenation, shown in Figures 2.19a and 2.19b, are consequences of dispersion, adsorption, and reaction after introduction of a MES waveform that is approximately a square wave of 4-minute duration. The periodicity of the TMC and MES concentrations is apparent. Adsorption of both TMC and MES on the Chromosorb 106 adsorbent can be described by a Langmuir isotherm. However, at the low concentrations of the experiment, the isotherm may be approximated as linear and no influence of the isotherm on the shapes of the observed concentration waves is expected. The effluent concentration history profile of TMC at port B is within the 30 second time resolution of the composition measurements, and the sharp cut at the end of each four minute period comes from rapidly switching the location of port B to the next column in line. The curved top of the TMC concentration wave can be attributed to variation of reaction rate with axial distance, in combination with the velocity of the TMC front.

Figure 2.19b shows that at port A the concentration of MES falls monotonically over each switching interval. The waveform seen here is the MES desorption tail as it is purged from the reactor. Axial dispersion is insufficient to account for the observed width, so adsorption is its most likely source. Independent experiments show that the adsorption of MES on $Al_2O_3$ can be described by a Langmuir isotherm (Fish and Carr, 1989). At mole fractions below 0.005, which correspond to the range in Figure 2.19b, the isotherm is approximately linear and dispersion of the trailing edge by the favorable isotherm is not expected to be important. This has been confirmed by numerical simulations using an axially dispersed plug flow model with a linear isotherm. The calculated concentration profiles did not have significant tail spreading due to dispersion or to the isotherm (Djoharie, 1995). The experimentally observed broadening can be explained if MES has a distribution of adsorption energies on the Chromosorb 106 adsorbent, such that a portion is not readily desorbed by the $N_2$ carrier gas. This would not be detected experimentally by the method of breakthrough curves that was used to determine the isotherm, but would be easily detected in desorption, such as in Figure 2.19b. It is most likely that what is being observed is desorption of MES from higher energy surface sites. If these amount to approximately 15–20% of the total concentration of sites, $N$, they can account for the observed loss of MES at port A, which is the principal reason that reactant conversion is less than the theoretical prediction of nearly unit conversion. Surface heterogeneity may be the principal reason for the discrepancy between theoretically predicted and observed conversions in all cases investigated thus far.

The steep adsorption fronts that move through the reactor sections present a challenge to numerical simulation of the SCMCR. An adaptive finite element method with uneven grid points was employed to simulate the multiple column MES hydrogenation reactor (Ray and Carr, 1995b), but difficulty in obtaining convergence was experienced in simulating the steep fronts typical of the experiments. Consequently, the reactor of Figure 2.18 was modified by moving port B one section ahead of the feed, and collecting only 15% of the flow at this point. The remainder was sent to the next section. This avoided large changes in flow rate and smoothed the concentration fronts, giving convergence of the solutions, at least for linear models. Finite difference methods have been found to be more satisfactory. An approach using adaptive gridding with four point quadratic upstream differencing for first derivatives, a centered difference scheme for second derivatives and second order Crank Nicholson integration (Sun and Meunier, 1991) has proved entirely satisfactory for dispersed plug flow modeling of a simulated countercurrent moving bed separator (Djoharie, 1995). In a model study of a SCMCR for oxidative coupling that utilized the dispersed plug flow model with a linear driving force approximation for mass transfer (Kruglov et al., 1996a), a third-order quadratic upstream difference scheme (Leonard, 1979) was used to approximate the convective term, and the system of differential algebraic equations was solved by the method of lines (Brenan et al., 1989).

### 2.7.6. Natural Gas Utilization

The understanding of SCMCR behavior gained from the MES hydrogenation studies have made it possible to consider other applications. An area of considerable current interest where the SCMCR can provide significant advantages over conventional chemical reactors is natural gas (methane) conversion to fuels and feedstocks. It is frequently pointed out that the extensive worldwide reserves of natural gas are an underutilized resource. Proven reserves are estimated to be capable of providing up to 100 years of transportation fuel (Smith, 1987). Utilization for purposes other than heat or energy from combustion would provide a feedstock for the chemical industry that is cleaner than coal or petroleum. However, much of the world's natural gas reserves are in remote regions. Accessing these by pipeline or by shipping liquefied natural gas requires significant capital investment. Natural gas from oil wells is frequently flared off as the only economical option, but this contributes to the atmospheric burden of the greenhouse gas, $CO_2$. Economically favorable on site conversion of natural gas to feedstocks and fuels that are transportable and more readily utilized would capture this valuable resource and also have positive environmental benefits.

The principal component of natural gas is methane, typically about 90%, with the balance being ethane, higher hydrocarbons and other minor constituents. Natural gas processing centers about methane chemistry; but the high C–H bond energy (437 kJ/mol) gives methane sufficient chemical stability that purely thermal processes leading to useful chemical products are endothermic, and may require high process temperatures that rule out commercial feasibility. Oxidative processes are normally exothermic, so lower temperature processing is possible in the presence of $O_2$. However, to limit complete oxidation to $CO_2$ and $H_2O$, and to maximize selectivity for the desired product, the quantity of $O_2$ must usually be much less than stoichiometric. With $O_2$ as the limiting reagent, $CH_4$ is incompletely converted and the desired product yield per pass through conventional reactors is low. This has inhibited commercialization of oxidative processes for methane conversion.

The conditions leading to poor yields in conventional reactors can be circumvented in the SCMCR. We have seen how the SCMCR improves the conversion of equilibrium limited reactions by separations that inhibit the rate of the reverse reaction. The adsorptive separation can be used to advantage in oxidative processes for methane conversion by rapidly removing reaction products, which are usually more reactive than methane and thus susceptible to further oxidation, from the oxidizing environment. Of greater significance is that the unconverted methane emerging from one SCMCR section can then be combined with additional $O_2$ and makeup methane and passed to the next section for further reaction. This can be repeated as long as desired, retaining methane in the SCMCR for improved conversion, while maintaining high $CH_4/O_2$ for good selectivity. In this way, the SCMCR can be applied to low conversion per pass reactions in order to improve performance. Two methane conversion reactions, oxidative coupling and partial oxidation, are discussed below.

### 2.7.6.1. Oxidative Coupling of Methane

Methane reacts with oxygen in the presence of many metal oxide catalysts at temperatures in the vicinity of 1000°K to form ethane and ethylene in a process called oxidative coupling of methane (OCM).

$$2CH_4 + \tfrac{1}{2}O_2 \Leftrightarrow C_2H_6 + H_2O$$

$$2CH_4 + O_2 \Leftrightarrow C_2H_4 + 2H_2O$$

A parallel path in which $CH_4$ is completely oxidized also occurs.

$$CH_4 + 2O_2 \Leftrightarrow CO_2 + 2H_2O$$

These reactions have been extensively investigated as a method for conversion of methane into ethylene (Hutchings et al., 1989; Lunsford, 1995) in order to utilize natural gas as a feedstock for this, the largest production volume organic chemical in the world. While OCM is not equilibrium limited, it is a low conversion per pass reaction if complete conversion to $CO_2$ and $H_2O$ is to be minimized. With some catalysts selectivity of the $C_2$ products reaches 80–90% at $CH_4/O_2$ around 50 or more. This severely limits the $CH_4$ conversion and the yield of $C_2H_4$ and $C_2H_6$. In conventional fixed-bed or fluidized bed reactors it has not proved possible to obtain $C_2$ yields much greater than about 20%. It has been estimated that a commercially viable process would require $C_2$ yields of at least 30% (Matherne and Culp, 1992; Kuo, 1992).

It is possible to substantially improve the yield of $C_2$'s by carrying out OCM in a SCMCR (Tonkovich et al., 1993), but to do so requires modification of the reactor in order to obtain satisfactory separations. At the high reaction temperature required for OCM most adsorbents will thermally degrade or will not provide enough adsorption selectivity to separate reactants and products. Consequently, it is not possible to mix catalyst and adsorbent as was the case for the MES hydrogenation, which occurs at about 200°C. To achieve good separations, Tonkovich et al. (1993) designed a SCMCR in which each reactor section consisted of a quartz tubular reactor packed with catalyst and maintained at high temperature, followed by a lower temperature column packed with adsorbent, as shown in Figure 2.21. Each reactor–adsorber combination is one section of the SCMCR, corresponding to a section of the integrated reactor–adsorber shown in Figure 2.18. It performs the same functions as a single column packed with a mixture of catalyst and adsorbent.

The reactor operates by switching the feed, which consists of a mixture of $CH_4$ and $O_2$, with $N_2$ as carrier gas, sequentially from one section to the next. Unreacted methane remains in the system, flowing from the reactor through the adsorber, where it elutes in advance of the more strongly adsorbed products and

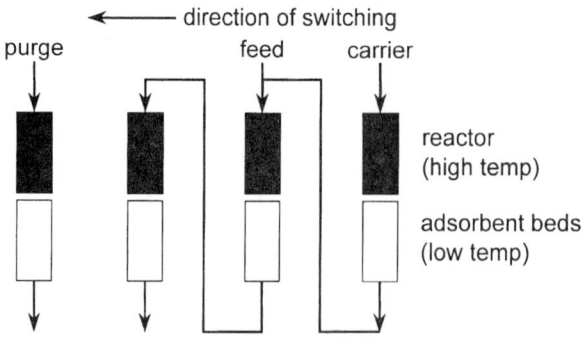

**Figure 2.21.** Multiple section SCMCR for the oxidative coupling of methane. Each section consists of a reactor and an adsorber.

passes on to the next section. The reaction products are rapidly removed by the adsorbent beds, thus suppressing over oxidation of $C_2H_4$. The feed consists of make up $CH_4$ and $O_2$, to replace the amounts reacted, and make up carrier. The $CH_4$ to $O_2$ ratio in the reactor is maintained at about 20:1 or larger for good $C_2$ selectivity and to minimize oxidation of $C_2H_4$. However, $CH_4/O_2$ in the make up stream should have the stoichiometric ratio of about 2:3 since it replaces reactants that have been converted in the stoichiometric ratio. Viewing the SCMCR as a black box, the reactants are fed at the stoichiometric ratio, but converted to products at conditions for high $C_2$ selectivity. In principle, $CH_4$ can be contacted to extinction since it is not deliberately removed from the SCMCR. In practice there are $CH_4$ losses, for much the same reasons discussed above for MES, which limit performance. Nevertheless, high $CH_4$ conversions and good $C_2$ yields have been obtained.

Conventional microreactor experiments with powdered $Sm_2O_3$ catalyst revealed that for high $CH_4/O_2$, in the range of 20 to 50, and at 1000°K, $C_2$ selectivities were greater than 95% but the $C_2$ yields were less than 10% due to low per pass $CH_4$ conversions (Tonkovich and Carr, 1994). However, in the SCMCR of Figure 2.21 with activated charcoal adsorbent and similar reaction conditions, these authors found $C_2$ yields that ranged to more than 50%. The SCMCR accomplishes this by maintaining high $C_2$ selectivity, greater than 90% at some conditions, and $CH_4$ conversion that is as much as 0.65. The effect of temperature, switching interval, and $CH_4/O_2$ were investigated and all found to have significant influence on reactor performance.

A mathematical model of the SCMCR for the oxidative coupling reaction was recently reported (Kruglov et al., 1996a). Input data for carrying out numerical simulations of SCMCR performance were obtained from experiments with a microreactor and adsorber column in series. Tests with three catalysts and three adsorbents showed that the best performance, obtained with a $YBa_2Zr_3O_5$ catalyst and activated charcoal, was a predicted 55% yield of $C_2$ products. SCMCR experiments with this combination of catalyst and adsorbent corroborated the model study (Kruglov et al., 1996b). Yields of $C_2$ products greater than those observed have not been obtained because of $CH_4$ losses. Minimization or prevention of $CH_4$ loss in the purge stream is an issue that remains as a challenge for further development and improvement of the SCMCR.

In the design of Kruglov et al. (1996a), it was decided to use only a single fixed-bed reactor and to direct the reactor effluent to a simulated countercurrent moving bed separator. This arrangement is conceptually no different from the four reactor-adsorber configuration above since at any given time feed enters the reactor, which is followed by an adsorbent bed acting as one section of the SCMCR. The reaction product stream is sent sequentially to the adsorbers, which are handled in very much the same manner as in the previous work (Tonkovich and Carr, 1994), and the overall reactor performance is unchanged. However, this arrangement is better than associating a separate fixed-bed reactor with each adsorbent bed. It uses less catalyst, requires a lower heat load, and avoids differences in fixed-bed reactor performance that might arise from differing rates of catalyst deactivation.

### 2.7.6.2. Partial Oxidation of Methane to Methanol

Methanol, in demand as a fuel oxygenate, is being investigated as an alternative fuel for both conventional auto engines and diesel powered vehicles, and it is an intermediate in the manufacture of a number of commodity chemicals. It is also used as a feedstock for methyl tert butyl ether production, a fuel oxygenate used in reformulated gasoline. Reformulated gasoline reduces CO, $NO_x$, volatile organics, and benzene emissions from automobiles, as required by the Clean Air Act.

Current technology for methanol production is a complex and costly two step process, consisting of steam reforming of $CH_4$ to produce CO and $H_2$, and methanol formation by passing CO and $H_2$ over a metal oxide catalyst. The overall chemistry is endothermic by about 125 kJ/mol, requiring significant energy consumption. One step conversion of $CH_4$ to $CH_3OH$ by partial oxidation is energetically attractive, since it is exothermic by 126 kJ/mol. However, to minimize formation of $CO_2$ and $H_2O$, it is necessary to carry out the reaction at high $CH_4/O_2$ ratios. With $O_2$ as the limiting reagent the $CH_4$ conversion, and hence the methanol yield, is limited. Literature reports of per pass methanol yields from methane partial oxidation in empty or fixed-bed tubular reactors are usually less than 10% (Gesser and Hunter, 1992). Methanol yields have thus been discouragingly low for commercial development. The SCMCR provides an approach to direct methanol synthesis with higher methane conversions and higher methanol yields for the same reasons that enhanced performance was realized in OCM. The overall reaction has the deceptively simple appearance of the insertion of an O atom into methane.

$$CH_4 + \tfrac{1}{2}O_2 \Leftrightarrow CH_3OH$$

This would be a desirable alternative to the two step synthesis gas route since a single reaction would reduce the complexity and cost of a methanol production facility. In particular, the absence of the energy intensive syngas step would make for a more energy efficient process. The ease with which methanol can be further oxidized is a potential difficulty, but this is circumvented with the SCMCR since rapid separation of methanol from oxygen in the adsorbers minimizes this unwanted secondary reaction. The SCMCR also permits contacting the reactants

## 2. Adsorption Technologies    113

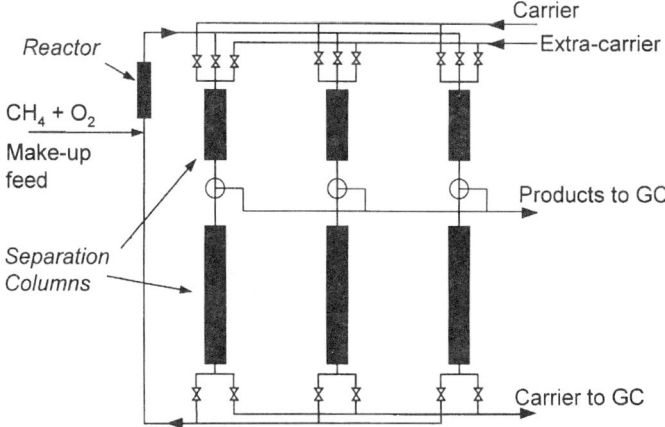

**Figure 2.22.** Single-reactor–multiple-adsorber modification of the multiple section SCMCR for the oxidative coupling of methane.

at high $CH_4/O_2$, where methanol selectivity is high, while feeding the reactor at a much lower makeup ratio as for OCM.

The methanol SCMCR consists of a quartz tubular reactor and three adsorbent beds, as shown in Figure 2.22. The single tubular reactor linked with each adsorption column, in turn as the column to column flow switching operation is carried out, follows the approach of the single fixed-bed reactor SCMCR for the OCM reaction, and was adopted for the capital and energy savings that would be realized in a practical process. The efficacy of solid catalysts for partial oxidation of methane has been questioned (Burch et al., 1989; Hunter et al., 1990). There is evidence that some catalysts are ineffective and that a substantial fraction of the reaction occurs homogeneously. In empty tubular reactors methanol selectivity in the 70% to 80% range has been reported for the homogeneous reaction carried out at elevated pressure (30 to 70 atm) and 400 to 500°C (Yarlagadda et al., 1988; Feng et al., 1994).

Recent experiments at the University of Minnesota have confirmed that good results can be obtained from the homogeneous reaction at elevated pressure. In the empty tubular reactor, 60% methanol selectivity has been obtained at 75 atm and 400°C, but with the $CH_4/O_2$ ratio of 25 the methane conversion was 5–8%, and the yield of methanol was only 3–4%. When the tubular reactor was interfaced with the adsorbers and operated in the SCMCR mode the yield of $CH_3OH$ was 17% demonstrating again that the SCMCR is effective for improving the performance of low conversion per pass reactions (Bjorklund and Carr, 1996). At this writing the methanol SCMCR has not been optimized, and it is expected that ongoing experimentation will lead to $CH_3OH$ yields significantly in excess of 17%.

### 2.7.6.3. Methanol from Synthesis Gas

An alternative to partial oxidation is $CH_3OH$ synthesis from CO and $H_2$. Numerical modeling of a simulated countercurrent moving bed for this process has been reported (Kruglov, 1994). A range of operating conditions were found for this reaction, which is equilibrium limited when carried out in nonseparative reactors, which gave 96–99% conversions in a single pass. This was attributed to removal of methanol from the reaction by adsorption onto the solid. Two different reactor configurations were considered. In one arrangement catalyst and adsorbent were mixed and the reactor was operated adiabatically, while in the other, which was isothermal, the catalyst and adsorbent were separated. In the isothermal reactor, only the adsorbent beds were handled by simulated countercurrency. The catalyst beds, which were connected in series with the adsorbent beds, were not in the flow-switching scheme. Reactor performance was improved by incorporating a stripping section immediately behind the feed section. This served to blow unconverted reactants into the reaction zone, preventing their loss with the purge gas when methanol was being stripped from the adsorbent. The axially dispersed plug flow model used the linear driving force approximation to describe adsorption. Experimental data for reaction kinetics over $Cu/ZnO/Al_2O_3$ catalyst, and adsorption on silica-alumina were taken from the literature. The evolution of axial concentration and temperature profiles in the reactor were presented. These give insight into how the reaction and separation proceed.

Dandekar and Funk (1995) have filed a patent application for methanol production via a SMB process. In this process either low-temperature methane or one of the feed components (CO, $H_2$) is used as the desorbent in order to minimize the reverse reaction of methanol during desorption. Complete processes for producing methanol from methane were considered, that is, the SMB units were combined with different processes for syngas generation including steam methane reforming, secondary reforming, and partial oxidation. The advantages of the SMB-based processes are ultimately due to the increased reaction conversion attributable to the adsorption of methanol. This enables operation at lower reactor pressure and with reduced or even eliminated recycle requirements.

## 2.7.7. The Pressure Swing Reactor

When chemical reactions are carried out in pressure swing adsorbers (PSA), reaction and adsorptive separation occur in concert, and the PSA may be termed a pressure swing reactor (PSR). This concept was introduced into the literature by Vaporciyan and Kadlec (1987), who investigated an isothermal, 1-D, dispersionless reaction/adsorption equilibrium model, termed the ideal model, for a single column packed with adsorbent and catalyst. This rapid periodic separating reactor was operated with a three step cycle consisting of high pressure vapor feed, delay with depressurization, and low pressure exhaust. Three solid catalyzed reactions, $A = 2C$, $2A = C$, and $A = \frac{1}{2}B + C$ were considered. All chemical species followed a linear adsorption isotherm, and local chemical and adsorption equilibrium was maintained. Numerical solution of the model equations revealed separations much like those from (nonreactive) PSA, but conversions beyond chemical equilibrium were not observed because the imposition of local chemical equilibrium obviated enhancement of conversions and yields.

In their subsequent work Vaporciyan and Kadlec (1989) relaxed the chemical equilibrium constraint for model studies of several reversible and irreversible reactions in a PSR operated in the same manner as the 1987 work. One of these, the irreversible oxidation of CO to $CO_2$ was investigated experimentally as well as by model calculations. The mathematical model differed from the previous work only in that finite reaction rates were employed. Separation and enhanced conversion were predicted by the models, and observed for the experiments. Since the feed $CO/O_2$ was diluted with excess $N_2$, though, both the effluent and delivery streams contained carbon oxide levels less than 5%. In the reversible reaction cases it was found that equilibrium conversions expected for nonseparative reactors could be shifted to favor reaction products, giving yields that were greater than equilibrium. Also, the experimentally observed $CO_2$ production rates were higher from PSR experiments than from a tubular flow reactor at the same inlet concentrations and pressures. This was attributed to inhibition of the reaction rate by CO. The inhibition was less in the PSR primarily because of the removal of separated CO in the exhaust stream, whereas the lack of separation in the nonseparative case leads to lower reaction rates. In addition, separation reversals were predicted and observed. An unusual feature observed in this work is separation inversion. That is, the adsorbable species is in the product stream, in contrast with simple cycle PSA where the negligibly adsorbed species is expected to be in the product stream. This was attributed to the fact that the concentration gradients created by the reaction and by the separation are in opposition. Model studies of multiple reversible reactions predicted improvements in selectivity and conversion over steady state operation. A patent based on the above work was obtained by Kadlec and Vaporciyan (1993) for the single bed PSR technology.

An interesting approach to PSR optimization via the effects of nonuniform axial distributions of adsorbent and catalyst on reactor performance has been advanced (Lee and Kadlec, 1988). If the adsorption selectivity for the components of the reaction mixture is varied with distance by charging the bed with different adsorbents at different positions, it is possible to manipulate the product stream composition. Also, if the axial distribution of catalyst is varied, it is possible to manipulate the reaction rate along the bed. These propositions have been investigated by a modeling study for the set of parallel reactions, $A = B$ and $A = C$. The optimal distributions of adsorbent capacity and catalyst activity to maximize the separation of B from C, and the selectivity of B, respectively, were calculated by a gradient search method.

A general model for nonisothermal adsorption and chemical reaction in a rapid PSA has been developed, and several numerical methods for solution of the model have been compared (Alpay et al., 1993). Solutions by orthogonal collocation, orthogonal collocation on finite elements, double orthogonal collocation on finite elements, and cells in series were compared. Orthogonal collocation on finite elements was found to be the most computationally efficient of these. The model was applied to the nonreactive separation of air and to the reactive separation of $A \Leftrightarrow B + 3C$, with B the only adsorbing species.

A two bed PSR with a first order irreversible reaction, $A \Rightarrow B$, in which only B is adsorbed, has been investigated (Kirkby and Morgan, 1994). An isothermal, 1-D, dispersionless model was formulated for a dilute feed of A in inert gas and solved by the method of characteristics. Three methods of operation typical of PSA were tested; simple cycles, purge cycles and backfill cycles. The simple cycle consisted of pressurization by feed, product release and depressurization. In this case the product stream was predicted to be highly enriched in B, a separation inversion corroborating the above cited work (Vaporciyan and Kadlec, 1989). When cycles incorporating purge and backfill were investigated, the concentration of B in the product stream was very low, but the waste stream from both was enriched in B. The results suggested that for $A \Rightarrow B$ it is not possible to obtain a stream that is enriched in B and also contains little or no A.

The isothermal, 1-D mathematical model of the two bed PSR was extended by Alpay et al. (1994) to some reversible reactions; isomerizations ($A \Leftrightarrow B$), dissociations/disproportionations ($2A \Leftrightarrow B + C$), and dehydrogenations ($A \Leftrightarrow B + nH_2$). In all cases, the reactant A and product C ($H_2$ during dehydrogenation) were assumed to be nonadsorbed, and the feed gas (and therefore the product gases) was diluted with inert and nonadsorbing carrier gas. Cycles included a simple sequence of feed pressurization, product release, and depressurization, as well as a sequence including a countercurrent low pressure product purge step. For all reactions studied, the purge cycle led to improved product yields relative to the simple cycle. In cases where there was no increase in mole number upon reaction, the model calculations predicted product yields exceeding the equilibrium limit. For dehydrogenation reactions the increase in moles due to the reaction causes equilibrium conversions to decrease with increasing pressure, and the predicted PSR performance was not as good. For example, the dehydrogenation of methylcyclohexane (MCH) to toluene (T) was modeled using kinetics data from microcatalytic reactor studies, and adsorption parameters from the pulse input chromatography method. In this reaction, elevated pressure depresses the equilibrium conversion and limits PSR performance because this reaction produces three moles of $H_2$ per mole of reactant. It was found that toluene yields exceed those from an equivalent plug flow reactor (PFR) only at PFR operating pressures greater than 7 bar. For dehydrogenations where only one mole of $H_2$ is eliminated from the reactant, the situation is more favorable, with PSR performance exceeding that of an equivalent PFR over a wider range of operating conditions.

A model study of $2A \Leftrightarrow B + C$ in a two bed PSR with both simple cycle and purge cycle has been reported (Chatsiriwech et al., 1994). The equilibrium conversion is independent of pressure, affording an opportunity to test the PSR against conventional reactors without interference from the effect of pressurization on the reaction itself. With B as the only adsorbed component, comparison with PSR performance showed that the yield of B exceeded chemical equilibrium, the maximum expected of the PFR, by almost twofold. As part of this work, three numerical approaches to solution of the model equations were tested. They are the method of characteristics, cells in series, and orthogonal collocation of finite elements. These all compared favorably. From a practical point of view, these authors pointed out that PSR units could potentially reduce operating temperatures for endothermic equilibrium controlled-reactions, thereby leading to lower energy costs or byproduct formation.

Prediction of greater than equilibrium conversion for $A = B + C$ has been obtained from numerical simulation of a three step (feed, delay, purge), single bed, pressure swing reactor (Lu and Rodrigues, 1994). The adsorption equilibrium model incorporated nonlinear, multicomponent isotherms, and axially dispersed plug flow. Since the reaction is favored by low pressure, the outlet of the PSR was maintained at atmospheric pressure. The model was solved by orthogonal collocation of finite elements. When B was considered to be the only adsorbed species the PSR model predicted 10–20% higher conversion than the equilibrium conversion that could be attained in a plug flow reactor. When both the reactant, A, and the product, B, are adsorbed, the PSR conversion exceeded the equilibrium limit by 10%.

Imai et al. (1985) experimentally illustrated the potential of hydrogen-absorbing alloys for enhancing the conversion of dehydrogenation reactions. They studied the dehydrogenation of cyclohexane in a flow reactor packed with a mixture of a $Pt/Al_2O_3$ catalyst and various metallic alloys. The initial reaction conversion of 11% at 150°C in the presence of $CaNi_5$ alloy was nearly double the conversion measured without the alloy. The conversion slowly declined with onstream time as the alloy became saturated with $H_2$, but it was found that it could be regenerated by purging with helium. A series of alloys were tested and their effectiveness in improving the reaction conversion was directly correlated with the initial rate of hydrogen absorption. By increasing the amount of catalyst and alloy, decreasing the feed flow rate, and using a prolonged activation treatment, it was demonstrated that cyclohexane conversions as high as 98.7% were possible with this system.

Goto et al. (1992) expanded upon this early work and proposed a cyclic PSA process based on the $CaNi_5$ hydrogen occlusion alloy. Individual reaction step experiments indicated that cyclohexane conversions up to 100% could be obtained in the presence of the alloy even when the catalyst-only system yielded only 30% conversion. The conversion increased with temperature, but was not sensitive to the range of feed gas flow rates studied. Regeneration of the alloy was determined to be more effective at higher temperature (requiring a shorter regeneration time and lower purge volume). A PSA cycle consisting of a reaction step and an inert gas purge step was studied with the experimental apparatus and was found to produce reproducible effluent streams and nearly 100% cyclohexane conversion.

The Thermally Coupled Pressure Swing Adsorption (TCPSA) process, which combines an adsorptive reactor system with the thermodynamic Stirling cycle, has been described by Keefer (1989;1990). Catalyst and adsorbent are contained in separate vessels and are connected to pistons that reverse gas flow and modify system pressure in cyclic fashion. The process provides for improved conversion of reactants by separating them from the products, and also directly recovers the heat of reaction (for an exothermic reaction) as mechanical energy, which is used in the process. A laboratory unit was built to demonstrate the TCPSA approach for the synthesis of ammonia. The TCPSA system was found to effectively trap $H_2$ in the reactor vessel until reacted, and produce a product stream containing enhanced levels of ammonia. The unit operated at a higher ammonia productivity than an analogous PFR system. The advantages of the TCPSA ammonia process are more effective integration of the process flowsheet and reduction of the process operating pressure. In general, hydrogenation and dehydrogenation reactions are expected to be particularly amenable to this approach due to the favorable fluid and thermal properties of $H_2$ in the Stirling cycle.

A pressure swing reactor system called the sorption enhanced reaction process (SERP) has been developed (Carvill et al., 1996; Anand et al., 1995). It consists of two or more fixed beds, each packed with a mixture of catalyst and adsorbent and alternately cycled through five steps; (i) high pressure sorption reaction (ii) depressurization (iii) purge with suitable purge gas (iv) purge with product gas, and (iv) pressurization with product gas. The purge gas in step (iii) should be cheap, nonadsorbed, and nonreactive with the catalyst, e.g., nitrogen, methane. The above process cycle has been designed for reactions that produce a desired product and a more strongly adsorbed byproduct.

A bench-scale SERP unit was built and used for evaluation of the reverse water gas shift reaction; $CO_2 + H_2 = H_2O + CO$. A zeolitic adsorbent was used to adsorb $H_2O$ and shift the reaction toward higher conversion to the desired product CO. High product purity, produced at feed gas pressure and enhanced conversion with respect to a PFR, was demonstrated. For example, the SERP conversion at 250°C was 36%, and the purity of the CO product was 99+%. At 250°C the equilibrium conversion in a nonseparative reactor would be only 9.8%. To obtain 36% conversion for this endothermic reaction in a PFR would require a temperature of at least 565°C, the temperature at which the equilibrium conversion is 36%, and the effluent composition would be 22% CO and $H_2O$, and 39% $CO_2$ and $H_2$. The lower temperature required for obtaining a given conversion, and the lessened downstream separation requirement, make application of the SERP concept attractive for reducing process costs and complexity.

This basic process cycle has also been considered for the production of $H_2$ via the steam methane reforming reaction (Anand et al., 1996; Hufton et al., 1997; Nataraj et al., 1996). The potential advantages of this process include (1) operation at a much lower temperature (350–500°C) than conventional reformers (800–900°C), yielding cheaper reactor construction, energy savings, and minimization of coking reactions, and (2) production of hydrogen product at feed pressure (lower compression costs) and at relatively high purity (minimize downstream separations).

The process utilizes reactors packed with a mixture of conventional nickel-based reforming catalyst and a selective $CO_2$ adsorbent. The adsorbent must be capable of reversibly adsorbing and desorbing $CO_2$, in pressure swing mode, at constant operating temperatures between 350 and 500°C and in the presence of up to 15 atm of steam. This represents quite a challenge to most materials, and only after an extensive materials development program has such an adsorbent been discovered (Anand et al., 1996).

The SERP concept for $H_2$ production was verified in a laboratory-scale reactor system (Hufton et al., 1997). The reactor was packed with a 3:1 mixture of adsorbent and catalyst. At 450°C and 55 psig, a feed containing 11% $CH_4$ and 89% $CH_4$ was converted to an effluent product of 95% $H_2$ (reaction step average). The methane conversion was found to be 68%. The thermodynamic limit for the conversion and hydrogen purity of a conventional system operated at these conditions is only 34% and 57%, respectively.

Preliminary process designs based on the SERP concept have been developed and compared with conventional SMR units. Economic analyses suggest that the SERP system can produce $H_2$ at a significantly lower cost than the conventional approach (2.5 MMSCFD basis).

A fixed-bed pressure swing reactor process was described by Dandekar et al. (1996) for the production of methanol from synthesis gas containing hydrogen and carbon dioxide. In this system the primary product, methanol, is the most strongly adsorbed component, and is produced during the low pressure regeneration step of the cycle. A preferred process sequence consists of an adsorption / reaction step to convert feed $CO_2$ and $H_2$ to methanol, a hydrogen purge step at reaction pressure, counter- or co-current depressurization and $H_2$ purge to produce a methanol-enriched product stream, and repressurization with $H_2$ to the reaction pressure. The high pressure $H_2$ purge is included to increase the conversion of residual carbon oxides in the reactor, while the low pressure $H_2$ purge prevents the reverse methanol decomposition reaction during the regeneration steps. Numerical studies of this system showed that the inclusion of the latter step could increase the reaction conversion from 27 to 63%. Pressurization with feed gas, a common separative PSA step, was found to be unattractive, while cocurrent depressurization with $H_2$ purge could potentially increase reactor conversion. Three process designs incorporating the pressure swing reactor concept were developed and compared with a conventional methanol production process. The PSR processes yielded higher carbon oxide conversions (by a factor of 2.3) and 5–28% lower compression costs than the conventional approach.

### 2.7.8. The Gas–Solid–Solid Trickle Flow Reactor

The gas–solid–solid trickle flow reactor (GSSTFR) is a three phase adsorptive chemical reactor for gas–solid contacting. In this approach to reactive separations, catalyst pellets are stationary in a packed bed, a chemically reacting gas flows upward through the bed, and a powder, which is the adsorbent, slowly flows downward over the catalyst countercurrent to the gas. If reaction products are selectively adsorbed on the powder, they can be removed from the reactor, and from the reacting gases, by the flowing adsorbent. If an equilibrium limited reaction can be run in this arrangement, then not only can reaction and separation occur simultaneously, but unfavorable equilibria can be shifted toward enhanced product formation. Likewise, reduction of the product mole fraction in the gas stream can yield higher overall reaction rates throughout the reactor since the reverse reaction rate is reduced. The continuous flow of initially cool adsorbent particles through the reactor can also help remove the heat of reaction for exothermic reactions (e.g., methanol or ammonia synthesis). It is apparent that the gas–solid–solid trickle flow reactor has similarities with the countercurrent moving bed chromatographic reactor.

Solids trickle flow has previously been studied for mass transfer operations and for heat transfer, but only relatively recently has the concept been applied to chemical reactions. The theoretical background has been presented, and a one dimensional steady state reactor model for methanol synthesis from CO and $H_2$ has been investigated. The model predicts that complete reactant conversion is attainable in single pass operation of a trickle flow reactor when appropriate gas residence time and adsorbent to feed gas ratio are chosen (Westerterp and Kuczynski, 1987). This is an important prediction, since in conventional fixed-bed catalytic reactors the unfavorable chemical equilibrium of the CO + $H_2$ reaction limits conversions, necessitating separation of methanol and recycling of the reactants, even when high pressure is applied to shift the equilibrium. The inefficiency of this reaction contributes to the complexity and high energy consumption of the commercial methane to methanol process. The theoretical model also suggests that superior process economics will be achieved with adiabatic reactors combined with interstage coolers in order to maintain lower reactor temperatures and the lowest possible solids flow rate.

An experimental investigation of the synthesis of methanol from CO and $H_2$ in a GSSTFR has confirmed the model predictions (Kuczynski et al., 1987). An amorphous silica alumina powder was found to selectively adsorb methanol in the operating temperature range of 498–523°K (Kuczynski et al., 1986). In experiments with the stoichiometric ratio of CO to $H_2$, and at total pressures in the range 5.0 to 6.3 MPa, conditions where unit conversion could be obtained were observed. Westerterp (1988) filed a patent based on this work.

A GSSTFR process design was developed by Westerterp et al. (1988) in order to assess the potential benefits of this approach for methanol synthesis. The reactor design was based on laboratory-scale data. Although an isothermal system was conceptualized, a combination of four adiabatic beds with cooling after each bed was preferred due to solid distribution issues. The reactor system was coupled with a series of five storage tanks that collected the spent adsorbent solids and recovered the adsorbed methanol. The adsorbent in each tank was regenerated by first purging with methanol, depressurizing to produce the pure methanol product, and then repressurizing with syngas to the process pressure. The solid was pneumatically transported to the top of the reactor with feed gas.

The economics of the proposed GSSTFR system were evaluated and compared with a conventional low pressure Lurgi system. For production of 1000 tpd methanol, the GSSTFR system has the potential to reduce cooling water consumption by 50%, recirculation energy by 70%, and catalyst amount by 70%. An energy credit can also be taken for high-pressure export steam, which is produced at 0.25 ton per ton of methanol product. Raw materials consumption is reduced by 12%. Investment costs were expected to be similar for the Lurgi and GSSTFR system because the savings realized from elimination of the recycle equipment was offset by the need for solids handling systems (storage tanks, high pressure pneumatic transport). At the current level of analysis, it appears that this approach has the potential to improve the economics of methanol production.

### 2.7.9. The Temperature Swing Reactor

The concept of shifting an equilibrium controlled reaction by removing a byproduct with a solid adsorbent is not new. Gluud et al. (1927) described such an approach for enhancing the conversion of the water gas shift reaction, $CO + H_2O = H_2 + CO_2$. They suggested the use of lime or dolomite for capturing carbon dioxide from the reaction mixture. Since the interaction between the adsorbent and $CO_2$ is chemical in nature (forms carbonates) and very strong, simple depressurization cannot be used to regenerate the adsorbent, even at reaction temperatures of 300–600°C. Instead, the solid material is heated to 900–1000°C by in situ burning of fuel that decomposes the carbonate and rejuvenates the adsorbent.

The approach taken by Gluud et al. is rather unique since the goal is to use the adsorbent to permit operation at higher temperatures without paying the penalty of poor conversion for this exothermic reaction. In a more conventional approach, specially formulated catalyst are used to produce acceptable reaction rates at more thermodynamically favorable lower temperatures. The inclusion of the adsorbent increases the conversion to acceptable levels at temperatures where catalysts are no longer necessary.

Han and Harrison (1994) have recently pursued this same approach to replace conventional water gas shift reactors and $CO_2$ separation units in modern $H_2$ production systems. Experimental fixed-bed reaction experiments were carried out with dolomite at 450–650°C, 15 atm, and with feed consisting of CO, $H_2O$, $CO_2$ and $H_2$ diluted in an excess of $N_2$. The shift and carbonation reactions were found to approach equilibrium at the conditions tested, and 99.5 to 99.8% of the carbon oxides were successfully removed from the reaction gas before breakthrough of the $CO_2$ adsorption front.

A later study (Han and Harrison, 1997) was focused on the cyclic adsorption/regeneration performance of the dolomite adsorbent. Repetitive adsorption (typically 550°C, 15 atm, 5% CO) and regeneration/calcination (750°C, 3.3 atm, 100% $N_2$) steps were performed. The adsorbent $CO_2$ capacity and carbonation rate both decreased with continued cycle exposure, especially when higher adsorption and regeneration temperatures were used.

Harrison et al. (1994) fashioned a conceptual process design for producing hydrogen from coal gas via the dolomite temperature swing approach and compared it with a conventional shift/PSA process and a high-temperature membrane separation process. Although the dolomite-based process was feasible, it suffered from lower hydrogen product purity and recovery. The major obstacle for this technology is the substantial regeneration energy requirement; 50% of the coal gas feed was needed as fuel for the regenerator, and this obviously reduces hydrogen recovery. Efficient solids transfer between the high-pressure reactor and the low pressure regenerator was also identified as a technical challenge which must be addressed for further development of this technology.

A treated form of dolomite was proposed by Brun-Tsekhovoi et al. (1988) to remove $CO_2$ from the reaction products of the steam methane reforming reaction. In this process, feed gas containing steam and methane is passed through a fluidized bed of aluminonickel catalyst and dolomite. Fresh or regenerated dolomite is added at the top of the fluidized bed. After reacting with $CO_2$ to form $CaCO_3$, the dolomite is separated from the smaller catalyst particles (based on pellet size/density) and then pneumatically transported to a fuel fired thermal regenerator. Since the regenerator must operate at the reactor pressure, a turbine is needed after the regenerator to recover energy from the effluent gas. Heat transfer aspects of this process are unique since the exothermic reaction between $CO_2$ and dolomite, combined with the exothermic water gas shift reaction, can provide enough energy to exceed the energy needed for the endothermic reforming reaction. Thus, the energy needed for this process is actually provided in the regenerator and not the fluidized bed.

The feasibility of this approach was tested out in laboratory and pilot-scale units (Brun-Tsekhovoi et al., 1988; Kurdyumov et al., 1996). These experiments indicated that the addition of fresh dolomite to a fluidized bed of catalyst could substantially increase the methane conversion and produce a relatively pure hydrogen product (94–98% dry basis) with very low carbon oxide impurities. The latter advantage has the potential to eliminate the need for conventional CO-shift, $CO_2$ removal, and methanation steps. The process also has the potential to reduce fuel consumption and eliminate costly high-temperature reactor tubes of the conventional reforming process.

A final approach for using $CO_2$ removal in the production of hydrogen is given by Lyon (1996). This process uses multiple fixed beds that contain a mixture of a suitable metal/metal oxide catalyst (e.g., Ni/NiO) and $CO_2$ acceptor (e.g., CaO). With the reactor at a temperature of 600–800°C, a feed stream of hydrocarbon (e.g., natural gas, fuel oil) and steam is passed through the reactor where it forms $H_2$ and $CO_2$. The energy needed for this reaction is provided by the simultaneous exothermic reaction between $CO_2$ and CaO to form $CaCO_3$, and the removal of $CO_2$ effectively drives the reforming reaction to high conversion. A secondary reactor stage operating at lower temperature can be included to decrease the $CO_2$ partial pressure even further, and subsequently increase the reforming reaction conversion. Once the $CO_2$ capacity is saturated, the reactor is regenerated by passing air through the bed. Oxygen reacts with the Ni catalyst to produce NiO. The heat liberated during this reaction be used to decompose the $CaCO_3$ to CaO, thereby regenerating the $CO_2$ acceptor. Conceptually, this approach can enhance the conversion of the reforming reaction to produce higher purity hydrogen product and also provide means for regenerating the $CO_2$ acceptor without cumbersome solids handling issues.

## 2.8. References for Adsorptive Reactors

Alpay, E., Kenney, C. N., Scott, D. M. (1993) Simulation of rapid pressure swing adsorption and reaction processes. *Chem. Eng. Sci.*, 48, 3173–3186.

Alpay, E., Chatsiriwech, D., Kershenbaum, L. S., Hull, C. P., and Kirkby, N. F. (1994) Combined reaction and separation in pressure swing processes. *Chem. Eng. Sci.*, 49, 5845–5864.

Anand, M., Carvill, B. T., Sircar, S. (1995) Process for Operating Equilibrium-Controlled Reactions. USA Patent Pending.

Anand, M., Hufton, J. R., Mayorga, S. G., Nataraj, S., Sircar, S., and Gaffney, T. R. (1996) Sorption enhanced reaction process (SERP) for the production of hydrogen. *Proceedings of the 1996 US DOE Hydrogen Program Review*, Vol. II, 537–552.

Bassett, D. W., and Habgood, H. W. (1960) A gas chromatographic study of the catalytic isomerization of cyclopropane. *J. Phys. Chem.*, 64, 769-773.

Bergh, C. (1946) Hypersorption process for the separation of light gases, *Trans. AIChE*, 42, 665.

Beroza, M., and Coad, R. A. (1967) Reaction gas chromatography, in *The Practice of Gas Chromatography*, L. S. Ettre and A. Zlatkis, eds., Interscience, pp. 461–510.

Bjorklund, M. C., and Carr, R. W. (1996) presented at AICHE National Meeting, Chicago, IL, paper 195a.

Brenan, K. E., Campbell, S. L., Petzold, L. R. (1989) *Numerical Solutions of Initial-Value Problems in Differential-Algebraic Equations*, Elsevier, New York.

Bridges, S., and Barker, P. E. (1993) Continuous cross current chromatographic reactors, in *Preparative and Production Scale Chromatography*, P. Ganetsos and P. E. Barker, eds., Marcel Dekker, New York, pp. 113–126.

Broughton, D. B. (1984) Production scale adsorptive separations of liquid mixtures by simulated moving bed technology. *Sep. Sci. Tech.*, 19, 723–736.

Brun-Tsekhovoi, A. R., Zadorin, A. N., Katsobashvili, Y. R., and Kourdyumov, S. S. (1988) The process of catalytic steam reforming of hydrocarbons in the presence of a carbon dioxide acceptor. Proceedings of the 7th World Hydrogen Energy Conference 2, 885–900.

Burch, R., Squire, G. D., and Tsang, S. C. (1989) Direct conversion of methane into methanol. *J. Chem. Soc. Faraday Trans. I*, 85, 3561–3568.

Carr, R. W. (1993) Continuous reaction chromatography. In *Preparative and Production Scale Chromatography,* P. Ganetsos and P. E. Barker, eds., Marcel Dekker, New York, pp. 421–447.

Carvill, B. T., Hufton, J. R., Anand, M., and Sircar, S. (1996) Sorption enhanced reaction process. *AIChE Journal,* 42, 2765-2772.

Chatsiriwech, D., Alpay, E., Kershenbaum, L. S., Hull, C. P., and Kirkby, N. F. (1994) Enhancement of catalytic reaction by pressure swing adsorption. *Catalysis Today,* 20, 351–366.

Cho, B. K., Carr, R. W., and Aris, R. (1980) A new continuous flow reactor for simultaneous reaction and separation. *Sep. Sci. Technol.,* 15, 679–696.

Cho, B. K., Aris, R., and Carr, R. W. (1982) Mathematical theory of a countercurrent catalytic reactor. *Proc. Roy. Soc. (London)* A383, 147.

Coca, J., Adrio, G., Jeng, C-Y., and Langer, S. H. (1993) Gas and liquid chromatographic reactors. in *Preparative and Production Scale Chromatography,* P. Ganetsos and P. E. Barker, eds., Marcel Dekker, N. Y., pp. 449–475.

Dandekar, H. W., and Funk, G. A. (1995) Process for Methanol Production Using Simulated Moving Bed Reactive Chromatography. USA Patent 5,449,696, issued 12 September 1995.

Dandekar, H. W., Funk, G. A., Swift, J. D., and Maurer, R. T. (1996) PSA Process with Reaction for Reversible Reactions. USA Patent 5,523,326, issued 4 June 1996.

Djoharie, H. (1995) Dynamics of Adsorption/Desorption in a Simulated Countercurrent Moving Bed Separator. M. S. Thesis, University of Minnesota.

Feng, W., Knopf, F. C., and Dooley, K. M. (1994) Effects of pressure, third bodies, and temperature profiling on the noncatalytic partial oxidation of methane. *Energy and Fuels,* 8, 815–822.

Fish, B. B., and Carr, R. W. (1989) An experimental investigation of the countercurrent moving bed chromatographic reactor. *Chem. Eng. Sci.,* 44, 1773–1783.

Fish, B. B., Carr, R. W., and Aris, R. (1993) Design and performance of a simulated countercurrent moving bed separator. *AIChE Journal,* 35, 737-745.

Ganetsos, G., and Barker, P. E., eds. (1993) *Preparative and production scale chromatography,* Chromatographic Science Series, 61, Section IIIB.

Gesser, H. D., and Hunter, N. R. (1992) Direct conversion of methane to methanol (DMTM). in *Methane Conversion by Oxidative Processes,* E. E. Wolf, ed., Van Nostrand Reinhold, New York pp. 403–425.

Giddings, J. C. (1962) Theoretical basis for a continuous, large capacity gas chromatographic apparatus. *Anal. Chem.,* 34, 37.

Gluud, W., Keller, K., Schoenfelder, R., and Klempt, W. (1931) Production of Hydrogen. USA Patent 1816523, issued 28 July 1931.

Goto, S., Tagawa, T., Omiya, T. (1993) Dehydrogenation of cyclohexane in a PSA reactor using a hydrogen occlusion alloy. *Chem. Eng. Essays* 19(6), 1–12.

Goto, S., and Takahashi, Y. (1993) Continuous rotating annular chromatography. In *Preparative and Production Scale Chromatography.* P. Ganetsos and P. E. Barker, eds., Marcel Dekker, New York 1993, p. 127

Han, C., and Harrison, D. P. (1994) Simultaneous shift reaction and carbon dioxide separation for the direct production of hydrogen. *Chem. Eng. Sci.* 49(24b), 5875–5883.

Han, C., and Harrison, D. P. (1997) Multicycle performance of a single step process for $H_2$ production. *Sep. Sci. Tech.* 32(1–4), 681–697 (b).

Harrison, D. P., Han, C., and Lee, G. (1994) A Calcium Oxide Sorbent Process for Bulk Separation of Carbon Dioxide.VI., DOE Report, Project DE-AC21-89MC26366.

Hufton, J. R., Mayorga, S. G., Gaffney, T. R., Nataraj, S., and Sircar, S. (1997) Sorption enhanced reaction process (SERP) for hydrogen production. *Proceedings of the 1997 US DOE Hydrogen Program Review*, 179–194.

Hunter, N. R., Gesser, H. D., Morton, L. A., Yarlagadda, P. S., and Fung, D. P. C. (1990) Methanol formation at high pressure by the catalyzed oxidation of natural gas and by the sensitized oxidation of methane. *Appl. Catal.*, 57, 45–54 (1990).

Hutchings, G. J., and Scurrrell, M. S. (1992) in *Methane Conversion by Oxidative Processes,* E. E. Wolf, ed., Van Nostrand Reinhold, N. Y. pp. 201–258.

Imai, H., Tagawa, T., and Kuraisbi, M. (1985) Acceleration effect of hydrogen storage alloys for the catalytic dehydrogenation of cyclohexane. *Mat. Res. Bull.* 20, 511–516.

Kadlec, R. H., and Vaporciyan, G. G. (1993) Periodic Chemical Processing System. USA Patent 5254368, issued 19 Oct 1993.

Keefer, B. G. (1989) Gas Phase Chemical Reactor. USA Patent 4,816,121, issued 28 March 1989.

Keefer, B. G. (1990) Demonstration of the TCPSA cycle chemical reactor for hydrogen processes. *Int. J. Hydrogen Energy* 15(7), 463–471.

Kirkby, N. F., Morgan, J. E. P. (1994) A theoretical investigation of pressure swing reaction. *Trans. Inst. Chem. Eng.*, 72, 541–550.

Kokes, R. J., Tobin, H., Emmett, P. H. (1955) New microcatalytic chromatographic technique for studying catalytic reactions. *J. Am. Chem. Soc.*, 77, 5860.

Kruglov, A. V. (1994) Methanol synthesis in a simulated countercurrent moving bed adsorptive catalytic reactor. *Chem. Eng. Sci.*, 49, 4699–4716.

Kruglov, A. V., Bjorklund M. C., Carr, R. W. (1996a) Optimization of the simulated countercurrent moving bed chromatographic reactor for the oxidative coupling of methane. *Chem. Eng. Sci.*, 51, 2495–2950.

Kruglov, A. V., Bjorklund, M. C., and Carr, R. W. (1996b) presented at 14th International Symposium on Chemical Reaction Engineering, Brussels, Belgium.

Kuczynski, M., Ooteghem, A., and Westerterp, K. R. (1986) Methanol adsorption by amorphous silica alumina in the critical temperature range. *Colloid Polymer Sci.* 264, 362–367.

Kuczynski, M., Oyevaar, M. H., Pieters, R. T., and Westerterp, K. R. (1987) Methanol synthesis in a countercurrent-gas-solid-solid trickle flow reactor. An experimental study. *Chem. Eng. Sci.*, **42,** 1887-1898.

Kuo, J. C. W. (1992) Engineering evaluation of direct methane conversion processes. In *Methane Conversion by Oxidative Processes,* E. E. Wolf, ed., Van Nostrand Reinhold, New York, pp 483–526.

Kurdyumov, S. S., Brun-Tsekhovoi, A. R., and Rozental, A. L. (1996) Steam conversion of methane in the presence of a carbon dioxide acceptor. *Petroleum Chemistry* 36(2), 139-143.

Lee, I. D., Kadlec, R. H. (1988) Effects of adsorbent and catalyst distributions in pressure swing reactors. *AIChE Symp. Ser.*, 84, 167-176.

Leonard, B. P. (1979) A stable and accurate convective modelling procedure based on quadratic upstream interpolation. *Comput. Methods Appl. Mech. Eng.*, 19, 59.

Lu, Z. P., and Rodrigues, A. E. (1994) Pressure swing adsorption reactors: Simulation of three step one bed process. *AIChE Journal*, 40, 1118-1137.

Lunsford, J. H. (1995) The catalytic oxidative coupling of methane. *Angew. Chem. Int. Ed., Engl.*, 34, 970-980.

Lyon, R. K. (1996) Methods and Systems for Heat Transfer by Unmixed Combustion. USA Patent PCT/US96/03694, filed 18 March 96.

Martin, A. J. P. (1949) Summarizing paper. *Discuss. Faraday Soc.*, 7, 332.

Matherne, J. P., and Culp, G. L. (1992) Direct conversion of methane to $C_2$'s and liquid fuels: Process economics. In *Methane Conversion by Oxidative Processes*, E. E. Wolf, ed., Van Nostrand Reinhold, New York pp 463-482.

Mazzotti, M., Kruglov, A., Neri, B., Gelosa, D., and Morbidelli, M. (1996) A continuous chromatographic reactor—SMBR. *Chem. Eng. Sci.*, 51, 1827-1836.

Nataraj, S., Carvill, B. T., Hufton, J. R., Mayorga, S. G., Gaffney, T. R., Brzozowski, J. R. (1996) Process for Operating Equilibrium-Controlled Reactions. USA Patent, Pending.

Paryjczak, T (1986) *Gas Chromatography in Adsorption and Catalysis*, Wiley, New York

Petroulas, T., Aris, R., and Carr, R. W. (1985) Analysis of the countercurrent moving bed chromatographic reactor. *Comp. and Maths. with Appls.*, 11, 5 34.

Ray, A., Tonkovich, A. L., Aris, R., and Carr, R. W. (1990) The simulated countercurrent moving bed chromatographic reactor. *Chem. Eng. Sci.*, 45, 2431-2437.

Ray, A. K., Carr, R. W., and Aris, R. (1994) The simulated countercurrent moving bed chromatographic reactor: A novel reactor separator. *Chem. Eng. Sci.*, 49, 469-480.

Ray, A. K., and Carr, R. W. (1995a) Experimental study of a laboratory scale simulated countercurrent moving bed chromatographic reactor. *Chem. Eng. Sci.*, 50, 2198-2202.

Ray, A. K., and Carr, R. W. (1995b) Numerical simulation of a simulated countercurrent moving bed chromatographic reactor. *Chem. Eng. Sci.*, 50, 3033-3041.

Rhee, H. K., Aris, R., and Amundson, N. R. (1986). *First Order Partial Differential Equations*, vol. 1, *Theory and Application of Single Equations*. Prentice Hall, Englewood Cliffs, N.J.

Ruthven, D. M. (1984) *Principles of Adsorption and Adsorption Processes*. Wiley, New York

Smith, D. H. J. (1987) The gas conversion challenge. *Pet. Tech.*, 332, 10.

Tonkovich, A. L. Y., Carr, R. W., and Aris, R. (1993) Enhanced $C_2$ yields from methane oxidative coupling by means of a separative chemical reactor. *Science*, 262, 221-223.

Tonkovich, A. L. Y., and Carr, R. W. (1994) A simulated countercurrent moving bed chromatographic reactor for the oxidative coupling of methane: Experimental results. *Chem. Eng. Sci.,* 49, 4647–4656.

Viswanathan, S., and Aris, R. (1974) Countercurrent moving bed chromatographic reactors. *Adv. Chem. Ser.,* 133, 191.

Vaporciyan, G. G., and Kadlec, R. H. (1987) Equilibrium limited periodic separating reactors. *AIChE Journal,* 33, 1334–1343.

Vaporciyan, G. G., and Kadlec, R. H. (1989) Periodic separating reactors: Experiments and theory. *AIChE Journal,* 35, 831–844.

Wardwell, A. W., Carr, R. W., and Aris, R. (1982) Continuous reaction chromatography: The dehydrogenation of cyclohexane over $Pt/Al_2O_3$. In *Chemical Reaction Engineering Boston,* Wei and Georgakis, eds., *ACS Symp. Ser.,* 196, pp. 297–306.

Westerterp, K. R. (1988) Process for Carrying out a Chemical Equilibrium Reaction. USA Patent 4731387, issued 15 March 1988.

Westerterp, K. R., Bodewes, T. N., Vrijland, M. S. A., and Kuczynski, M. (1988) Two new methanol converters. *Hydrocarbon Processing* (November), 69–73.

Westerterp, K. R., and Kuczynski, M. (1987) A model for a countercurrent gas–solid–solid trickle flow reactor for equilibrium reactions. The methanol synthesis. *Chem. Eng. Sci.,* 42, 1871–1885.

Yang, R. T. (1987) *Gas Separation by Adsorption Processes.* Butterworth, Stoneham, MA.

## 2.9. General Texts on Adsorption and Adsorption Processes

Barton, P.I., Modeling and Simulation of Combined Discrete/Continuous Processes, Ph.D. Thesis, Imperial College, London, May, 1992.

Basmadjian, D., *The Little Adsorption Book,* CRC Press, Boca Raton, FL (1997).

Ganestos, G. and Barker, P. (eds.), *Preparative and Production Scale Chromatography,* Dekker, NY (1993).

International Conferences on Fundamentals and Applications of Adsorption (Proceedings).

Jarvis, R.B., Robust Dynamic Simulation of Chemical Engineering Processes, Ph.D. Thesis, Imperial College, London, May, 1993.

LeVan, M. D. ed., *Fundamentals of Adsorption* (Fifth Int. Conference on Adsorption, Asilomar), Kluwer, Boston (1996).

Li, N. and Strathmann, H. edS., *Separations Technology,* Engineering Foundation, New York (1988).

Liapis, A. I. ed., *Fundamentals of Adsorption* (Second Intl. Conference on Adsorption, Santa Barbara), Engineering Foundation, New York (1987).

Mersmann, A. B., and Scholl, S.E. eds., *Fundamentals of Adsorption* (Third Intl. Conference on Adsorption, Sonthofen), Engineering Foundation, New York (1991)

Myers, A. L., and Belfort G., eds., *Fundamentals of Adsorption* (First Intl. Conference on Adsorption, Schloss Elmau), Engineering Foundation, New York (1984).

Rodrigues, A. E., and Tondeur, D. eds., *Percolation Processes,* Proc. of NATO ASI, Espinho, Portugal (1978). Sijthoff and Noordhoff, Alphenaan den rijn, Holland (1981).

Ruthven, D.M., Farooq, S., and Knaebel, K. *Pressure Swing Adsorption,* VCH, New York (1994).

Ruthven, D. M., *Principles of Adsorption and Adsorption Processes,* John Wiley, New York (1984).

Suzuki, M., *Adsorption Engineering,* Kodansha-Elsevier, Tokyo (1990).

Suzuki M., ed., *Fundamentals of Adsorption* (Fourth Int. Conference on Adsorption), Kodansha, Tokyo (1993).

Vansant, E. F., and Dewolfs, R. eds., *Separations Technology,* Proc. of Int. Symp. on Gas Separation Technology, Antwerp, Sept. 1989.

Vansant, E. F. ed., *Separations Technology,* Proc. of Third Int. Symp. On Separation Technology, Antwerp, Aug., 1993.

Wakao, H., and K. Kaguei, *Heat and Mass Transfer in Packed Beds,* Gordon and Breach, New York (1982).

Wankat, P. C., *Large Scale Adsorption and Chromatography,* CRC Press, Boca Raton, FL (1986).

Yang, R. T., *Gas Separation by Adsorption Processes,* Butterworth, Stoneham, MA (1987).

For information on adsorbents see also the Proceedings of the International Zeolite Conferences (eleven conferences spanning the period 1967–1996). For a useful compilation of isotherm data see

Valenzuela, D. P., and A.L. Myers, *Adsorption Equilibrium Handbook,* Prentice Hall, Englewood Cliffs, NJ (1989).

# 3

# Membrane Technologies

**PRIMARY AUTHORS**

Kamalesh Sirkar
New Jersey Institute of Technology

Pushpinder Puri
Air Products and Chemicals, Inc.

Anna Lee Y. Tonkovich
Battelle Pacific Northwest National Laboratory

**CONTRIBUTING AUTHORS**

Richard W. Baker
Membrane Technology and Research, Inc.

Dibakar Bhattacharyya
University of Kentucky

Douglas E. Fain
Lockheed Martin Energy Systems

Mike Harold
DuPont

Jamie Hestekin
University of Kentucky

Jerry Lin
University of Cincinnati

Hans Wijmans
Membrane Technology and Research, Inc.

**EDITOR**

David R. Shonnard
Michigan Technological University

## 3.1. Membrane Technology Overview

This section provides an overview of membrane technologies relevant to this monograph. The first part of this section will provide a general overview in terms of the existing membrane separation processes, commonly employed membrane materials, membrane module types, and membrane selection. For an extended treatment, see Ho and Sirkar (1992), Mulder (1996), and Rautenbach and Albrecht (1989). The second part will discuss a variety of engineering, economic, environmental, and energy considerations intrinsic to the introduction of membrane separation technologies into chemical processes and systems.

### 3.1.1. General Overview

A membrane is interposed between two bulk phases in a membrane separation process. The membrane is a separate phase that is immiscible with and distinct from both bulk phases. The bulk phases are generally either liquid or gaseous or both; but they are rarely solid. The most common membrane phase is that of a solid which is either porous, microporous, or nonporous. The pores may contain a gas, a liquid, a gel, or other fluid phase. A liquid phase can also act as a membrane. The membrane phase may or may not have fixed electrical charges. The membrane thickness is always very small. The most common membrane material is polymeric, while inorganic membranes that are of ceramic (and/or metallic) origin are also being increasingly employed.

The feed gas or liquid mixture to be separated flows on one side of the membrane (Figure 3.1) in the separation device. A second bulk phase, gas or liquid, flows on the other side of the membrane and is variously called the *permeate*, the *sweep*, the *strip stream*, the *receiving phase*, the *dialyzate*, etc. In some membrane processes, the second stream is produced from the feed stream; no external stream is introduced into the membrane device from the outside (Figure 3.1a). In others, an external stream is introduced into the permeate side (Figure 3.1b). Conditions are created (in general) such that the partial pressure $p_i$ or concentration $c_i$ of any species $i$ in the feed or the total pressure $P$ of the feed stream is greater than the value of the corresponding quantity in the permeate stream. (Such conditions ensure that the gradient of chemical potential $\mu_i$ of any species $i$ across the membrane is nonzero, that is, $\nabla \mu_i \neq 0$ since the driving force is defined as $-\nabla \mu_i$). The difference in $\mu_i$ across the membrane drives the transport of any species $i$ from the feed stream to the permeate side. The membrane, however, selectively allows one of the species to go through the membrane much faster than the other species. Consequently, the permeate stream is enriched in the species which moves through the membrane much faster. The fraction of the feed remaining on

## 3. Membrane Technologies

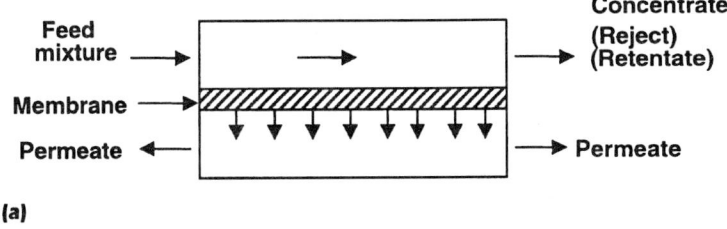

(a)

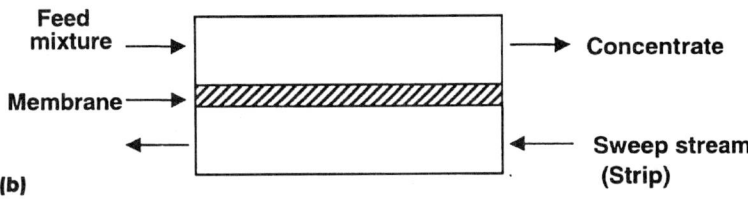

(b)

**Figure 3.1.** (a) Membrane separation device in which the permeate stream is obtained from the feed stream only. (b) Membrane separation device in which a separate stream is introduced from outside into the permeate side of the device.

the feed side and exiting the device is called the *reject* or the *retentate* or the *concentrate* (Figure 3.1a). This stream is highly enriched in the species which is either rejected by the membrane or is transported much more slowly through the membrane. [*Exception:* In the electrodialysis process (see Section 3.1.1.1.6), the feed stream leaving the membrane device is call the diluate whereas the strip (wash) stream introduced from outside to the permeate side is called the concentrate.]

The mechanism by which a species is rejected or favored by the membrane can be due to a variety of phenomena characteristic of individual membrane separation processes. In general, the concentration of a species in the membrane at the feed–membrane interface is different from that in the feed phase at the same interface. The extent of this partitioning or distribution of the species between the feed phase and the membrane phase is often quite important for separation. A species that is favored by the membrane phase will have a higher membrane-phase concentration than another species. This concentration will drive the diffusion of the species (or will be convected) through the membrane. The diffusivity or the convective characteristic of the transport of a species is, in addition, often crucial for separation. Higher diffusivity can lead to a higher transport rate. Finally, the species has to be partitioned at the membrane–permeate interface into the permeate (or the sweep or the strip or the receiving fluid phase). The overall transport rate of any species is directly influenced by the species partition coefficients at these two interfaces between the membrane and the two surrounding bulk fluid phases as well as by the species transport coefficients (e.g., diffusion coefficients, etc.) in the membrane phase.

The rate at which a species $i$ is transported through the membrane from the feed to the permeate side is in many cases proportional to its partial pressure $(p_i' - p_i'')$ or its concentration difference $(c_i' - c_i'')$ across the membrane. The flux of species $i$ through a membrane of effective thickness $\delta_m$ may be described by

$$\text{Flux} = \left(\frac{Q_i}{\delta_m}\right)\{(p_i' - p_i'') \text{ or } (c_i' - c_i'')\} \frac{\text{moles}}{\text{cm}^2 - \text{sec}} \quad (3.1)$$

Here superscripts "/" and "//" refer, respectively, to the feed side and the permeate side. The quantity $Q_i$ is called the *permeability coefficient* of species $i$ and $(Q_i/\delta_m)$ the *permeance* of species $i$. The quantity $Q_i$ for any species may be represented for illustrative purposes as a product of the species diffusion coefficient in the membrane and the species partition coefficient between the membrane and the fluid phases. It is desirable to have a high value of $(Q_i/\delta_m)$ for a species needed to be transported selectively through the membrane from the feed mixture. The species which is to be retained preferentially by the membrane should have as low a value of the permeance as possible. For a given membrane material, the membrane thickness $\delta_m$ is generally reduced to as low as a level as possible ($<0.1\ \mu\text{m}$). Such a low thickness of the effective membrane layer has to be supported on a much more porous and rigid substrate of much larger thickness. Practical membranes have, therefore, most often a *composite* structure; they are called *thin film composites* (TFCs).

Commonly, the term *solute rejection* or *solute retention* is employed in liquid separations to quantify the extent of retention of a give species by the membrane. These terms are defined by

$$R = \frac{c_i''}{c_i'} \quad (3.2)$$

A value of $R = 1$ implies $c_i'' = 0$; that is, the solute $i$ is completely retained by the membrane. Correspondingly, $R = 0$ implies that the solute concentration in the permeate stream on the other side of the membrane is the same as that of the feed. Usually processes where solvent is to be purified of a given solute by permeating the solvent through the membrane or a solute is to be concentrated by removing the solvent through the membrane employ such a definition. A more general definition that follows can be used to identify the extent of membrane preference for species $i$ over $j$. For two species $i$ and $j$ in the feed stream being separated by the

## 3. Membrane Technologies

membrane, an indicator of the extent of enrichment of the permeate stream in species $i$ vis-à-vis species $j$ is the *separation factor* $\alpha_{ij}$:

$$\alpha_{ij} = \frac{c_i''/c_j''}{c_i'/c_j'} \tag{3.3}$$

$$\alpha_{ij} = \frac{p_i''/p_j''}{p_i'/p_j'} \tag{3.4}$$

Under appropriate conditions (e.g., very low permeate side partial pressure or concentration compared to that on the feed side), this separation factor is called the *ideal separation factor* $\alpha_{ij}^*$ and is obtained from the following relation

$$\alpha_{ij}^* = \frac{Q_i/\delta_m}{Q_j/\delta_m} \tag{3.5}$$

To produce a permeate stream highly enriched in species $i$, the value of $\alpha_{ij}^*$ should be as high as possible. In addition, the magnitude of $(Q_i/\delta_m)$ should be large to reduce the membrane surface area needed to transfer a certain amount of species $i$ to the permeate stream. Further, this membrane surface area should be packed in a compact piece of equipment. Membrane separation devices, therefore, should possess the following characteristics: a membrane that is highly selective to species $i$ over $j$ (high $\alpha_{ij}^*$), has a high value of $Q_i$, very low value of $\delta_m$, and a high membrane surface area per unit equipment volume.

The membrane device design should also be such as to minimize the flow pressure drop in the phases flowing in the device. The device design should also ensure a high species transfer coefficient in the fluid phases in the direction of transport normal to the membrane surface. Lastly, the membrane performance and the device must be stable in the physical and chemical environment for an extended period (preferably three years or more). The above-mentioned requirements have to be satisfied for the membrane separation technology to succeed in the chosen application (unless economics indicate otherwise).

The transmembrane transport of a species $i$ considered so far was driven by a difference in partial pressure or concentration of species $i$ or a difference in total pressure across the membrane. In particular membrane separation processes, an *electrical potential difference* is applied by means of a cathode and an anode on two different sides of a membrane. In such a case, the membrane most often has fixed ionic charges. Only ionic species driven by the electrical field can move through such charged membranes (except electroosmotically driven water). Different rates of transport of different ionic species through such membranes lead to sepa-

separation due to the intrinsically different rates of transport of different species through a particular membrane. These processes are, therefore, characterized as *rate-governed membrane separation processes*.

Recent innovations have, however, led to membranes also being used in *equilibrium-based separation processes* like gas absorption/stripping, solvent extraction, distillation, etc. (Ho and Sirkar, 1992). In such processes, the two bulk phases present on two sides of the membrane contact each other through the pores of the membrane. Separation achieved relies on the difference in the composition of the two contacting bulk phases at equilibrium. The microporous/porous membrane does not *usually* participate in such a separation except to achieve nondispersive operation and other operational advantages like absence of flooding/loading, etc. Conventional analysis of equilibrium separation processes identified above are often adopted for such membrane-based equilibrium separation processes.

### 3.1.1.1. Existing Membrane Separation Processes

This subsection will present a brief introduction to different membrane separation processes employed in industrial practice. However, we provide first a compact basis for gross operational characterization of the membrane processes based on the nature of the feed stream to be separated, the separation goal, the driving force for separation, and the nature of the permeate stream.

The feed streams to be separated are either liquid solutions/suspensions or gaseous mixtures with or without particulates. For *feed liquid solutions/suspensions*, the following membrane-based separations processes may be relevant: *microfiltration, ultrafiltration, nanofiltration, reverse osmosis, dialysis, electrodialysis, emulsion liquid membranes, membrane-based solvent extraction, pervaporation, membrane-based stripping*. For separation of *gaseous mixtures, membrane gas permeation, membrane-based gas absorption,* and *vapor permeation processes* are employed. To remove particulates from gaseous feed streams, *microfiltration* is often adopted.

When a feed liquid mixture to be separated contains particles larger than 10 $\mu$m, conventional filtration and other separation methods are adopted to remove the particles. When the feed liquid suspension contains colloidal or fine particles in the size range of 10 $\mu$m to 0.1 $\mu$m, *microfiltration* is employed to retain these particulates as the liquid solution is forced through the membrane by hydrostatic pressure and appears as the filtrate. For macrosolutes in the range of 500 daltons to 1000 Å, *ultrafiltration* is frequently selected to provide an essentially macrosolute-free permeate solution. *Nanofiltration* is increasingly the membrane process of choice to remove/recover solutes of molecular weight between 150 and 500 daltons from the feed liquid solution as the solvent and smaller solutes go through the membrane. For separations of feed solutions having microsolutes of molecular weight <150 daltons, any one of the following membrane processes may be selected: *reverse osmosis, dialysis, electrodialysis, emulsion liquid membranes, membrane-based solvent extraction, pervaporation, membrane-based stripping*. Of these processes, the last two can be used only if the solutes to be removed from the solution are volatile.

The following processes employ high to moderate pressures in the feed solution to drive the solvent present in the feed selectively through the membrane to the lower pressure permeate side in preference to solutes/particles. They are, therefore, identified often as *pressure-driven membrane separation processes, and include reverse osmosis, nanofiltration, ultrafiltration, microfiltration*. The process goal is to recover either a purer solvent as a permeate or a concentrated solution of the microsolute/macrosolute/particles. A high rate of flow of the solvent through a nonporous/porous membrane is characteristic of these processes.

When microsolutes are removed selectively through the membrane instead of the solvent to either purify the solvent or recover the microsolute or both, the following processes are relevant: *dialysis, emulsion liquid membranes, membrane-based solvent extraction, membrane-based stripping, electrodialysis*. In all of these processes except electrodialysis, chemical potential differences between the feed and the receiving phases in the form of concentration differences (or effective concentration or partial pressure differences in the case of membrane-based stripping) drives the solute flux which is always diffusive in nature. In electrodialysis, however, an electric field is applied to the separation device through an anode and a cathode to selectively transport ionic species obtained from electrolytic solutes through ion-exchange membranes; the solvent is purified of the electrolytes and a concentrated solution of electrolytes is obtained on the permeate side.

In the liquid separation processes identified so far (except membrane-based stripping), both phases on the two sides of the membrane are liquid. In the *membrane-based stripping process*, volatile species from the liquid feed are removed by stripping through a membrane into a gas or vapor stream on the other side of the membrane. The amount of volatile species present in the feed liquid is generally small. In the process of *pervaporation*-based separation of volatile liquid mixtures, the permeate side is gaseous; the feed liquid is generally at atmospheric pressure. The permeate side is generally under considerable vacuum. Volatile solutes from the feed permeate through the membrane under an effective concentration or partial pressure difference; the volatile permeated solutes are recovered by condensation. The species permeating through the membrane may be the solvent or microsolutes in the feed or both.

For the separation of a gas mixture by *membrane gas permeation* and the separation of a gas–vapor mixture or vapor–vapor mixture by vapor permeation, the driving force across the membrane is due to partial pressure difference of the transported species and the process is diffusive. The two phases on two sides of the membrane are gaseous in nature. In the process of *membrane-based gas absorption*, the feed gas mixture is contacted via a membrane with an absorbent liquid which preferentially absorbs the solute gas or vapor species; the driving force is due to the effective partial pressure difference of the species between the two phases.

This and other principal characteristics of commercialized membrane separation processes are provided in Table 3.1 (adopted from Ho and Sirkar (1992)). Each of these individual membrane separation processes is described very briefly now.

### 3.1.1.1.1. Reverse Osmosis (RO)

In the reverse osmosis process, the pressure of the feed liquid solution, $P_f$, is raised to a high level above that of the permeate side $(P_p)$ to permeate the solvent through the membrane, while the microsolute is rejected by the membrane. The excess pressure of the feed solution over that of the permeate solution $\Delta P = (P_f - P_p)$ must be significantly larger than the difference in osmotic pressure of the feed and the permeate, $\Delta \pi = (\pi_f - \pi_p)$ (Figure 3.2). The membrane does have a small solute permeability in most practical cases so that the permeate is not completely solute-free. The solvent is driven by the difference $(\Delta P - \Delta \pi)$ across the membrane. A simplified expression for the solvent flux through the membrane in RO is

$$N_{solvent} = \left(\frac{Q_{solv}}{\delta_m}\right)(\Delta P - \Delta \pi) \qquad (3.6)$$

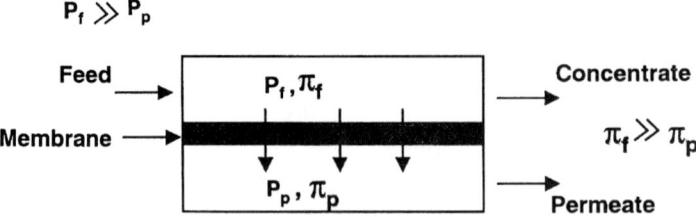

**Figure 3.2.** Reverse Osmosis-based separation of the solvent from the microsolute.

**Table 3.1** Principal Characteristics of Commercialized Membrane Separation Processes*

| Separation Process | Separation Goal | Nature of Species Retained (Size) | Nature of Species Transported through Membrane | Minor/Major Species Transported | Driving Force | Mechanism for Transport/Selectivity | Phase of Feed and Permeate Streams |
|---|---|---|---|---|---|---|---|
| Gas permeation | Stream/streams enriched or depleted in a particular species | Larger species retained unless highly soluble | Gaseous. Smaller species/more soluble species | Either | Concentration gradient (partial pressure difference) | Solution-diffusion | Gaseous |
| Pervaporation | Same as above | Same as above | More soluble/smaller/ more volatile nonelectrolytes | Preferably minor species | Concentration gradient, temperature gradient | Solution-diffusion | Liquid feed, gaseous permeate |
| Dialysis | Macrosolute solution free of microsolute, microsolute solution free of macrosolute | >0.02 μm retained, >0.005 μm retained in hemodialysis | Microsolute, smaller solute | Minor species. Solvent transported under osmotic unbalance | Concentration gradient | Sieving, hindered diffusion in microporous membranes | Liquid |
| Electrodialysis | Solution free of microions, concentrated solution of microions, fractionation of microions | Co-ions, macroions[a] and water retained | Microionic species | Minor ionic species, small amounts of water by electroosmosis | Electrical potential gradient, electro-osmosis (minor amount) | Counter-ion transport via ion-exchange membranes | Liquid |

[a] Macromolecules with charge
* Adapted from Ho and Sirkar (1992) with permission.

**Table 3.1** (continued)

| Separation Process | Separation Goal | Nature of Species Retained (Size) | Nature of Species Transported through Membrane | Minor/Major Species Transported | Driving Force | Mechanism for Transport/Selectivity | Phase of Feed and Permeate Streams |
|---|---|---|---|---|---|---|---|
| Reverse osmosis | Solvent free of all solutes, concentrated solution of microsolutes | 1 to 10 Å microsolute species | Solvent. Species retained may be electrolytic or volatile | Major species solvent | Hydrostatic pressure gradient vs. osmotic pressure gradient | Preferential sorption/capillary flow (solution–diffusion–imperfection) | Liquid |
| Ultrafiltration | Solution free of macrosolute, macrosolute solution free of microsolute, macrosolute fractionation | 10 to 200 Å macrosolute species | Solution of microsolutes | Major solvent, minor microsolutes | Hydrostatic pressure gradient vs. small osmotic pressure gradient | Sieving | Liquid |
| Microfiltration | Solution free of particles, gas free of particles | 0.02 to 10 μm particles | Solution/gas free of particles | Major solvent, minor microsolutes/macrosolutes | Hydrostatic pressure gradient | Sieving | Liquid or gas |
| Emulsion liquid membrane | Stream/streams enriched or depleted in a particular species | Generally not size-selective except in host–guest chemistry | Species with high solubility in liquid membranes | Minor species. Can be major species in organic mixture separation | Concentration gradient, pH gradient | Solution–diffusion, facilitated transport | Generally liquid feed, emulsion containing permeate |

| Process | Products | Species retained | Species transferred | Role of membrane | Driving force | Basis for separation | Phases |
|---|---|---|---|---|---|---|---|
| Membrane solvent extraction | Solute-free feed solution, solute-enriched extractant solution | Microsolute and macrosolute not extracted | Species preferentially extracted, nonionic or ionic | Minor. Membrane as phase interface stabilizer | Concentration gradient, different solubilities in extractant | Selective partitioning between extract and feed phases | Liquid feed and extract, both immiscible |
| Membrane gas absorption | Solute-free feed gas, solute-enriched absorbent | Gaseous species not absorbed | Gaseous species preferentially absorbed | Usually minor. Membrane as phase barrier for absorbent | Concentration gradient, different solubilities in absorbent | Selective absorption between absorbent/feed | Gaseous feed, liquid absorbent, both immiscible |
| Membrane stripping | Solute-free feed liquid, solute enriched in stripping gas | Liquid species not stripped | Volatile species stripped from feed | Usually minor. Membrane as phase barrier for feed | Concentration gradient, different volatilities | Higher volatility between feed/stripping gas | Liquid feed, stripping gas, both immiscible |
| Vapor permeation | Stream/streams enriched or depleted in a particular species | Less soluble species retained | Vapor. More soluble species | Either | Concentration gradient (partial pressure difference) | Solution–diffusion | Gaseous |

The membranes and systems currently available can treat dilute aqueous feed solutions of inorganic/organic electrolytes as well as organic nonelectrolytic microsolutes. Reverse osmosis membranes currently available are not suitable for organic solvents. The range of pressures of feed solutions varies from 5.6–10.5 MPa for seawater desalination (where NaCl rejection values are upward of 0.99) to 1.4–4.2 MPa for brackish water desalination (NaCl rejection are in the range of 0.95–0.98) (Bhattacharyya et al., 1992). The following *short* but illustrative list of solutes or collections of solutes are rejected well by current RO membranes: NaCl, $Na_2SO_4$, phenols, heavy metals, many pesticides, COD (chemical oxygen demand), TOC (total organic carbon), many organics, especially of somewhat higher molecular weight. Many currently available membranes have poorer rejections of hydrogen-bonding solutes like urea, inorganic mineral acids like HCl, $HNO_3$, HF, etc. Extensive details of the performance of various RO membranes are provided in Williams et al. (1992).

Depending on the membrane material, the allowable feed solution pH range varies from 4–6 (cellulose acetate) to 3–11 (thin film composite membranes). These membranes have generally almost no chlorine tolerance. The feed solutions have to be pretreated to remove suspended solids/particulates, colloids, scale-forming salts, metal oxides, biological/organic foulants. Spiral-wound membrane modules as well as hollow-fiber membrane modules are almost exclusively used for clean feeds; tubular and plate-and-frame devices are employed for highly fouling feeds. The chapters by Bhattacharyya et al. (1992) provide an excellent introduction/overview of such aspects. Recent notable developments are identified in Sirkar (1997).

*3.1.1.1.2. Nanofiltration (NF)*
Nanofiltration, sometimes called loose RO, is employed commercially to reject divalent anions and the corresponding heavy metals from aqueous solutions. The rejection values of $SO_4^{2-}$-containing salts are around 0.98, where as, the rejection of monovalent anions like $Cl^-$ is lower, around 0.1 to 0.4. Thus high osmotic pressure feeds (e.g., high NaCl concentration) may be treated without requiring the feed to have a high pressure (0.3 to 1.4 MPa). Such rejections of divalent ions are achieved by a negative fixed charge on the skin layer of the thin film composite membrane. Uncharged nanofiltration membranes are also being used to retain larger molecular weight inorganic/organic (especially organic) compounds in the molecular weight range of 150–500 daltons. The solvent is exclusively water. Details of NF membrane and their applications, especially in drinking water treatment, are available in Bhattacharyya et al. (1992) and Williams et al. (1992).

New organic-solvent-resistant nanofiltration membranes have become available from Kiryat Weizmann, Rehovot, Israe (through Koch Membrane Systems,

Wilmington, MA). These membranes may be employed to retain organic molecules of larger molecular weights (200–700 daltons) as the solvent is allowed to pass through the membrane under pressure. Both types of NF membranes are generally thin film composites.

*3.1.1.1.3 Ultrafiltration (UF)*
In the ultrafiltration process, solvent from aqueous solutions of colloids, proteins, macromolecules, latices, viruses, etc. is driven by hydrostatic pressure through the membrane while the macrosolutes are retained by the membrane. The solvent is almost always water; the feed solution pressure can vary from 2 to 10 atmospheres. The UF membrane pore sizes can vary from 10 to 1000 Å; correspondingly the macrosolutes retained by the membrane have molecular weights of 500 daltons and upward. The driving force for the solvent flux is proportional to ($\Delta P - \Delta \pi$) as in RO (Figure 3.2). The macrosolutes rejected/retained by the membrane have a pronounced tendency to form a deposit on the feed side of the membrane surface. Such a deposit creates considerable additional resistance to solvent transport (the macrosolute retention is also affected). The design of ultrafiltration membrane device and the operating conditions (e.g., backpulsing or high shear) can reduce the solvent flux-limiting role of this deposit.

As the solvent is removed from the feed solution through the UF membrane, microsolutes (e.g., salt) are removed as well. Sometimes UF is used to purify the solution of macrosolutes by removing microsolute impurities. This is conventionally done by continuously adding the solvent (i.e., water) to the feed solution as UF progresses and removes the microsolute impurity in the filtrate. This process is called *diafiltration* and is frequently practiced for protein purification. Comprehensive introductions to UF and UF technology are available in Cheryan (1986) and Kulkarni et al. (1992). There are numerous applications of UF in food industry, biotechnology and pharmaceutical industries, water treatment, process industries, pulp and paper industries. Table 3.2 illustrates a few of these (Kulkarni et al., 1992).

*3.1.1.1.4. Microfiltration (MF)*
Pressures of 1 psig to 50 psig (0.11–0.45 MPa) are employed to remove solvents through microfiltration membranes which retain particles in the size range of 0.01 to 10 $\mu$m. Microfiltration membranes are used routinely to treat both aqueous and organic solutions. Microfiltration is used to remove particles from gases as well. Particulate materials removed via MF include bacteria, whole cells, large macromolecules (MW $\geq 10^6$), aerosols, dust, pigment particles, catalyst particles, etc. When particles are retained on the membrane surface and solvent passes

through the membrane for particle sizes larger than membrane pore sizes, the mechanism is identified as *surface filtration*. If the membrane pore size is somewhat larger than that of the particles and yet the particles are captured inside the membrane thickness, *depth filtration* is said to be active; in fact, a significant part of the microfiltration industry employs fibrous/granular filter medium to do just that as the filtered solvent/gases flow through. Such filter media are sometimes radically different from conventional membranes which are quite thin.

Surface filtration-based MF is operated either as a *deadend filtration* or as *crossflow filtration*. Deadend filtration is necessarily a periodic process; the filtration is stopped every so often to remove a cake of particles from the filter top or to

**Table 3.2.** List of Some Ultrafiltration Applications*

| Process | Separation |
|---|---|
| Electrophoretic paint | Process rinse water, recycle paint to dip tank, allow reuse of rinse water |
| Cheese whey | Concentrate/fractionate proteins from lactose and inorganics |
| Juice clarification | Remove haze components from apple juice |
| Textile sizing agents | Recover polyvinyl alcohol after scouring of woven goods |
| Wine clarification | Remove haze components from red and white wines |
| Oil/water emulsions | Metal cutting oils (lubricants) concentrated from wastewater for incineration |
| Polymer latex | Latex emulsions concentrated from wastewater |
| Dewaxing | Separation of wax components from lower paraffins |
| Deasphalting | Solvent recovery/recycle for deasphalting of heavy crudes |
| Egg-white preconcentration | Partial dewatering before spray drying |
| Fermentation broth | Separate low molecular weight organics/therapeutic agents from cells or cell debris |
| Kaolin concentration | Partial dewatering of clay slurry before centrifugation |
| Water treatment | Concentration before sludge dewatering |
| Affinity membranes | Retain ligand complex from noncomplexed proteins |
| Reverse osmosis pretreatment | Retain colloidal silica, bacteria |

* From Ho and Sirkar (1992) with permission.

replace the filter. In crossflow filtration, the liquid flows tangentially over the membrane surface at a high velocity to keep the layer of rejected particles on the membrane very thin and achieve a high solvent flux. Often a backpulse of filtrate is used to breakup this deposited particle layer in a periodic fashion. Such backpulsing is being increasingly employed in crossflow liquid filtration. Successful application of such backpulsing is appearing in gas filtration as well (see Sirkar, 1997).

The efficiency of the microfiltration membrane/medium in removing particles from a liquid or gas stream is usually expressed either by a removal efficiency defined by $(1 - P)$ where

$$P = \frac{\phi_{out}}{\phi_{in}} \qquad (3.7)$$

or by the log reduction value (LRV)

$$\text{LRV} = \log_{10}\left(\frac{\phi_{in}}{\phi_{out}}\right) \qquad (3.8)$$

Here $\phi$ represents the particle volume fraction in the fluid under consideration. LRV indicates the number of orders of magnitudes by which the particle concentration is reduced in the treated fluid ($\phi_{out}$) from that in the feed fluid ($\phi_{in}$). In many applications, the LRV required may be 8–10 or even higher. This is particularly true for removal of bacteria, pyrogens, viruses, etc. in biological/sterile applications. Particle removal needs in semiconductor manufacture are also very stringent. Such stringent requirements have led to frequent use of membranes that are not asymmetric and are homogeneous unlike RO and UF where asymmetry is a requirement for high solvent flux. Successful applications of such microfiltration membranes are integral to the development of modern pharmaceutical industry, semiconductor manufacture, beverage industry and laboratory-scale analysis. A comprehensive introduction to the fundamentals, design and application of microfiltration has been provided by Davis (1992), Goel et al. (1992) and Mir et al. (1992) in Chapters 31–35 of Ho and Sirkar (1992).

*3.1.1.1.5. Dialysis*
In this process, microsolutes (salt, urea, creatinine, sulfuric acid, caustic soda, etc.) diffuse through a microporous membrane from the feed solution to the permeate stream (the dialyzate) driven by its concentration gradient. The feed, the dialyzate and the membrane pores contain the same solvent. Macrosolutes in the feed (protein, hemicellulose, etc.) are retained by the membrane and are, therefore, purified of the microsolutes. The pressures on the two sides of the membrane are essentially the same. For the feed solution to be purified/separated in a

continuous manner, fresh dialyzate solution has to be supplied continuously to the membrane device so that a high concentration difference can be maintained (Figure 3.1b). (If the two solutions, feed solution and the dialyzate, are kept on two sides of the membrane without replenishing the dialyzate solution, both chambers will soon have the same concentration of the microsolute.) An inevitable result of the dialysis process is a dialyzate solution that is more dilute in the microsolute than the feed solution. If the microsolute is a pollutant, use of dialysis is questionable. If the microsolute is an acid or an alkali having commercial value, dialysis is useful in recovering a usable chemical, although in a more dilute form.

Dialysis is used extensively for treatment of end-stage renal disease, for example, waste metabolites in blood are removed by dialysis into a dialyzate stream. Additional applications include reduction of alcohol from beer using water as the dialyzate stream. Kessler and Klein (1992) have provided an extensive treatment of dialysis as a membrane process.

### 3.1.1.1.6 Electrodialysis (ED)

The process of electrodialysis uses ion-exchange membranes and an electrical field applied through an anode and a cathode in a cell. Two types of ion-exchange membranes are simultaneously used in conventional ED units: cation-exchange membrane and anion-exchange membrane. Due to the requirement of electroneutrality (i.e., no net charge anywhere), a membrane having fixed positively charged group allows negatively charged ions (*counterions*) to enter the membrane immersed in water. Such membranes are called *anion-exchange membranes*. These membranes reject positively charged ions (*coions*). A membrane having fixed negatively charged groups, on the other hand, allows positively charged ions (*counterions*) to enter the membrane immersed in water. Such membranes are identified as *cation-exchange membranes*. Both kinds of membranes reject *coions* (ions having charge similar to the fixed charge in the membrane). Any such ion-exchange membrane interposed between an anode and cathode in an electrical cell with an externally applied voltage will allow only specific ions to go through it. For example, a cation-exchange membrane will allow only a cation to move toward the cathode through the membrane just as an anion-exchange membrane will allow only an anion to move through the membrane toward the anode.

These basic phenomena are utilized in an electrodialysis unit having multiple sets of alternating cation-exchange membrane (C) and anion-exchange membrane (A) to remove electrolytes from an aqueous solution. Consider Figure 3.3, where an aqueous solution containing an electrolyt (e.g., NaCl) has to be purified. Consider the solution between an anion-exchange membrane and a cation-exchange membrane identified as the *diluate*. Cations (e.g., Na⁺ in this solution) will move through the cation-exchange membrane toward the cathode; anions (e.g., Cl⁻ in this solution) will move in the opposite direction toward the anode through the anion-exchange membrane. Thus, this diluate solution is purified of NaCl; the solutions on the other sides of the two ion-exchange membranes, however, get concentrated in NaCl due to the same phenomena taking place through the two other membranes. The aqueous solution in these two adjacent chambers are called the concentrate. Electrical potential is, thus, used in ED to remove electrolytic solutes from water; simultaneously a concentrated electrolytic solution is produced. One anion-exchange membrane, one cation-exchange membrane and the diluate and concentrate solution in adjacent chambers constitute one unit cell in such a process (Figure 3.3). A stack of, say, 100 such unit cells (up to 500) are put together in one device for commercial operation.

This separation process is economically efficient for electrolytic solutions that are neither very concentrated nor very dilute. When the electrolytic solution becomes concentrated, the selectivity of the ion-exchange membranes for the counterions decreases; the coions are not rejected as well. When the elecrolytic solution becomes too dilute, not enough ions are there to conduct electricity. A method called *electrodeionization* (briefly described below) is then practiced to produce ultrapure water.

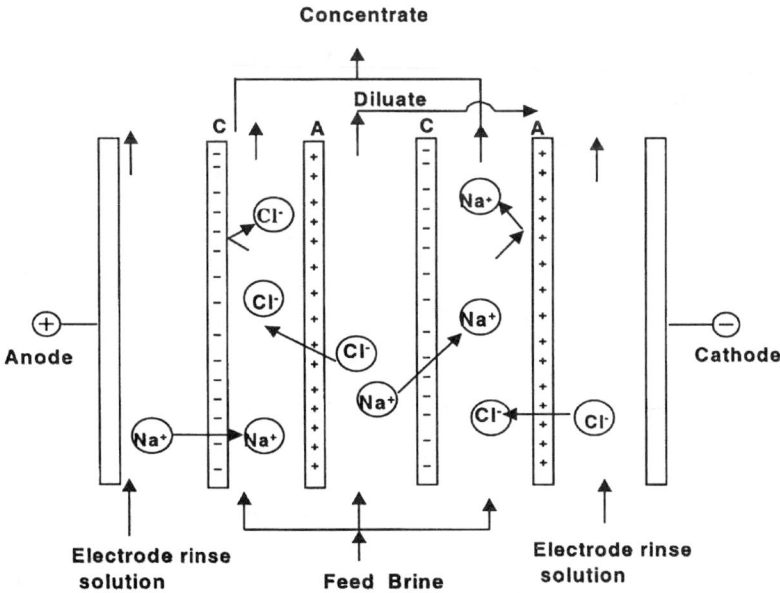

**Figure 3.3** Schematic of feed brine purification by the electrodialysis process.

In industrial practice, ED employs a stack of thin flat ion-exchange membranes (0.5–2 mm thickness). As many as 200–1000 ion-exchange membrane sheets exist between the two electrodes. The adjacent membranes, cation-exchange membrane and anion-exchange membrane, are separated by a thin plastic spacer/gasket. These spacers are designed to distribute the flow of the diluate fluid/concentrate fluid efficiently with high mass transfer coefficient to ensure that species are supplied efficiently from the bulk solution to the boundary layers (next to the membrane) which get depleted of ionic species due to selective transport through the membrane. In the *electrodeionization process* (Ganzi, 1988) these spacers are designed to hold a bed of mixed ion-exchange resin beads which create conditions for conduction of the current when the bulk concentration of the electrolyte is drastically reduced. This technique produces high purity water without requiring extra chemicals needed to regenerate ion-exchange beds in a conventional ion-exchange bed-based demineralization process. This process is highly useful for pollution prevention.

Another application of electrodialysis process is based on what is called "*water-splitting*" technology. This technology converts a salt stream (e.g., NaCl in water) into two streams, one containing NaOH and the other containing HCl. In many chemical processes, salts are produced via acid–base neutralization needed in the process. To revert back to the acid and the base, *bipolar membranes* (Figure 3.4) are used. Consider a cation-exchange and an anion-exchange membrane placed together with almost no gap between them. This assembly of a bipolar membrane is put between two electrodes and a voltage applied. Water present in between the two membranes is split to produce $H^+$ and $OH^-$ ions which conduct the current. However, $H^+$ ions are conducted away through the cation-exchange membrane toward the cathode just as $OH^-$ ions are removed through the anion-exchange membrane toward the anode. If this bipolar membrane-based cell incorporates also monopolar membranes used in conventional ED, such that the cell location where $H^+$ ions come is also supplied with $Cl^-$ ions and the cell location where $OH^-$ ions appear is also supplied with $Na^+$, we obtain two separate streams, one containing NaOH and the other containing HCl. Thus, the solution of NaCl fed into particular chambers of this combined stack of bipolar and monopolar membranes can be converted into two separate streams of the acid HCl and the base NaOH. Comprehensive treatments of electrodialysis are available in Strathmann (1992) and Schaffer and Mintz (1980).

A widely commercialized technique that uses ion-exchange membranes of only one type in an electrodialysis stack without any cathode or anode (so no applied voltage) is called *diffusion dialysis*. Consider a stack of anion-exchange membranes designed to allow free acid (e.g., HCl) to diffuse rapidly into fresh water (dialysate) while metallic cations are rejected. One can, therefore, recover

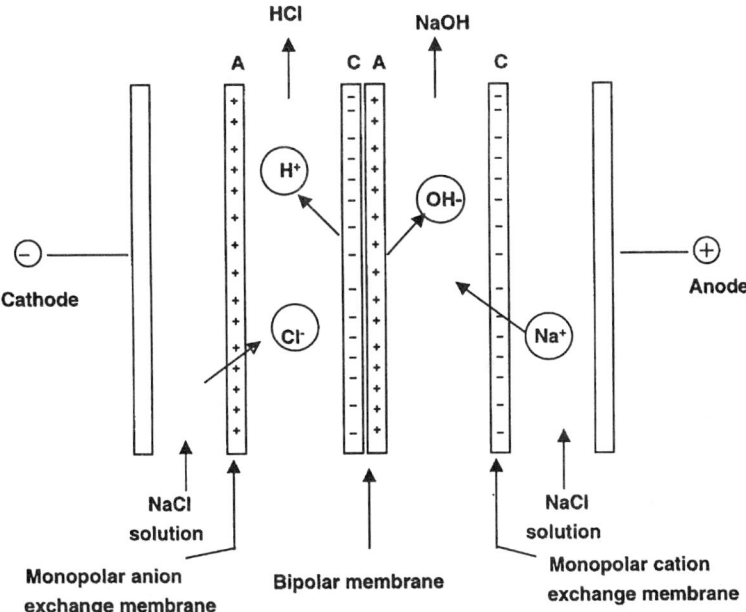

**Figure 3.4.** Bipolar membrane in an electrodialysis cell containing monopolar ion exchange membranes to produce the acid HCl and the base NaOH from a NaCl solution.

HCl, $HNO_3$, HF, $H_2SO_4$, etc. from spent pickling baths, metal finishing baths, battery wastes, etc. Similarly, using a stack of cation-exchange membranes, one can easily recover caustic from metal finishing baths, photographic baths, etchant streams, etc. These techniques satisfy the need for resource recovery and waste minimization. In either types of stack, the feed solution to be treated and fresh water are introduced in adjacent chambers between any pair of similar ion-exchange membranes. The fresh water streams (the dialysate) pick up either the acid of the alkali as the feed solution is stripped of the acid or the alkali (Kobuchi et al., 1987).

Another technique employing a single ion-exchange membrane without any applied potential difference in a stack of such membranes is called *Donnan Dialysis* (Kessler and Klein, 1992). Unlike all other processes and techniques discussed so far, Donnan dialysis relies for separation on differences in concentration at *equilibrium* between two solutions on two sides of an ion-exchange membrane. Consider a dilute solution 2 of a salt MX separated by a cation-exchange membrane from a concentrated solution 1 of an acid HX having a common anion between the two solutions. The anion cannot be exchanged between the two solutions. However, cations can be. Since the $H^+$ ion concentration in the acid solution 1 is high, it diffuses to the salt solution side 2. To maintain electroneutrality (no net charge anywhere), $M^+$ ions get concentrated in side 1 to a level far higher than in the original salt solution (side 2) to satisfy the Donnan equilibrium relation:

$$\frac{[M^+]_1}{[M^+]_2} = \frac{[H^+]_1}{[H^+]_2} = K \text{ (a constant independent of the species)} \quad (3.9)$$

where subscript 1 is the acid side of the membrane having a very high $[H^+]$ and subscript 2 is the salt side having a low $[H^+]$. Therefore, the salt side $M^+$ concentration is highly reduced. A similar process may be carried out with anion-exchange membranes whereby anions will be exchanged between the two solutions on two sides. For example, $OH^-$ ions from a concentrated solution of NaOH are used to transfer citrate ions from citrus juices to the caustic solution and achieve sweetening of citrus juices.

*3.1.1.1.7. Emulsion Liquid Membranes (ELM)*

In liquid membrane processes, a thin layer of a liquid is created between the feed phase and the strip/receiving phase. This thin liquid layer acts as the membrane and is immiscible with both the feed as well as the receiving phases. In the emulsion liquid membranes, both the feed and the strip phases are generally liquid. When the feed solution is aqueous, the strip solution is also aqueous but the liquid membrane is an oil phase immiscible with both. For an organic feed solution and an organic strip solution, the liquid membrane is an aqueous phase immiscible with the two organic phases.

The special characteristics of ELMs consists in their being essentially double emulsions: for an aqueous feed they are water/oil/water (W/O/W) type whereas for an organic feed they are oil/water/oil (O/W/O) type. Such a double emulsion is achieved by first creating, for example, a W/O emulsion and then dispersing this emulsion in a continuous water phase. Correspondingly, for a feed oil phase, the initial emulsion is O/W which is then dispersed in a continuous oil phase. In reality, the membrane phase appears as the continuous phase in the emulsion globules (100–2000 $\mu$m in diameter) which contain 1–3-$\mu$m droplets of the internal phase while the emulsion globules themselves are dispersed in an external continuous phase (Figure 3.5).

Unlike other membranes made of solid material (organic or inorganic), ELMs are made for treating a certain amount of feed phase or for a certain time and then they are destroyed. The liquid membrane phase is usually recirculated to participate in a cycle of membrane making, membrane-based separation and membrane destruction via emulsion breaking. Figure 3.6 shows the process of emulsification, then formation of the double emulsion and removal of the desired species from the feed to the receiving phase and finally destruction of the ELMs.

3. Membrane Technologies

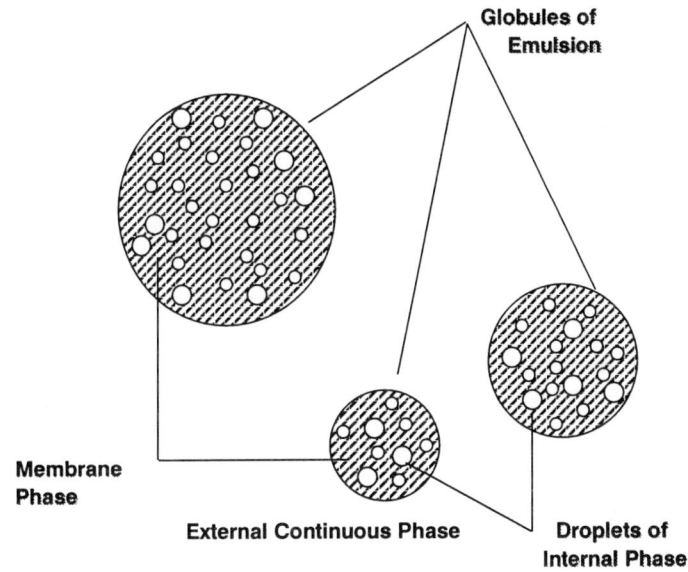

**Figure 3.5.** Schematic of an emulsion liquid membrane system.

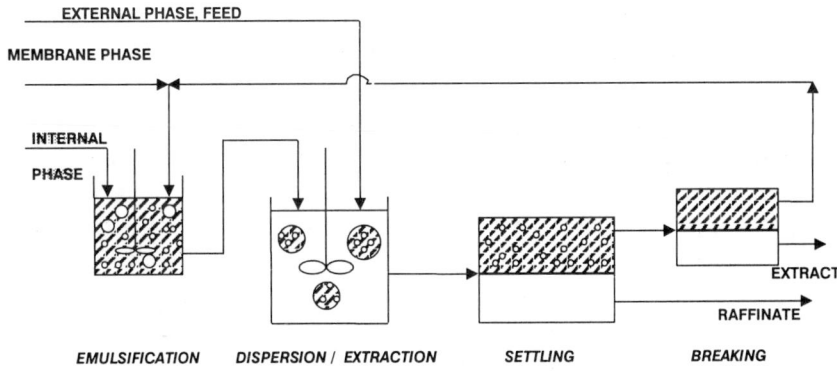

**Figure 3.6.** Schematic of a continuous emulsion liquid membrane process [From Ho and Li (1992) with permission.]

This is followed by collection of the enriched and the depleted phases and recirculation of the liquid membrane material for renewed membrane formation.

This technique is generally applied to two basic types of selective transport of a species in a feed solution. *In the first type*, a species in the feed solution (e.g., phenol in water) preferentially partitions into an oil membrane, diffuses to the strip side and partitions into the strip phase which, however, contains a reagent (e.g., NaOH) that produces sodium phenolate which cannot diffuse back to the feed. This is a case of diffusion with preferably an almost zero concentration of the species in the receiving phase. *In the second type*, a species in the feed solution (e.g., a heavy metal ion) undergoes an interfacial complexation reaction with a complexing agent in the liquid membrane. The complex diffuses through the liquid membrane to the strip side where the conditions facilitate dissociation of the complex and release of the feed species into the strip solution. This is a form of *facilitated transport*. These transport processes are implemented very rapidly in an ELM process due to the extraordinarily large surface area created in a double emulsion structure.

Emulsion liquid membranes have been studied extensively in laboratories for removing heavy metals, alkali metal cations, anions (e.g., chloride, sulfate, chromates), ammonia, phenol, organic and inorganic acids, amino acids from aqueous solutions as well as fractionation of hydrocarbons. There are three large scale commercial applications: removal of $Zn^{2+}$ from wastewater in the viscose fiber industry, removal of phenol from wastewater, and removal of cyanide from waste liquors in gold processing. Mixer-settlers or column extractors are generally employed in large-scale applications of the dispersion/extraction and settling part of the ELM process shown in Figure 3.6. A comprehensive introduction to the fundamentals and theory of ELM processes are provided in Ho and Li (1992). Additional treatments on design and applications of ELM processes are provided in Chapters 38–40 of Ho and Sirkar (1992).

*3.1.1.1.8. Membrane-Based Solvent Extraction*
In conventional solvent extraction, a solvent is employed to selectively extract a solute species from another immiscible liquid phase by dispersing one phase as drops in the other phase in a mixer. After the extraction is over, the dispersed phase is separated from the continuous phase via coalescence in a settler. In membrane-based solvent extraction (Figure 3.7) neither phase is dispersed in the other; rather the two immiscible phases flow on two sides of a porous/microporous membrane. One phase preferentially wets the membrane pores and contacts the other phase at each pore mouth on the other side of the membrane. By maintaining the phase excluded from the membrane pores at a pressure equal to or higher than that of the phase in the membrane pores, the phase interfaces are immobilized at each pore mouth on the side of the membrane having the phase excluded from the pores. At the phase interface, solute partitioning, extraction or back extraction goes on nondispersively.

Compared to conventional solvent extraction in mixer-settlers and agitated columns, this technique has a number of advantages: (a) there is no need to have a density difference between the two phases; (b) there is no flooding or loading in

## 3. Membrane Technologies

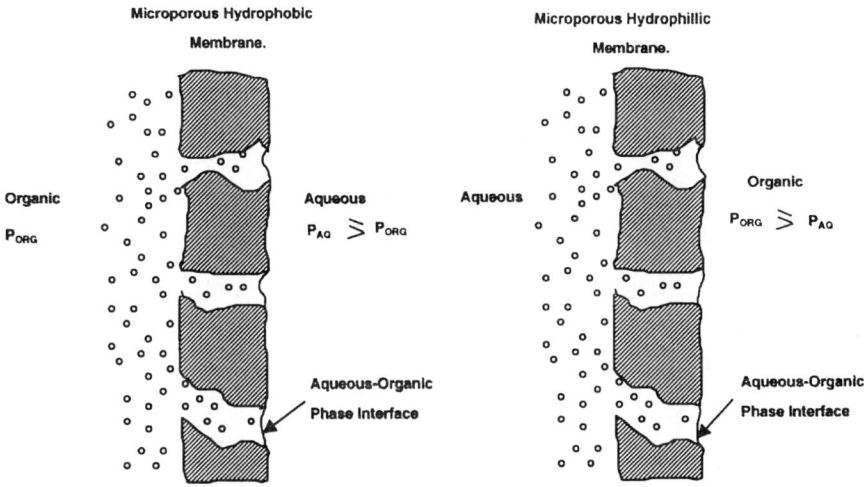

**Figure 3.7.** Nondispersive solvent extraction concept employing either a microporous hydrophobic membrane of a microporous hydrophilic membrane.

the column; (c) there is no emulsion formation; (d) the interfacial area is known and is independent of the interfacial tension; (e) loss of solvent is reduced to a minimum; and (f) when the membrane is in the hollow fine fiber form, the surface area/volume can be very high. As a result the height of a transfer unit in a membrane-based solvent extraction device can be very small. To achieve such advantages of nondispersive extraction, the excess pressure of the phase not in the pores must not exceed a breakthrough pressure.

Such a technique has been studied in the laboratory and pilot plant scale for a variety of applications. These include organic pollutant extraction from wastewater, metal extraction and back extraction, pharmaceutical extractions, fermentation product extraction and aromatics extraction. The technique is also being employed in extractive bioreactors as well as enzymatic resolution of racemic mixtures. The last one is being used in one industry in a large scale (Sirkar, 1997).

The rate of extraction of a species from the feed to the solvent being used for extraction is primarily controlled by the mass-transfer coefficients in the feed liquid phase, the membrane pores and the extracting solvent phase. The rate is also determined by the interfacial area between the two phases in a given location of membrane extractor. Maximization of the mass-transfer rate is achieved by increasing the value of each transfer coefficient and the amount of interfacial area. Large-scale membrane solvent extraction devices, therefore, employ hollow fine fibers which can provide values of interfacial area/equipment volume that are 5–30 times larger than conventional extraction devices. The mass transfer coefficient on the shell-side of such hollow-fiber extractors are increased in the latest devices by locally having transverse flow of the shell-side liquid instead of parallel flow. These recent devices achieve countercurrent flow over the module by means of a shell-side baffle. The microporous hollow fine fibers are either hydrophobic (primarily of polypropylene) or hydrophilic. An introduction to the technique, methods of analysis, design and equipment details is provided in Prasad and Sirkar (1992), Reed et al. (1995) and Sirkar (1995).

*3.1.1.1.9. Pervaporation*

This process is employed to selectively remove one of the volatile species from a liquid solution of at least two volatile species flowing at atmospheric pressure on one side of a nonporous membrane (Figure 3.8). The temperature of the liquid solution can be as high as the boiling point of the species desired to be removed, but they are generally somewhat lower. On the other side (permeate side) of the membrane, a reasonably high vacuum is maintained in most common forms of pervaporation. Sometimes a sweep gas is passed instead of maintaining a vacuum. This creates conditions for permselective evaporation of the liquid solute species through the membrane. Invariably, the gaseous permeate stream obtained through the membrane by means of a vacuum is condensed. If the membrane is sufficiently selective to the desired species and if the desired species is also more volatile than the others, then the condensate can be an almost pure liquid of the desired species. To successfully implement such a process, the partial

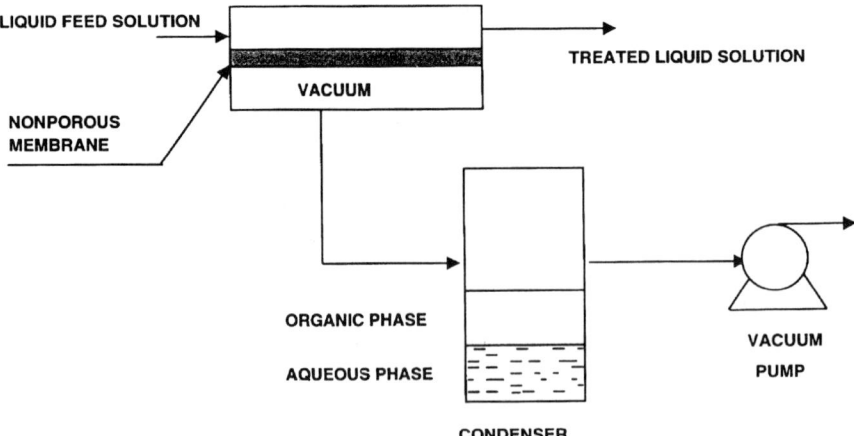

**Figure 3.8.** Pervaporation (PV) process schematic for separating an aqueous-organic feed solution.

pressure of the desired species in the permeate vapor stream must be considerably lower than the partial pressure of the species in a hypothetical gas phase in equilibrium with the feed solution. This partial pressure difference provides the driving force for permeation.

Large-scale commercial use of pervaporation has started. The membranes are polymeric and generally are in the form of a thin film composite. Flat membrane modules by GFT are the most common. Spiral-wound membranes as well as hollow-fiber membranes have been commercially employed. There are three application areas: (a) removal of water from an organic stream; (b) removal of organics from an aqueous stream, (c) separation of organic–organic mixtures.

Ethanol dehydration is practiced in a large scale to obtain a 99.5% + ethanol stream on the feed side as water is selectively removed through a membrane of polyvinyl alcohol (GFT) by vacuum from a feed solution containing 5–20 wt% water. Selectivities for water over alcohol may be as high as 2000–3000. Removal of water from other organics (e.g., xylene, methanol, methylethylketone, dichloroethylene, $n$-butanol, etc.) is also practiced commercially. Selectivities for water over the organics can be very high (the less polar the organic the higher the selectivity). Molecular sieve membranes are also being evaluated for such solvent drying applications. In the second application category, volatile organic compounds (VOCs) are being removed from their dilute solutions in water by rubbery membranes (commonly of polydimethylsiloxane) prepared as a composite on a porous glassy polymeric support. Three important examples are: (a) removal (and recovery) of VOC pollutants from water; (b) alcohol reduction in beer by removing alcohol through the membrane; and (c) recovery of aromas from food processing streams.

Potentially the application area of largest importance to chemical/petrochemical industry is the separation of organic–organic mixtures. This application area has not yet been blessed with a large-scale commercial application. Potential systems of great importance where the relative volatilities are low and traditional separation processes are costly are: (a) aromatics/aliphatics; (b) branched hydrocarbons from n-paraffins; (c) purification of dilute mixtures of isopropyl alcohol from hexane/heptane; and (d) mixtures of isomers. Fleming and Slater (1992) in a sequence of four chapters have provided a good introduction to pervaporation as well as other application details. Huang (1993) should be consulted for additional treatments. Sirkar (1997) has identified the large-scale tests done at Exxon by Ho on aromatics/saturates separation.

*3.1.1.1.10. Membrane-based Stripping*
In membrane-based stripping, microporous/porous membranes are employed in industrial practice to achieve stripping-based removal of volatile species from a

liquid (which is mostly water). In conventional stripping processes not employing membranes, air or $N_2$ or steam or other gases/vapors are bubbled through a liquid (e.g., an aqueous solution) to remove the dissolved gases or volatile species from the liquid. Vacuum may also be employed to strip the volatile species. Membrane-based stripping for the same gas–liquid system employs a porous/microporous membrane between the liquid and the stripping gas/vapor phase.

The membrane material and the operating conditions are such that the porous/microporous membrane pores are gas-filled and are not wetted by the liquid. For aqueous solutions, hydrophobic membranes are needed. Further, the liquid pressure (higher than that of the stripping gas) must not exceed that of the stripping gas by a critical value. This ensures that the resistance to transfer of the stripped species through the membrane pores is much less than that in the liquid phase. The difference between the partial pressure of the stripped species in equilibrium with the feed solution and that in the stripping gas drives the transport process. The resistance to transport for a volatile species not too soluble in the feed liquid is primarily controlled by the liquid phase. The membrane module design thus enhances the liquid phase transfer coefficient.

Membrane-based stripping is commercially used most often for removing volatile species from water. For deoxygenation of water (e.g., for boiler or semiconductor processing) vacuum (and sometimes $N_2$ purge) is pulled on the permeate side of the porous membrane. For stripping VOCs from water, air is employed. Decarbonation of aqueous process streams is an additional application. Hollow-fiber modules containing microporous hydrophobic polypropylene fibers are being used in large scale for the last 3–4 years; they employ either vacuum or a stripping gas ($N_2$/air) or both. Hollow-fiber membrane modules available now are not yet suitable for steam stripping. Consult Sirkar (1992) and Sengupta et al. (1995) for an introduction to and greater details on various aspects of membrane-based stripping.

For those liquids which may wet the pores of the microporous fibers, it is necessary to have a thin nonporous polymeric coating on the fiber internal diameter or outside diameter; the liquid should flow on the side that has the coating with a vacuum or a stripping gas. This last type of membrane-based stripping (Poddar et al., 1996a) is not yet employed in large scale.

*3.1.1.1.11. Membrane Gas Permeation*
The feed fluid phase in all membrane separation processes considered so far was liquid. In the three processes to be considered now, the feed fluid phase is gaseous. The mixture may be of permanent gases, gases and vapors, or vapors only. In this subsection, we consider membrane gas permeation for separation of gas mixtures. Although no special principles are involved when vapors are present

and are to be separated, selective permeation of vapors will be considered under *vapor permeation*.

In the membrane gas permeation process, the feed gas mixture flows on one side of a *nonporous membrane* (Figure 3.1a) and the gas mixture that permeates through the membrane flows on the other side. Sometimes, a sweep gas or a purge gas stream is introduced from an external source into the permeate side. It sweeps away the permeated gas mixture (Figure 3.1b) and reduces the partial pressure of the permeated gas species. The permeation process of any species is driven by a positive partial pressure difference between the feed side and the permeate side, that is, $(p_i' - p_i'')$. This is achieved under conditions of no sweep or purge gas stream by maintaining the feed gas pressure $(P)$ considerably higher than the permeate gas pressure $(p)$. The smaller the pressure ratio, $\gamma\ (=p/P)$, the higher the extent of separation. In fact, when $\gamma \to 0$, the highest separation is achieved and the value of $\alpha_{ij}$ becomes $\alpha_{ij}^*$ [see Eq. (3.5)]. For a given pressure ratio, the separation achieved is determined by the ratio of the species permeabilities $(Q_{im}/Q_{jm})$ for the gas-pair $i$-$j$. Almost pure $i$th species is obtained in the permeate stream only when $(Q_{im}/Q_{jm}) \to \infty$ and $\gamma \to 0$. Since $(Q_{im}/Q_{jm})$ varies between 2–100 for most systems, highly purified $i$th species is rarely obtained by a gas permeation process; membrane gas separation is thus generally considered suitable for bulk gas separation.

The feed gas pressure in membrane gas separation can vary from atmospheric to as high as 900–1100 psig. The permeate gas pressure is lower. The temperature of operation varies from ambient to 60–70°C. Gas mixtures successfully separated in large-scale in industry are: $O_2$-$N_2$; $CO_2$-$CH_4$; $H_2$-$CO$; $H_2$-$N_2$; $H_2$-$CH_4$; $He$-$CH_4$; $H_2S$-$CH_4$ (the first species preferentially goes through the membrane and is enriched in the permeate). Very large units are in operation throughout the world. The membrane modules are either in spiral-wound form or hollow fine fiber form. The membranes are generally polymeric in nature and the membrane structure may be asymmetric or composite. But the selective layer is nonporous for polymeric materials.

The fraction of the feed gas appearing as the permeate in a separator (Figure 3.1a) is called the *cut*. The larger is the membrane area for a given feed gas flow rate, the higher is the value of the cut. Usually, one is interested in a particular species in the mixture. This species may appear in the concentrate (reject) stream or in the permeate stream. In the production of $N_2$-enriched air (NEA), for example by an $O_2$-selective membrane, the higher the fraction of the feed $N_2$ appearing in the concentrate, the lower is the concentration of $N_2$ in this concentrate. If we look at the same separation as an example of producing $O_2$-enriched air in the permeate, the higher the $O_2$ concentration in the permeate, the lower is the $N_2$ concentration in concentrate (reject), and the lower is the value of the cut.

For a given membrane device under given operating conditions, the highest permeate composition is obtained at the lowest cut and at the lowest pressure ratio.

The nonporous polymeric membranes used for permanent gas separation are most often glassy since they provide much higher selectivity. A rubbery membrane may have a much higher value of the permeability coefficient $Q_{im}$ for any species $i$ but the selectivities are usually quite low except for a system like $H_2S/CH_4$ ( $\cong 30$) or $SO_2/N_2$ ( $= 10$). The permeability coefficient of a permanent gas through a rubbery membrane does not vary with the partial pressure of the gas. For a glassy membrane there is some dependence. If the species $i$ is a vapor, then the permeability coefficient is likely to be a strong function of the vapor partial pressure, especially in a rubbery membrane. An extensive introduction to the membrane gas permeation processes, design of gas permeation systems, their applications and economics is provided in the chapters by Zolandz and Fleming (1992) in the Membrane Handbook. Additional accounts of progress are available in Koros and Fleming (1993), Stern (1994) and Puri (1996).

Porous/microporous membranes are not used in commercial gas separation processes since they can provide only a limited selectivity (generated in Knudsen flow). Microporous membranes used in membrane reactors will provide a selectivity of $\alpha_{ij} = (M_j/M_i)^{1/2}$ for two gases $i$ and $j$ (where $M_i$ and $M_j$ are the molecular weights of species $i$ and $j$, respectively, and $M_i < M_j$) if the gases flow through the membrane is under *Knudsen diffusion* conditions. If the pore sizes and conditions are such that there is *Poiseuille flow*, there will be no separation.

In microporous membranes with pore sizes in the range of 5–10 Å, very high selectivity can, however, be developed if one of the species in the gas mixture is strongly adsorbed in the pore surface. This species can be transported by surface diffusion along the pore; simultaneously, other species not adsorbed on the pore surface may be blocked from being transported through the pores. Such a process is being developed to remove hydrocarbon vapors (strongly adsorbed on carbonized surfaces) from hydrogen via nanoporous carbon membranes. The membrane transport mechanism is identified as selective surface flow (SSF) (Rao and Sircar, 1993).

*3.1.1.1.12. Vapor Permeation*
When the feed gas mixture has some condensable species that are being separated by permeation through a membrane, the gas permeation process is specifically identified as *vapor permeation* process. When separation of a vapor mixture is involved, obviously each component of the feed is undergoing vapor permeation and the separation process involves *vapor permeation* only. Separation of vapor mixtures like water/ethanol, benzene/cyclohexane, and methanol/$n$-propanol involves vapor permeation just as separations of volatile organic compounds

(VOCs) like toluene, octane, trichloroethane, methanol, etc. from air/$N_2$ etc. are well-known examples of the vapor permeation process.

The vapor permeation process may be carried out with the feed mixture around atmospheric pressure and the permeate side under vacuum. The permeate is highly enriched in the more permeable and highly condensable vapor. Condensation and compression is employed on the permeate stream. *Alternately,* the feed mixture may be compressed (and in many cases followed by condensation) and then sent to a vapor permeation unit where the permeate may be at atmospheric pressure.

There are a number of commercial units recovering VOCs from a variety of inert gaseous streams. Such units employ rubbery (most often silicone) membranes which selectively and rapidly permeate VOCs in strong preference over $N_2$/air or other gases. The permeated vapor species is recovered by condensation in the permeated stream under vacuum. Cen and Lichtenthaler (1995) provide a compact introduction to vapor permeation.

*3.1.1.1.13. Membrane-Based Gas Absorption*

In traditional gas absorption processes, the gas mixture is dispersed as bubbles in a liquid or the liquid is dispersed as drops in the gas mixture. A particular gas species is removed from the gas by being selectively absorbed in the liquid through the gas–liquid interfaces created by the bubbles or drops. In membrane-based gas absorption, the gas and the liquid develop a gas–liquid interface at every membrane pore mouth in a porous/microporous membrane (Figure 3.9). The absorption occurs through this interface. The gas mixture and the liquid stream flow on two sides of the membrane. The membrane material is generally hydrophobic and the conditions of operation are such that the pores are gas-filled. For aqueous absorbent solutions that do not spontaneously wet the pores of a hydrophobic membrane, the pressure of the absorbent liquid is maintained higher than that of the gas to prevent gas dispersion in the liquid but not high enough to displace the gas from the pores. From the bulk gas, gas species diffuse into the gas-filled pore and are then absorbed into the liquid. Absorption can then be carried out nondispersively without any problem of loading, flooding, foaming, or weeping.

The flux of a gas species $i$ being absorbed can be described by means of the product of an overall mass-transfer coefficient $K_l$ (or $K_g$) and the concentration difference (or the partial pressure difference) of species $i$ between the gas and the liquid:

$$N_i = K_l(c_i^* - c_{ilb}) = K_g(p_{igb} - p_i^*) \qquad (3.10)$$

Here $p_{igb}$ is the partial pressure of species $i$ in the bulk gas and $p_i^*$ is the hypothetical gas phase partial pressure of $i$ in equilibrium with the bulk liquid. This

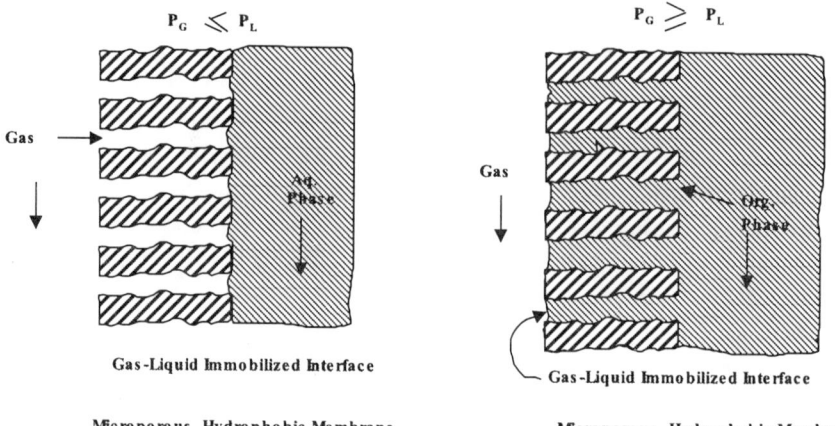

**Figure 3.9.** Nondispersive gas absorption concept employing a gas-filled or liquid-filled microporous hydrophobic membrane.

mode of description is identical to that used in conventional gas absorption processes where $K_l$ has contributions from the gas film and the liquid film. In membrane-based gas absorption, there is an additional resistance, that of the gas-filled pore. Usually this is negligible. The overall rate of transfer of species $i$ in a given absorber volume is obtained by

$$N_i a = W_i = K_l a (c_i^* - c_{ilb}) = K_g a (p_{igb} - p_i^*) \tag{3.11}$$

where $a$ is the interfacial area/volume. When the membrane absorber employs hollow fine fibers, the value of $a$ can be high anywhere from 25 to 100 cm²/cm³. Packed beds and plate towers, traditionally used for gas absorption however, have much lower values of $a$, sometimes by an order of magnitude. Thus, $K_l a$ or $K_g a$ is at least a few times higher in hollow-fiber membrane-based absorbers which are, therefore, very compact.

Many absorbents have some organic species which may slowly seep into the pore for extended time operation in membrane absorbers. Any liquid in the pore will increase the resistance to diffusion. To avoid such situations, sometimes the liquid side of the microporous membrane has a thin nonporous polymeric coating which is highly permeable to the species to be absorbed but is essentially impermeable to the liquid species, which could wet the pore. The condition for ensuring nondispersive operation (in terms of gas bubbling in the liquid) is that the liquid-side pressure should be higher than the gas-side pressure without the liquid appearing on the gas side (Poddar et al., 1996b).

Membrane-based gas absorption is being commercially employed for adding $CO_2$ to carbonate beverages, absorbing $CO_2$ from flue gases in amine solutions, supplying oxygen to fermentation broths, adding $O_3$ to purify water in semiconductor processing, oxygenation of blood, and others. This technique is likely to have some utility in gas–liquid reactors employed in chemical industry. Consult Sirkar (1992) and Reed et al. (1995) for fundamentals, design principles and applications of membrane-based gas absorption. Additional information on large-scale applications are provided in Sirkar (1995) and Sirkar (1997).

### 3.1.1.2. Membrane Materials

*3.1.1.2.1. Polymeric*
The very early polymeric membranes were made using cellulosic materials like cellulose nitrate, cellulose acetate and regenerated cellulose for reverse osmosis, ultrafiltration, gas separation and hemodialysis applications. Today, polymeric membranes are made essentially from all commercial polymers, including homopolymers, copolymers, polymer blends, and ionic polymers. As an example, the various polymers used in gas and liquid separation membranes are given in Table 3.3.

All polymeric materials used for membranes should have some amorphous structure. Heterogeneous blends are not desirable because they lack uniformity of the membrane performance and mechanical properties. Criteria for selecting a polymer for a given application are complex; however, they generally include properties such as flux, selectivity, mechanical integrity, and durability. Information on the physical and chemical requirements of the membranes and membrane durability is generally missing because it is either kept proprietary by the manufacturers or is not available at all. Membrane materials should have a modulus high enough so that in membrane form it has the mechanical strength to withstand operating pressure, pressure fluctuations, resistance to creep, and resistance to abrasion. It should have a glass transition temperature high enough to give it thermal stability and preservation of basic properties over a wide range of temperatures. It should be chemical resistant to bear the effect of a wide range of feed components and contaminants. Also, all polymers age with time at ambient conditions, and the aging phenomenon can be aggravated at membrane use conditions.

***Gas Separation Membrane Polymers.*** By far, polymeric gas separation membranes are most widely used in commercial applications. For gas separation applications, amorphous glassy polymers offer the best permselective properties. A polymeric membrane primarily transports gases through it by solution-diffusion mechanism. In this, the permeating gas first dissolves in the polymer and then is transported through the intersticial spaces in the polymer chains by diffusion. For a given gas pair, the ratio of the solubility of the two gases, sometimes referred to

**Table 3.3.** Membrane Polymers

| Polymer Class | Polymer Type |
|---|---|
| 1. Homopolymers | |
| 1.1. Glassy Polymers | |
| 1.1.1. Amorphous | Cellulose acetate (CA), polysulfone (PSF), polystyrene (PS), polypropylene (PP), polyimides (PI), polyarylates (PAR) |
| 1.1.2. Semi-crystalline Polyethylene | (PE), polytetrafluoroethylene (PTFE), polyvinyl alcohol (PVA), polyvinylidenefluoride (PVDF), polyamides (PA), polycarbonates (PC), polyphynelene oxide (PPO) |
| 1.1.3. Crystalline | None |
| 1.2. Elastomers | Polydimethysiloxane (PDMS), polyphosphazine |
| 2. Copolymers | |
| 2.1. Block Copolymers | Styrene-butadiene copolymer |
| 2.2. Random Copolymers | Nitrile-butadine rubber (NBR), styrene-butadiene rubber (SBR), acrylonitrile-butadiene-styrene rubber (ABS) |
| 2.3. Grafted Copolymers | PTFE, PVDF, etc., grafted with poly n-vinylpyrrolidone (PVP) or poly 4-vinylpyridine (P4VP) |
| 3. Polymer Blends | |
| 3.1. Homogeneous Blends | Nylons, PVP+PVA, PAN+PVP |
| 3.2. Heterogeneous Blends | None |
| 4. Ionic Polymers (Ion Exchange Membranes) | |
| 4.1. Mobile Counter Ion | $PSF(SO_3^-)H^+$, Nafion ($Li^+$, $Al^{+3}$ or $Na^+$) |
| 4.2. Fixed Counter Ion | Introduction of betaine group |
| 5. Conducting Polymers | Polyaniline (PA), polypyrrole, poly n-methyl pyrrole |

as the solubility selectivity, in the various polymers is generally a small number (1.2–1.8). The ratio of diffusion through a polymer matrix, referred as the diffusion selectivity, is however a strong function of the type of the polymer and polymeric membrane morphology, and is the major contributor to the overall gas transport and its selectivity. However, the permeability of a gas through a polymer bears an inverse relation to the selectivity of the gas pair. Thus causing a gas flux limitation through a highly selective polymeric membrane. This has led to the development of better membrane fabrication methods to achieve the thinnest possible defect-free barrier layer and the search for alternate membrane materials which follow different transport mechanisms.

The polymeric membranes for gas separation offer the advantages of cheaper raw material and ease of membrane and membrane module fabrication. However, they suffer from a broad resistance to contaminants and solvents. Furthermore, polymeric membranes can operate at only moderate temperatures (up to 100°C) and not very high pressures (~100 bar). Novel polymers are being developed which have better solvent resistance and can operate at higher temperatures.

Several remarkable developments have been made in recent years which permit the fabrication of mechanically sturdy membranes with separation layers only in hundreds of angstrom thicknesses. Almost all these novel methods are based on the fundamental concepts of asymmetric membrane structure formation, which has a dense skin integrally supported on a microporous support. The technique to make such a membrane was first taught by Loeb and Sourirajan (Loeb and Sourirajan, 1963). Their method consisted of making a polymer solution in a mixture of a volatile and nonvolatile solvent, from which the volatile solvent was partly evaporated prior to coagulation. More details of which can be found elsewhere (Kesting and Fritsche, 1993). Based on this concept, several researchers have developed novel commercial ways of making efficient gas separation membranes. Noteworthy examples are Kesting et al. (1989) who produced density gradient skinned membranes with skin thickness of 300–400 Å. These unitary membranes are made from one polymer and are economical only if the polymer used is relatively inexpensive. For specialty polymers, which offer superior separation properties, several novel methods have been proposed to make competitive membranes. These include the making of asymmetric coatings on commercially available microporous substrates (Puri, 1988; Puri et al., 1996) and the use of coextrusion methods to spin two polymers simultaneously (Ekiner, 1992), one providing the mechanical support while the other offering the separation properties. These methods are being used commercially to make gas separation membranes (e.g., L'air Liquide and Ube). These newer developments have virtually revolutionized the gas separation industry.

***Liquid Separation Membrane Polymers.*** In the earlier reverse osmosis (RO) applications, cellulose acetate membranes were used exclusively, however, currently more efficient polyamides are used. Polysulfone, PEI and PA have also been extensively used for the RO/UF (ultrafiltration) applications. Polymers with hydrophilic properties show a high degree of swelling in an aqueous environment, when not cross-linked. The flux of water through a hydrophilic membrane will be high. Generally flux and selectivity bear an inverse relationship for polymeric membranes. Thus, a balance has to be struck between the flux and selectivity by the choice of polymer for a given application. Unfortunately, most synthetic polymers are hydrophobic therefore, surface tension and contact angle (capillarity) are important measures. Most desalination membranes are made of modified polyamides or polyimides and are produced in-situ by interfacial polymerization (Cadolle, 1981) on a microporous polymer substrate (such as nonwoven PET or microporous PSF).

*3.1.1.2.2. Inorganic*
With the development of sol–gel, zeolite, and other methods, inorganic membranes can now be produced in composite construction, which produces much higher permeance and makes them commercially useful. In addition, there is quite an extensive list of other real advantages for inorganic membranes. The obvious advantages are the wider range of operating conditions and the long life cycle in very harsh environments. The membranes can be used in a very broad range of temperatures, from very low (cryogenic) temperatures to 600°C (maybe 1000°C). The operating pressures can be high vacuum to more than 1500 psid (pressure difference across the membrane).

Since membranes can be fabricated with a very broad range of inorganic materials, they can be made with materials compatible with almost any reactive environment. The most important advantage is the ability to design a membrane to provide very high separation factors for almost any separation process. The design advantage comes from being able to choose materials, pore size, and operating conditions to enhance one or more of the large number of different types of transport mechanisms, in order to maximize the separation factor for a given separation process. Since inorganic membranes are porous with large void fractions (free volumes), they can be made with very high permeance.

It is the current practice to compare membranes based on the cost per unit area of membrane. It would be more meaningful to compare membranes on the basis of cost per unit throughput per unit time per unit area. On a unit area basis, ceramic membranes may cost as much as 100 times or more that of polymer membranes. With spiral-wound polymer membrane rolls or hollow fibers, as much as 1000 times more membrane area can be packed into a unit volume than one can achieve with inorganic membranes. However, the permeance of the inorganic membranes may be 1000 to 10,000 times higher than the polymer membranes. Therefore, to achieve the same throughput only 1/1000 to 1/10,000 the area will be needed for the inorganic membranes, thus providing a cost advantage for the inorganic membranes. In addition, higher pressure drops can be achieved and the hydrodynamics can be much more favorable for the inorganic membrane.

For more than 20 years, expectation has been high that inorganic membranes would make major improvements in the economics of many industrial separation processes. Much research and development has been conducted during that time. A great deal of good experimental data has been accumulated, some of which is still not well understood. That work is showing the extensiveness of possibilities for inorganic membranes. These data give good evidence that inorganic membranes can be designed to accomplish a large number of industrial separations. Inorganic membranes have the potential for revolutionizing many industrial processes, even to the point of changing the fundamental processes for separations. That research and development (R&D) also shows that we do not yet have the means to manufacture most of these membranes in industrial configurations and at sufficiently low cost to make them economically feasible and generally available for more than a few niche markets. A review of the kinds of R&D being performed on inorganic membranes was given in the proceedings of The 1996 Fourteenth Annual Membrane Technology/Separation Planning Conference (Fain, 1996) conducted by Business Communications Co., Inc. The review was based on impressions from the papers given at the Fourth International Conference on Inorganic Membranes (ICIM$_4$-96). Current R&D is being directed mostly toward laboratory configurations for obtaining membranes with high separation factors, but not much concentration is given to achieving a low cost manufacturing process. However, there is good evidence that we will not have to wait much longer.

The composition and subject distribution of papers presented at The ICIM$_4$-96 showed some change from previous conferences. The papers indicate the work is becoming more sophisticated and producing more and better information about the possibilities for inorganic membranes. Commercial manufacture of these more sophisticated membranes is still lacking. A list of paper categories is given in Table 3.4.

**Table 3.4.** Fourth International Conference on Inorganic Membranes (ICIM$_4$-96) Papers

| Category | No. Papers | Category | No. Papers |
|---|---|---|---|
| Membrane Preparation and Performance | 18 | Membrane Reactors | 10 |
| Liquid Filtration | 16 | Membrane Characterization | 6 |
| Palladium Membranes | 15 | Electrolyte Transport | 5 |
| Dense Membranes | 10 | Zeolite Membranes | 4 |

Essentially all of the papers in the preparation and performance category represent modifications of sol–gel membranes. There is still a great deal of interest in liquid filtration, but it is more toward ultra and nano filtration. The biggest surprise is the large increase in papers on palladium and dense membranes. Essentially all of the palladium membranes are composites made by applying palladium layers to ceramic or metal supports by various means. These membranes are primarily for separating hydrogen. The dense membranes are composites with the top layer of various nonporous oxides or perovskites. These membranes are for high temperature oxygen separation or for fuel cells. The small number of papers on zeolites was also a surprise, perhaps indicating a decrease in interest. The papers given at ICIM$_4$-96 show that there is increasing interest in inorganic membranes and the research is becoming more sophisticated. There is still a serious need for concern about fabrication methods that can be adapted to economical manufacturing.

In the first plenary session of the First International Conference on Inorganic Membranes in 1989, Jean Charpin, rightly, gave the CEA (Commissariat á l'Energie Atomique) credit for the significant increase in interest in inorganic membranes. As a spin-off from the French gaseous diffusion program, the French government allowed one company, Société Céramiques Techniques (SCT), to manufacture a series of ceramic membranes for use as micro, ultra, and nano filters for commercial use. Clearly, this was the event that triggered so much activity in academia and industry in inorganic membranes. ALCOA purchased 85% of SCT in the late 1980s, apparently with the intent of marketing in the United States the membranes manufactured in France, and to develop additional types of membranes for a larger range of applications. With relatively little success in achieving additional new membranes, ALCOA decided to concentrate on its basic business and sold its interest in SCT to U.S. Filter Corporation. U.S. Filter is actively marketing an excellent, but expensive, series of ceramic membranes manufactured in France.

Excellent evidence of the commercialization of inorganic membranes is being demonstrated at the East Tennessee Technology Park (ETTP) (formerly the Oak Ridge Gaseous Diffusion Plant) by the Membrane Technology Division (MTD). This group of scientists represents the sole repository of the previous, more than one billion dollar expenditure (in current dollars), on research and development for the United States Gaseous Diffusion Program. While this technology is still classified confidential restricted data, these scientists are developing a whole new family of inorganic membranes for industrial and waste management applications which have the potential for being declared unclassified and available for commercial use. We are working with the Department of Energy (DOE) Office of Fossil Energy for hydrogen recovery in coal gasification and petroleum applications. We have seven classified Cooperative Research and Development Agreements (CRADAs), each for a different application, with two private industry companies, Pall Corporation and Coors Technical Ceramics.

MTDs major asset is the fundamental and practical understanding of the character and functioning of inorganic membranes that has been developed through sophisticated theoretical and experimental research in the transport of molecules through inorganic membranes. The ability to measure gas flow rates with precision up to about 0.005% over a wide range of pressure and temperature allows the detailed examination of the transport mechanisms. With such detail one can deconvolute (separate) the net flow into various different types of flow and understand the physical and chemical mechanisms for the transport. Understanding the functional character of these transport mechanisms allows one to design membranes to achieve maximum separation factors for any given fluid mixture. Transport is affected by the membrane material and pore size, the geometry effects of the size of the individual molecules, the effect of the thickness of the adsorption layer, the transport of molecules in the adsorbed layer, etc.

Perhaps the most convincing evidence of the impending potential for inorganic membranes is the interest being shown by industry. This interest can best be shown by the seven CRADAs that have been signed with industrial companies. Publicly released information on these CRADAs is listed below.

1. CRADA No. ORNL94-0266; Separate Hydrogen from Petroleum Refinery Gases
   Industry Partner: Coors Technical Ceramics
2. CRADA No. K-2595-0332; Photocatalytic Decomposition of VOCs in Groundwater
   Industry Partner: Coors Technical Ceramics
3. CRADA No. K-2595-0333; Separate Hydrogen from Olefin Production
   Industry Partner: Coors Technical Ceramics
4. CRADA No. K-2596-0448; Reverse Osmosis to Provide Ultrapure Water
   Industry Partner: Pall Corporation
5. CRADA No. K-2596-0449; Cleanable High Efficiency Particulate Air (HEPA) Filter
   Industry Partner: Pall Corporation
6. CRADA No. Y-1297-0462; Porous Stainless Steel Filters
   Industry Partner: Pall Corporation
7. CRADA No. K-2597-0486; Photocatalytic Decomposition of VOCs in Ultrapure Water
   Industry Partner: Pall Corporation

The effort on each of these CRADAs will be to design and develop the membrane and provide a test module for the specific application indicated above. After the membrane has been fabricated the DOE will examine the membrane and determine whether the membrane can be declared unclassified. If the membrane is declared unclassified, it can be manufactured in a secure facility and the membrane and/or a module with the membranes can be marketed commercially.

With the characterization and modeling tools available, the MTD is interested in working with industrial companies to evaluate the use of a membrane for any process of interest. If the probability is significant that such a membrane can be developed, we can develop a "statement of work" and initiate contract R&D to develop that membrane. The types of membranes and applications are virtually endless.

***Glass.*** Some serious effort has been made to produce membranes with different pore sizes from various types of porous glass. Such membranes could be made in the form of hollow fibers. If perfected, these membrane could have the advantage of being incorporated into modules with large amounts of frontal surface area. However, the hollow-fiber wall thickness is 20 $\mu$m or larger and the permeance is low. The pore size and structure is difficult to engineer. In addition, the modules are difficult to make, they have poor hydrodynamics, and the areas of application are limited.

***Zeolites.*** Zeolites represent a particularly interesting approach to ceramic membranes. Zeolites are attractive because they can be made with different very precise pore sizes. Considerable attention has been given to producing membranes with various zeolites. The basic problem is growing the zeolite particles into a membrane without defects and thin enough to have reasonable permeance. Many labs are attempting to grow zeolite membranes in place on a porous support in an appropriate environment. These membranes are ideal for acquiring important transport data. However, it is difficult to grow them thin enough without defects and most likely they will be expensive to fabricate. Without a major breakthrough, they are not likely to be of significant commercial interest.

***Ceramic.*** The membranes available from U.S. Filter Corporation are based on the sol–gel process. Most of the research in industry and academia has been and continues to be devoted to this process. The following is a very simplified explanation of how this type of membrane is fabricated. A porous ceramic tube (usually alpha alumina) with a pore diameter of about 15 $\mu$m is used as a substrate or support tube for the membrane. The support tubes can be single tubes or multiliths with many channels. The membrane is formed by slip casting a colloidal suspension or sol onto the inside surface of the tube. As the liquid capillary forces pull the sol into the pores of the support material, water is lost and a gel is formed, hence, sol–gel process. The gel is then carefully dried and then calcined. The drying and calcining processes are critical steps and require a well controlled and timed increase in the temperature of the drying and calcining environment. Significant shrinkage occurs during these processes and many defects and cracks can be formed. The membrane layer should be as thin as possible to achieve high permeance. Thin layers tend to have fewer cracks and defects when cast on a very smooth surface, but are much more troublesome when cast on a porous support tube.

To reduce the number of defects, the slip cast layers applied to the support material are made as thin as possible. Several layers are applied, each dried and calcined sequentially. The size and distribution of pore sizes in the layers are determined by the size and distribution of size of particles in the sol. All of the applied layers can have the same sol particle size or they can be progressively smaller. The first layer should produce the largest pore size and each subsequent layer produces a smaller pore size using a smaller particle size sol. The pore size of the final or top layer determines whether the membrane will be a micro, ultra, or nano filter. Many different types of material may be used for the membrane. Membranes have been made with different sizes of alumina, silica, titania, zirconia, and other materials which can be obtained in a finely divided colloidal suspension. Important variables are, formation of the colloidal suspension or sol, the formation of the membrane from the sol, and the application of the membrane to a porous support. About fifteen companies are manufacturing and selling membranes made with some variation of the sol–gel process.

The basic sol–gel ceramic membranes are primarily used for micro, ultra, and nano filters. The word filter here is used in the context of removing solids from fluids, either gases or liquids. The separation mechanism is one of exclusion. The pore size of the top layer should be about the size of the solids to be removed. The smallest pore size commercially available now is about 5 nm. Advantages of ceramic filters are that they do not foul as badly or quickly and can be cleaned more easily.

At the ETTP, we have made several ceramic membranes with mean pore diameters ranging between 5 and 0.5 nm. For a binary mixture, it is intuitive that molecules can be separated by their size if the pore diameter is smaller than the larger molecule but larger than the smaller molecule. That is the definition of a molecular sieve membrane. With such a membrane, the small molecules can go through the membrane, but the larger molecules can not. With such a membrane, a pure stream of the smaller molecules is obtained. The separation factor is infinite (the separation factor is the ratio of the velocity of the faster molecule to the slower molecule). We have developed a mathematical model and have shown experimentally that molecular sieving effects occur in a continuous way and approach infinity as the above condition is approached. Different molecules adsorb on different materials in different ways. This adsorption can effect the size of the pores and the adsorbed molecules can diffuse on the surface. By making a membrane with the right materials and pore size, one can separate small molecules from large ones or with the proper different types of material and pore size one can separate larger more adsorbable molecules from smaller less adsorbing molecules. This has also been demonstrated experimentally (Fain and Roettger, 1996, 1997).

Most worldwide research has been devoted to making membranes with smaller pore sizes for new applications, but apparently no such membranes have been made that are practical and useful or for commercialization. The primary approach is to modify existing membranes. The U.S. Filter membranes or similar membranes are being used as starting material for various modification treatments to decrease its pore size. A number of modification methods are being attempted. These methods range from various forms of chemical vapor deposition (CVD), to using templates for pore forming. These methods produce interesting membranes to learn more about transport and separation processes. These expectations open up a whole new set of opportunities for ceramic membranes. However, at best this makes an already expensive membrane more expensive and at worst an impractical manufacturing process. Development must be directed toward more effective and less expensive manufacturing methods.

***Carbon-Based.*** Some work has been directed toward carbon-based membranes. Some micro, ultra, and nano filters have been made with carbon. These include particulates, small fiber composites, and clothes. However, the market is still limited and small. The development with the most potential is the nano filters made with very high surface area activated carbons. These materials are capable of producing membranes with pore diameters of 0.5 nm or less. They have molecular sieving capability and very high permeance for hydrocarbons. These were first developed by Koresh and Soffer (1985) in Israel. There were expectations for commercial membranes, but they do not appear to be available as yet.

Rao and Sircar with Air Products (Anand, 1995) have conducted extensive development on such a membrane. Their membrane is made by applying an emulsion of PVDC (polyvinylidene chloride) in a sol–gel-like manner to an alumina support tube and then drying, pyrolysizing, and activating the remaining carbon. This membrane is referred to as an SSF (selective surface flow) membrane. Air Products is in the initial phase of marketing this membrane for purifying, recovering, and recycling hydrogen from mixtures with hydrocarbons. At

ETTP in Oak Ridge, one of our CRADAs is for separating (and recycling) high purity hydrogen from petroleum refinery gases. A carbon-based membrane is being considered along with others.

The extent of the applications of carbon-based membrane has not been adequately examined as yet, but it is likely to have significant and large application.

***Conclusions.*** All of these inorganic membranes represent significant new opportunities that organic polymer membranes cannot fulfill. The range of applications include not only micro, ultra, and nano filtration, but molecular separations in gases and liquids, desalination, photocatalysis, and membrane assisted electrical separation processes (and probably many not yet known). As attention is applied to developing more economical manufacturing methods, inorganic membranes will become available for reducing energy use and generally improving the economics of a large number of industrial processes and methods for waste management. A listing of many of the current manufacturers and suppliers of inorganic membranes is given in Table 3.5.

**Table 3.5.** Manufacturers and Suppliers of Inorganic Membranes

| Company | Ceramic | $Al_2O_3$ | Por. Metal or C/$ZrO_2$ | SiC | C | Glass $SiO_2$ | Metal Salts |
|---|---|---|---|---|---|---|---|
| Air Products | | | | | | | X |
| Agency Ind. Sci./Japan | X | | | | | | X |
| APV (distributor) | X | X | | | | | |
| Alcan/Anotec | | X | | | | | |
| Arco Chemical | X | | | | | | |
| Asahi Glass | | | | | | | X |
| Carre/Dupont | | | X | | | | |
| CEA/SFEC/Tech-Sep. | X | X | X | | | | |
| Coors Ceramics | X | X | | X | | | |
| Corning Glass (support) | X | | | | | | |
| CeraMem | X | X | | | | | |
| DuPont | X | X | | | | | |
| Fastek (dist.) | X | | | | | | |
| Gaston Cty./Union Carbide | | | X | | | | |
| Ciba-Geigy | | | | | | | X |
| GFT | | | | | X | | |
| General Motors | X | X | | | | | |

**Table 3.5.** (continued)

| Company | Ceramic | Al$_2$O$_3$ | Por. Metal* or C/ZrO$_2$ | SiC | C | Glass SiO$_2$ | Metal Salts |
|---|---|---|---|---|---|---|---|
| Mitsubishi Heavy Ind. | X | X | | | | | |
| Mitsubishi Jukogyo | X | | | | | | |
| NGK | X | X | | | | | X |
| Negev Nuclear Research | | | | | X | | |
| Nitto Electric | X | | | | | | |
| Norton | X | X | | | | | |
| NASA (U.S.) | | | X | | | | |
| Osmonics | X | X | | | | | |
| Pall | | | | | | | X |
| Schott Glaswerke | | | | | | X | |
| SFEC/RH. Poul./Tech-Sep | X | X | X | | | | |
| Stichting Energieon (Neth.) | | X | | | | | |
| Sumitomo | X | X | | | | | |
| Swiss Aluminum | | X | | X | | | |
| TDK | X | X | X | | | | |
| Teijin | | | | | | | X |
| Toray | | | | | X | | X |
| Toyota | X | | | | | | |
| Union Carbide | X | | X | | | | |
| U.S. Filter | X | X | | | | | |
| Weitzhann Inst. | X | | | | | | |
| Westinghouse | X | X | | | | | |

### 3.1.1.3 Membrane Modules and Systems

Membranes are manufactured in three geometric configurations: flat sheets, tubular, and hollow fibers. The aim of a membrane module design is to provide assemblies of the individual membranes in a usable form, such that feed stream can be separated into permeate and retentate stream. An assembly of membrane module(s) with auxiliary equipment such as compressors, pumps, heat exchanger, etc. constitutes a membrane system. A membrane system is what is used as an industrial unit operation for separations.

It should be recognized that in a membrane separation system the membrane module cost may only be a fraction of the total system cost. Compressors, pre treatment, heat exchangers, coalescers, filters, scrubbers, knockout tanks, instrumentation and controls may consume major capital. It is therefore desirable that membrane modules be highly environmentally resistant and/or can be operated over a wide range of temperature and pressure so that the need and cost associated with the peripheral equipment is minimal.

Detailed discussions of modules and systems is described elsewhere (Ho and Sircar, 1992; Koros and Fleming, 1993). The intent here is to bring out the key points that affect the design and performance of a module and system. Much of the discussion here is directed toward gas separation, however, the general considerations are applicable to liquids as well.

*Membrane Modules*. Membrane modules can, in general, be made by using membranes of either dense, asymmetric or thin film composite morphologies. A brief description of the three kinds of membranes is given below:

1. Dense membranes have walls of uniform density and essentially zero porosity. They are usually made by melt spinning or solution casting.
2. Asymmetric membranes have a thin dense skin (which constitutes the separating layer) imbedded in a wall with a gradation of porosity through its thickness. They are made by phase inversion processes.
3. Thin film composite membranes have a single or multiple coating of one or more polymers applied to the surface of a porous substrate that provides a support for the coating(s). Thin films are also made in-situ by interfacial polymerization.

Whereas dense and asymmetric membranes are usually made of a single polymeric material, thin film composites are generally made by applying a coating different from the material of the substrate. This results in weaker adhesion between the separating layer and the substrate. In this case the probability of rupture is high when feed pressure is applied to the side opposite to the coating.

The goal of an efficient module design is to provide maximum membrane surface area per unit volume of the module while maintaining high separation efficiency of the module. A good membrane module design also requires the upstream and downstream considerations to make it compatible with the process in which it is being used. The ultimate goal, however, is to provide a cost effective and mechanically robust separation unit which can withstand the process pressure, temperature and their upsets. The materials of construction must be chemically compatible with the process application. Standard chemical industry components (piping, valves, liquid knockouts, heat exchangers, controls and instrumentation) should be used in the system using the membrane modules for their ease of acceptance in the industry. Modularity, simplicity, and lightness of weight are also desirable in certain applications.

In an effective membrane module design, the pressure drop on the feed side should be minimized in order to avoid significant loss of driving force. Therefore, proper design of the fiber dimension (or gap in case of the flat sheets) and length, its packing density and packing configuration become important. The maximum length of a membrane leaf or that of a hollow fiber is determined by the allowable pressure drop and manufacturing considerations. Bypassing, channeling, and stagnation of the fluid stream should be avoided, as should any unused section of the membranes. Stagnant zones will locally concentrate nonpermeate in the feed and make membrane area ineffective. Channeling and by-passing of feed stream or permeate streams will adversely affect the module separation efficiency. Effects of flow patterns on the performance of a module are described by Antonson et. al. (1977). Another very important criteria on the module efficiency design which is generally overlooked by the module manufacturers is the impact of nonuniformity of membrane properties on the module efficiency. All theoretical models assume membranes of uniform properties. In reality, it is rare when membranes of identical performance are made or used in a given module. Therefore, there is always a gap between the actual and predicted performance of a membrane module.

One key difference in the design of a gas separation vs. a liquid separation module is the viscosity of the fluids to be separated. The lower viscosity of gases permits smaller fiber geometries at equivalent pressure differentials, permitting larger membrane area and hence productivity per unit volume of the module.

Membrane modules can be operated in one of the three following modes:

1. Feed stream is perpendicular to the axis of the membrane and permeate flows in the same direction but on the opposite side of the membrane. (crosscurrent flow).
2. Feed stream flows parallel to the axis of the membrane and permeate also flows parallel to the axis of the membrane and in the same direction (cocurrent flow).
3. Feed stream flow parallel to the axis of the membrane and permeate also flows parallel to the axis of the membrane, but in the opposite direction (countercurrent flow).

The mode of operation determines the nature of the membrane tube-sheet, design of the headers, and the number and location of the feed, product and reject ports. All three flow patterns are achievable in a tubular or hollow-fiber membrane module. However, in a flat sheet module, bulk feed and permeate flows may neither be cross-current nor cocurrent, but may be at right angles to each other. Countercurrent flow modules are the most efficient for the gas separation systems. The efficiency of the membrane module, however, depends on several other factors in addition to those described above.

## 3. Membrane Technologies

*Flat Membrane Modules.* The flat sheet membranes are formed into membrane modules of two common configurations. The first flat films are stacked with spacers in-between, in a similar fashion as in a plate-and-frame filter (Nielsen et al., 1980; Sourirajan and Matsura, 1985). This configuration is inefficient because it offers a very small surface area per unit volume of the module ($\sim$100–150 m$^2$/m$^3$) and hence is expensive. Plate and frame type modules have a very limited commercial use. In the second configuration, flat sheet membranes are assembled into high surface area separators (500–1000 m$^2$/m$^3$) by the use of spiral-wound membrane module design. A typical design of spiral-wound module is shown in Figure 3.10 and details are given elsewhere (Sourirajan and Matsura, 1985).

Although expensive and difficult to make, a spiral-wound module offers distinct advantages in very high pressure applications (>100 bar), especially when the membrane flux is very high. This is because wider flow channels offer low pressure drop and hence better module efficiencies. Several modifications of the basic spiral-wound module design have been proposed over the years such as multileaf spiral modules.

*Hollow-Fiber and Tubular Membrane Modules.* Hollow fibers or tubular membranes are generally stacked in a configuration similar to that of a shell and tube heat exchanger. Large numbers of fibers having similar length are grouped together in a pressurizable shell or housing in which the opposite ends of the fiber are potted and sealed in a material which serves to form a tube sheet at each end. The volume within the shell which is toward the exterior of the fiber (shell side) is effectively sealed by the tube sheets from the volume within the bores of the hollow fibers. A typical hollow-fiber module design is shown in the Figure 3.11,

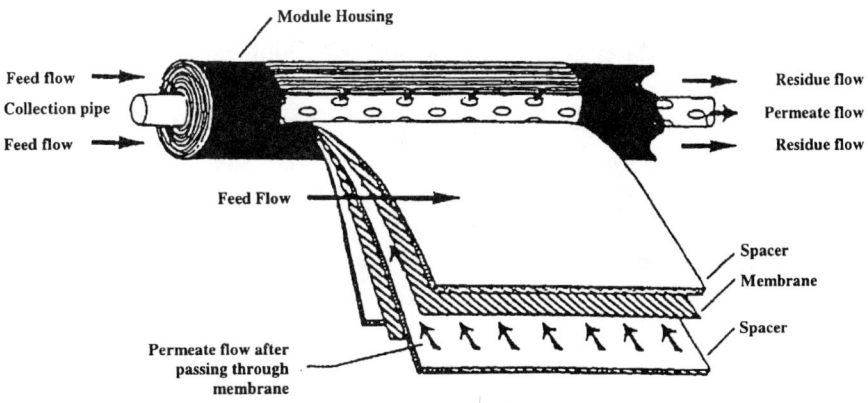

**Figure 3.10.** Spiral-wound membrane module.

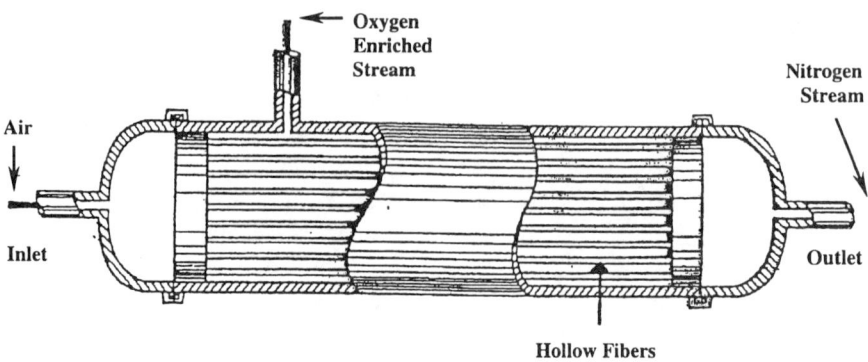

**Figure 3.11.** Hollow-fiber membrane module.

and details can be found elsewhere (Kesling and Fritzsche, 1993). This module can be operated with feed either on the shellside or on the bore side of the fibers or tubes with permeate coming on the opposite side.

A typical, tubular hollow-fiber membrane module can accommodate 1000–5000 $m^2$ area per cubic meter volume of the module, depending on the fiber external diameter and its packing density in the module. Patent literature is full of the various hollow-fiber membrane module designs with the key emphasis on the improvement of manufacturing costs and module separation efficiency. When pressurized feed is on the shell side, the hollow fibers of a given material and dimension collapse when the shear yield stress is reached in the fiber wall. With feed on the bore side, fibers burst when ultimate shear stress is reached in the fiber wall. Since, for a given fine hollow tube, collapse pressure is much higher than the burst pressure (typically 2–4 times), a shell side feed is generally preferred in the high pressure (>60 bar) applications. nonuniformity of the feed flow distribution and deviation from countercurrent flow are valid concerns in modules adopted for shell side feed. The flow patterns in these modules can be countercurrent or radial cross flow. Radial cross flow can also be achieved by a typical double ended module design with a central feed distribution tube with permeate removal from the side ports (Johnson,1988). As another example of a module design, Caskey et al. (1990, 1991) discuss a module design with the feed on the bore side of the fibers. Concentric or helical baffles, within the fiber bundle and on the outside of the casing are arranged to channel the permeate so that it flows countercurrently to the feed.

For low pressure applications (~10 barg), such as nitrogen production, bore side feed is more desired because it prevents channeling of the feed stream. Axial flow in both shell side and bore side are desired to maximize the countercurrent nature of feed to permeate streams to produce high purity product gases. Since in a bore fed module, fibers themselves act as a pressure vessel, a module design has been proposed which uses an outer shell to house the membranes (Edward, 1989). Such a module, obviously can't provide countercurrent flow between the feed and the permeate streams, and hence, is not efficient in all applications.

While many problems in gas separation efficiency can be appropriately addressed by simply using multiple modules connected in series, this is a costly solution because of increased number of pressure vessels which are required. Module designs with multiple stage operation within a single module have also been claimed to obtain better flow efficiencies and costs (Puri and Kalthod, 1993; Elhaus and Mäkel, 1991).

An alternative to parallel stacked membranes are helically wound fiber modules, which are claimed to be more structured and to prevent channeling. In the earlier version (Coplan, 1987), the potted ends of the fibers were cut differently at each end so that fiber openings at one end are closed at the other, and vice versa. Two concentric fiber bundles are thus formed in one module to simulate two modules in series. This design was used for shell feed configuration. However, in recent modifications, such modules have been proposed for bore side feed with countercurrent feed to permeate flow (Bikson and Giglia, 1992). This module design claims to promote complete radial mixing of the permeate stream on the shell side while minimizing axial mixing on both the shell and bore sides. Perrin (1989) discloses a hollow-fiber membrane module design having two different types of fibers with different separation characteristics. The two types of fibers are helically wound on a mandrel, either intertwined or in alternating layers, but spaced so that only one of each fiber extends to each end of the bundle. When each end of the bundle is potted in a tube sheet and cut, only the type of fiber extending to that end of the bundle is severed and opened so that permeate can exit the bores of the fibers.

A true crosscurrent hollow-fiber membrane design has been patented independently by Akzo and Standard Oil using a very similar concept (Nicholas, 1990). In this design, short lengths of the fibers are used with feed stream perpendicular to the fibers. The modules are made by stacking individual layers of fiber mats perpendicular to each other and then potting the stack around its periphery. The fiber ends are opened by machining the potting resin. A module of any length can be made by stacking the various layers of the membrane.

*Membrane Systems.* A membrane separation system may operate using a single membrane module or several membrane modules. When several modules are connected in parallel, the net performance sums up the performance of the individual modules. However, when modules are used in series (i.e., two, three, or more), a variety of opportunities exist to manipulate the various feed, permeate, and reject streams to optimize the effectiveness of membranes of a given perm selective property. Agrawal and Xu (1996) have proposed a generalized scheme to determine system configurations using two-, three- and multi-modules. A list of various membrane module and system manufacturers is given in Table 3.6.

***Operating Schemes for Membrane Systems.*** The quality of separation for a membrane module is not only determined by the selectivity, but also by two process parameters listed below.

**Table 3.6.** Major Manufacturers of Membrane Modules/Systems

| Type | Manufacturer | Application |
|---|---|---|
| Plate-and-Frame | DDS—Denmark | R.O. |
|  | GKKS—Germany |  |
| Spiral-Wound Modules | Toray Industries | R.O. |
|  | UOP, Separex | Gas Sep. |
|  | UOP, Fluid Systems | R.O. |
|  | DOW, Film Tech. | R.O. |
|  | Nitto Denko | R.O. |
|  | Osmonics, Inc. | R.O. |
| Tubular Modules | Amicon, Althin | Medical Applications |
|  | Baxter | Ultrafiltration |
|  | Nitto-Denko - Japan | Water Purification |
|  | Osmonics | Ultrafiltration |
| Hollow-Fiber Modules | Permea/Air Products | Gas Sep. |
|  | L'air Liquide | Gas Sep. |
|  | Praxair | Gas Sep. |
|  | AG Technology | Gas Sep. |
|  | Ube | Ultrafiltration |
|  |  | Gas Sep. |
|  | Asatin Chem. | R.O. |
|  | DuPont | R.O. |
|  | Toyobo Co. | R.O. |

1. The ratio of the permeate flow to the feed flow, generally referred to as "stage cut." A higher ratio will reduce the quality of the separation as the partial pressure of the lesser-permeating component is raised along the membrane, resulting in its increased permeation.
2. The ratio of permeate pressure to the feed pressure for a given component of the feed stream. A high pressure ratio will give better separation. However, it may result in energy penalty as the permeate is available at lower pressure than the feed pressure.

In its simplest form, a membrane module can be used with the feed stream on the shell or bore side and permeate coming in opposite to the membrane with the flow of permeate and feed stream opposite to each other as shown in Figure 3.12a. Better efficiencies can be obtained by using two-stage modules. However, the complexities and costs increase with three- and more stage operations.

Using a two-stage module as shown in the schematic in Figure 3.12b will improve single-stage operation and are preferred in many applications. Two-stage system offers advantages over a single larger module, with better feed distribution on the shell side and a flow pattern approaching that of countercurrent. Also, any bypassing of the feed gas in the first module (e.g., due to leaks) can be corrected by remixing the feed in the second stage. The disadvantages, however, are obvious. First, it requires two modules and is therefore more expensive. Second, due to intermediate mixing, a single large module will be more efficient if it operates with ideal countercurrent flow.

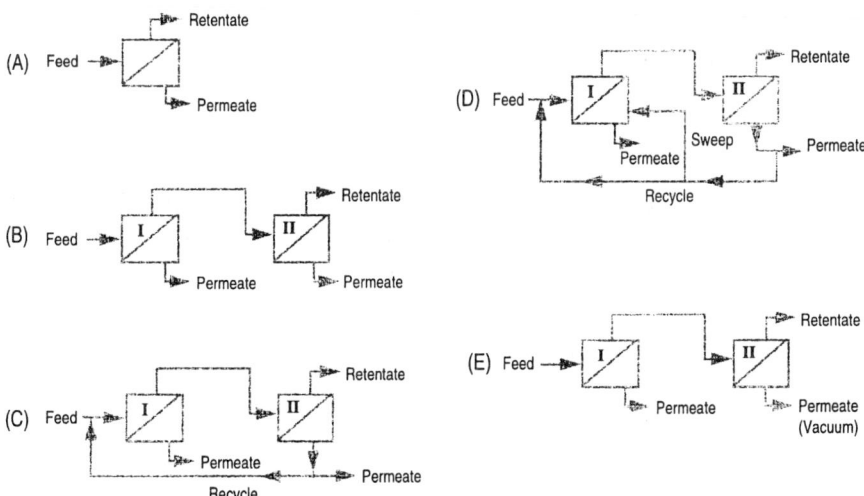

**Figure 3.12.** Membrane module process systems.

The two-stage system can also be operated in a mode where the permeate from stage II is recycled as shown in the schematic in Figure 3.12c (Doshi, 1987). For air separation operations, the recycle of the permeate from the second stage, which is nitrogen gas rich, offers better operational efficiency for the system.

The two-stage module can be operated in still another mode, in which a part or all of the permeate from stage II can be used as a sweep stream in stage I, to provide increased module performance efficiency. The use of permeate from the second stage to sweep the permeate from the first stage is also well documented (Antonson et al., 1977).

Another mode of operation for a two-stage module is one in which part of the permeate from stage II is used as a sweep stream in stage I and another part is used as a recycle stream, as shown in the Figure 3.12d. The two modules used in the above schematics can be of the same size or different sizes, as determined by optimization of system operation. The optimum depends on the performance properties of the membranes as well as the purity of the two streams needed. The two modules used in a system can also have membranes with different perm-selective properties. Depending on the needs of a given gas separation operation, the fibers of different materials or a separating barrier can be used in stages I and II without altering the cost of membrane module fabrication.

In another mode of operation of the two-stage system, permeate from stage II is evacuated by any suitable vacuum-producing device (e.g., steam or water ejectors, or vacuum pump), as shown in Figure 3.12e. Even moderately low pressure produces significant increases in both productivity and recovery. Though applying the vacuum to the stage II permeate is advantageous, both permeate streams may also be under vacuum.

Thus, a large number of system schemes can be developed to make optimum use of membranes of given perm-selective properties.

## 3.2. Selected Emerging Noneactive in-Process Source Reduction Membrane Applications

### 3.2.1. Membrane Gas Separation Opportunities in the Control of Greenhouse Effect

Introduction
The sun's radiation which reaches the earth's surface results in the heating of the ground surface, melting of the ice and snow, evaporation of water and plant photosynthesis. The earth's hot surface, in turn, acts as a radiator of energy in the infrared band. A majority of the outgoing infrared radiation is absorbed by a few

atmospheric gases, such as carbon dioxide, water vapor, methane, ozone, nitrous oxide, chlorofluorocarbons (CFCs), halons, etc., known as greenhouse gases. As the concentration of these gases increases in the earth's atmosphere, the retention of radiated energy also increases, possibly resulting in the warming of the earth's surface. Of the various gases listed above, $CO_2$, because of its high concentrations in the atmosphere, is responsible for half of the human contribution to global warming (Reck, 1995). The capture of $CO_2$, is therefore, the main topic of this paper. However, in principle the membrane technology described here is applicable to other greenhouse gases as well.

The amount of carbon dioxide in our atmosphere is steadily increasing and is expected to increase from the current 300–350 ppmv to 500–900 ppmv by year 2100. To curb the increase in $CO_2$ in the atmosphere, the 1992 UN Framework Convention on Climate Change set up a goal "to achieve stabilization of greenhouse gas concentration in the atmosphere at a level that would prevent dangerous anthropogenic interferences with the climate system." As a consequence, the atmospheric concentration of the $CO_2$ should probably not exceed 450 ppmv. One recommendation is for all the industrialized nations to reduce $CO_2$ emission to 20% below the 1990 levels by year 2010. The cost of doing this, using the existing technologies used in the power generation industry, will be in excess of $2.5 trillion (Richels, 1996). Therefore, attention needs to be given to technologies and ideas outside the normal thinking sphere. New and emerging technologies may offer a better solution than the existing practices. Gas separation membrane technology is one of those newer separation processes which may offer a low cost $CO_2$ mitigation path.

*Sources of Carbon Dioxide Emission*

Carbon dioxide is a naturally occurring chemical and exists in high concentrations in the various natural gas streams. Most natural gas streams contain 6–14% $CO_2$, for some special fields $CO_2$ concentration up to 71% has been reported (Natuna, Indonesia). The major sources of manmade carbon dioxide include: fossil fuel combustion for industry; transportation; space heating; electricity generation and cooking. Emissions from fossil fuel combustion account for about 65% of extra carbon dioxide now found in the atmosphere. The flue gas streams from the various fuel sources (i.e., combustion exhaust gases) contain $CO_2$, $N_2$, $H_2O$, $O_2$, CO, $SO_x$, $NO_x$ and particulate matter. These gas streams are generally at atmospheric pressure and at temperature $>100°C$. The flue gas discharge rates can vary from several hundred to several million cubic meters per day.

Typical composition of the flue gases from the various fuels combusted with air are given in Table 3.7.

**Table 3.7.** Composition of Flue Gases from Various Fuels (%)

| Fuel | $CO_2$ | $H_2O$ | $O_2$ | $N_2$ |
|---|---|---|---|---|
| Coal | 15.4 | 6.2 | 1.8 | 76.6 |
| Oil | 12.9 | 10.3 | 1.8 | 74.9 |
| Gas | 8.7 | 17.4 | 1.7 | 72.1 |

*Methods of Capturing Carbon Dioxide*
Currently, scrubbing of $CO_2$ with solvents (physical or chemical absorption) is exclusively used to capture $CO_2$ wheresoever possible or required. However, several other technologies such as adsorption, membrane and cryogenics can also be applied.

The conventional *absorption process* operates at elevated pressure and hence requires feed compression. This also necessitates removal of particulate matter from the feed gases. The scrubbing is typically done in packed beds using aqueous amine solutions. The process used is a steady state chemical process. The key issues associated with the scrubbing process are solvent stability, solvent concentration and regeneration. Much of the energy in this process is consumed in the desorber. The conventional solvent scrubbing technology was developed over 60 years ago as general nonselective solvent for removal of acid gases, namely $CO_2$ and $H_2S$ from natural gas streams. The application of $CO_2$ removal from flue gases requires modification of this process to incorporate inhibitors to resist solvent degradation and equipment corrosion by oxygen present in the gas as well as the upstream treatment of the flue gases to remove the species of both $SO_x$ and $NO_x$ present.

The presence of other acid gases and their concentration relative to the amount of $CO_2$ is likely to be a key factor in determining the optimum $CO_2$ recovery process. Chemical solvents tend to react with both $SO_x$ and $NO_x$ to form heat stable salts which are not easily recoverable. This can result in unacceptable solvent losses unless $SO_x$ and $NO_x$ are removed upstream. $SO_2$ is generally much more soluble in physical solvents than $CO_2$. For example $SO_2$ is 100 times more soluble than $CO_2$ in Selexol and is not easily recovered, hence $SO_2$ levels need to be reduced to 5–10 ppmv in the feed. $NO_x$ compounds are not particularly soluble in the physical solvents and in general are less a handicap to $CO_2$ recovery than $SO_x$.

The *adsorption process* is a nonsteady state process requiring a battery of adsorbers. The feed stream needs to be cleaned and compressed as in the absorption process. The adsorption process is well suited for small volumes. As for large

3. Membrane Technologies

streams, the process becomes cumbersome and inefficient. The specificity of the adsorbent changes over a period of time due to the presence of contaminants and, the temperature and pressure operating conditions.

The *cryogenic process* requires the feed stream to be free from both the particulate matter and moisture. The process operates on the principle of selective condensation of $CO_2$. Unfortunately, $CO_2$ has a propensity to solidify and hence cause problems in the process.

Finally, $CO_2$ can be removed by the use of *membranes*, which is the major topic of this paper.

### 3.2.1.1. Technology Description

*Separation of $CO_2$ by Membranes*

Membrane-based gas separation systems are now a widely accepted and employed unit operation in the industry. World-wide annual installed gas handling capacity of membranes since 1977, when the first commercial installation was done, has risen from a fraction of a million to over 10 million cubic meters/day in early nineties (Puri, 1996b). Numerous membrane systems are in operation today to: (a) recover hydrogen from purge gas and hydrocarbon streams; (b) for adjustment of the $CO/H_2$ ratio in syngas; (c) for removal of $CO_2$ from natural gas; (d) and for separation of air, and others. Lower cost, ease of operation, operational flexibility and portability are a few reasons membrane-based systems are chosen over adsorption and cryogenic-based separations in certain applications.

To capture carbon dioxide, membranes can be used in two forms, as gas selective membranes or as contactors between the gas and a scrubbing liquid. These are discussed in detail here.

*Selective Gas Separation Membranes*

The polymeric gas separation membranes are extensively used for carbon dioxide removal from natural gas and other gas streams. The ceramic membranes which have the potential to offer superior performance are still under development. Polymeric membranes which are selective to carbon dioxide are also selective to water transport. The feed stream needs to be compressed and the compressor can be coupled with an expander to recover energy. A schematic of a typical membrane installation for removal of carbon dioxide from flue gas stream is shown in Figure 3.13.

The flue gas is first cooled then compressed in three stages with three interstage coolers after which it is fed into the membrane module. The retentate stream, which is depleted in $CO_2$, is allowed to expand. Through this operation a part of the compression energy is recovered. The heat delivered by compression is

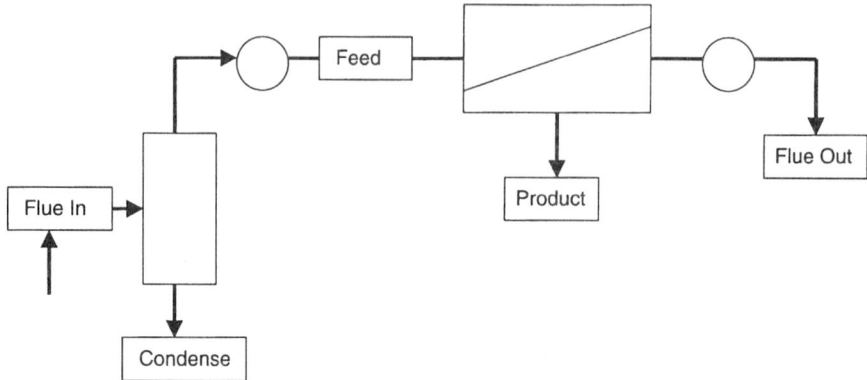

**Figure 3.13.** Typical Process Design of a Gas Separation Membrane Plant (Riemer, 1993).

more than sufficient to increase the temperature of the retentate prior to expansion. The resulting permeate stream is enriched in $CO_2$. More complex schemes which offer better efficiency or greater removal of carbon dioxide have also been proposed (Callahan, 1995).

*Polymeric Membranes*
A number of different polymers are now currently employed in commercial membranes which is discussed in Section 3.1.1.2. Typical examples of polymers used for carbon dioxide removal are polysulfone, cellulose acetate, polyphynelene oxide, polyimide, polyether imide, etc. These polymers have high glass transition temperature, high permeability for both water and carbon dioxide, and offer a higher $CO_2/N_2$ or $CO_2/CH_4$ selectivity. Polymeric membranes are easy to fabricate and are commercially available at competitive prices. A list of various membrane module manufacturers and the membrane materials is given in Table 3.8.

*Ceramic Membranes*
Membranes made from inorganic materials for removal of carbon dioxide from flue gases offer a special feature of their stability at elevated temperature (>100°C). For power plants, the flue gases may be at very high temperatures (400–600°C). Though inorganic membranes are not commercial for gas separation, they have been made in the lab by several methods such as chemical vapor deposition, sol–gel and hydrothermal synthesis. Starting from a variety of materials, zeolitic or ceramic membranes of controlled pore size (10–50 Å) and a very narrow pore size distribution have been reported, thus offering potential for ceramic membranes with both high flux and high selectivity. A large number of

**Table 3.8.** List of Membrane Manufacturers

| Company | Membrane Module | Applications |
|---|---|---|
| **Selective Membranes** | | |
| Air Products (Permea) | Hollow-Fiber | $O_2, N_2, H_2, H_2O, CO_2$ |
| Praxair | Hollow-Fiber | $N_2, H_2O$ |
| L'air Liquide (Medal) | Hollow-Fiber | $N_2, H_2, O_2$ |
| MG Industries (Generon) | Hollow-Fiber | $N_2$ |
| Aquillo | Hollow-Fiber | $N_2$ |
| Cynara | Hollow-Fiber | $CO_2$ |
| Separex (UOP) | Spiral-Wound | $CO_2, H_2$ |
| Ube | Hollow-Fiber | $H_2$ |
| GKKS, MTR | Spiral-Wound | $H_2O$, Organic vapors |
| Keverner (GMS) | Spiral-Wound | $CO_2$ |
| **Membrane Gas Absorbers** | | |
| TNO | Gortex/CORAL | $CO_2$ |
| Keverner | Gortex/MEA | $CO_2$ |

recent ceramic membrane developments for $CO_2$ separation were reported at the 1996 ICOM meeting (Proceedings, 1996). Most noteworthy was the work by Asaeda et al. (1996), who reported silica membranes with $CO_2/N_2$ separation factor of 32 at 50°C. They also observed that their membranes were very sensitive to the presence of water. For water saturated $CO_2/N_2$ feed gases, using the same membranes, a separation factor of 80 was reported at room temperature. The increase in the separation factor was attributed to hydration of oxide surfaces.

*Membrane Gas–Liquid Contactors*
Gas absorption membranes are membranes which are used as contacting device between a gas and a liquid. The separation is caused by the presence of an absorption liquid on one side of the membrane which selectively removes certain components of the stream. Hydrophobic microporous membranes can be used as gas–liquid contactors between the carbon dioxide-rich feed stream and a scrubbing liquid. The feed gas need not be compressed or dried for acid gas separation.

The feed gas is generally on the bore side of the hollow fibers with the absorbing liquid (such as carbonate or amine solutions) on the shell side. Carbon dioxide diffuses through the membrane pores and reacts with the absorbing liquid. Gas absorption membranes have inherently superior selectivity to polymeric membranes due to the absorption process. A properly designed membrane gas–liquid contactor offers an order of magnitude higher mass transfer coefficient (0.01–0.1 m$^2$/sec) than conventional packed beds. Membrane absorbers coupled with membrane desorbers can make an integrated process which can be highly energy efficient.

Membrane gas–liquid contactors as gas absorption/desorption devices have been identified as promising technology for the recovery of $CO_2$ from flue gas streams in the power industry. The hollow-fiber membrane contactors offer a compact and low weight contacting unit. Other advantages offered by the membrane gas absorbers are: (a) independent control of flow rates of the gas and liquid streams; (b) no entrainment, flooding channeling or foaming; (c) not influenced by the tilt of the offshore platform; (d) low liquid pumping power needed, and (e) flexibility. However these advantages are over-shadowed by the newness of this technology, difficulty in fault diagnosis (e.g., broken fibers), piping complexity, linear scalability and chemical compatibility of the cheaper polymeric membrane materials.

Further developments of superior gas–liquid membrane contactor membranes and modules and superior absorption liquids, which can be used at higher concentrations at lower circulation rates, will make this technology very attractive. Energy considerations for desorption are also very important. For monoethanolamine (MEA) as a sorbent one requires about 180 KJ/mole $CO_2$, consisting of 60 KJ/mole for desorption, 80 KJ for evaporation and 40KJ for reheating. Monoethanolamine can be used in higher concentration sorbents which require lower energy for desorption and can help reduce the overall energy need.

Two commercial-size gas absorbers using membranes for the removal of carbon dioxide from flue gases have been reported. In one, TNO in the Netherlands (Feron and Jansen, 1996) uses membranes made from a Teflon type of material and a proprietary scrubbing liquid, named CORAL. The intended applications for the $CO_2$ are in the horticultural industry. The second unit has been reported by Kvaerner Water Systems in Norway (Falk-Pedersen and Dannstrom, 1996). This membrane scrubber uses Gore-Tex membranes in a unique module design. The carbon dioxide recovery system has been installed on an offshore power generation platform. This system offers low space requirements; weight, maintenance, and energy consumption. Membranes are used for both the absorption and desorption steps. The desorption design is a lot more complex than the absorption design. The system weighs about 20 tons in comparison with a conventional solvent scrubber which weighed 80 tons.

### 3.2.1.2. Commercial Applications and Their Economics

Removal of carbon dioxide from flue gases and from natural gas streams are probably two of the largest applications of carbon dioxide removal. The current costs of carbon dioxide recovery from various streams range from $35–114/ton $CO_2$ depending on the nature of the feed stream, chemicals used and location of the separation plant (Falk-Pedersen and Dannstrom, 1996). The estimated selling price of crude $CO_2$ is about $6–12/ton $CO_2$ depending on volume. Thus there is little economic incentive to recover carbon dioxide. Therefore, the major driver for removal of carbon dioxide from flue gas and other $CO_2$-rich streams has to be the greenhouse effect. The motivation for further developments of membrane processes therefore is not the absolute economics of carbon dioxide recovery but the relative economics of the various carbon dioxide removal technologies.

Carbon dioxide removal from natural gas has been commercially used for several decades and economics of this process are well known. However, processes for carbon dioxide recovery from flue gases are of recent origin and hence the economics are not well known. Both of these membrane process applications are compared with amine absorption processes. These data are used for guidance only because commercially available polymeric membranes are mostly used for bulk removal of carbon dioxide. A combination of membranes with other separation processes such as cryogenic, physical/chemical absorption and adsorption processes may improve overall economics for larger systems. Also, membrane systems for carbon dioxide separation can be used in parallel to the existing scrubber to debottleneck them and avoid expensive capital costs. One specific feature of polymeric membrane is that they are capable of cocapture of $SO_x$ and $NO_x$ with $CO_2$ which offers an edge over conventional scrubbing methods.

A few examples of the economic evaluation of carbon dioxide removal by membranes are given below. Detailed description of these can be obtained from the original references cited.

*Carbon Dioxide Removal in Power Generation Industry*

The key technical problem in the removal of carbon dioxide from the power industry exhaust gases is the separation of nitrogen from carbon dioxide at elevated temperatures. The contaminants in the gas stream are unused oxygen, water, unburned hydrocarbons, $NO_x$ and $SO_x$. Detailed composition of the flue gases generated from the various combustion cycles are given in Table 3.9. When combustion is done with $O_2$ and $CO_2$ is used as a dilutent gas, $NO_x$ formation can

**Table 3.9.** Composition of Flue Gas Streams (mole %) (Riemer, 1993)

|  | PCFP | GFCC | IGCC | CFCA |
|---|---|---|---|---|
| $N_2$ | 71.0 | 75.0 | 75.3 | 0.6 |
| $CO_2$ | 12.0 | 3.4 | 7.0 | 62.6 |
| $H_2O$ | 11.1 | 6.9 | 4.9 | 31.5 |
| $O_2$ | 4.4 | 13.8 | 11.9 | 4.5 |
| Ar | 0.8 | 0.9 | 0.9 | 0.3 |
| $SO_2$ | 190 mg/m³ (STP) | — | 2 ppm | 0.3 |
| $NO_x$ | 666 mg/m³ (STP) | 25 ppm | — | 0.07 |

| PCFP | Pulverized Coal Fired Power plant |
|---|---|
| GFCC | Gas Fired Combined Cycle |
| IGCC | Integrated Gasification Combined Cycle |
| CFCA | Coal fired plant with Combustion in CO2-rich Atmosphere |

be minimized. Here the separation of $CO_2$ from recycle gases is easier to remove due to its high concentration in the feed stream.

In the power industry the economics of $CO_2$ recovery process are measured in their impact on the overall efficiency of the power generation unit. A comparison of 80% $CO_2$ removal from the flue gas discharged from the various combustion cycles using polymeric membranes is given in Table 3.10 (Riemer, 1993). The efficiency of the power generation unit is calculated taking into account the amount of compression energy required and the amount of compression energy recovered in the expansion. The cost of avoided $CO_2$ was calculated by the following formula.

$$\text{Avoided } CO_2 \text{ cost} = \frac{\begin{bmatrix} \text{Power cost with} \\ CO_2 \text{ removal} \end{bmatrix} - \begin{bmatrix} \text{Power cost without} \\ CO_2 \text{ removal} \end{bmatrix}}{\begin{bmatrix} CO_2 \text{ emission} \\ \text{without removal} \end{bmatrix} - \begin{bmatrix} CO_2 \text{ emission} \\ \text{with removal} \end{bmatrix}}$$

These calculations are for a 500-MW power plant using polyphynelene oxide membranes for $CO_2$ recovery. The cost of membrane assumed is $150/m², which constitutes only a minor fraction of the total $CO_2$ recovery system. The data show that the use of membrane gas separation leads to a sizable reduction in power plant output. This is due to large energy requirements associated with the compression of gas feed to the membrane unit. The cost of carbon dioxide avoidance

## 3. Membrane Technologies

**Table 3.10.** Power Generation Efficiency and Avoided $CO_2$ Cost for Various Power Generation Cycles Using a Membrane-Based Gas Separation

| Power Generation Cycle | Efficiency | | Avoided $CO_2$ Cost $/tonne |
|---|---|---|---|
| | No $CO_2$ Removal | With $CO_2$ Removal | |
| PCFP | 39.9 | 31.1 | 47 |
| GFCC | 52.0 | 31.0 | 335 |
| IGCC | 41.7 | 25.6 | 125 |
| CFCA | 32.8 | 30.8 | 9 |

is also very high. The most cost effective option is combustion in a $CO_2$-rich atmosphere.

The performance data for the same application with a membrane absorber using monoethanolamine (MEA) solvent are given in Table 3.11. The membrane absorber cost is assumed to be $60/m² of membrane area in the module. For all the combustion cycles, the power generation efficiency significantly decreases due to the incorporation of the $CO_2$ absorption process. However, avoided $CO_2$ cost is lower in all cases in comparison to selective membranes. Data from Tables 3.10 and 3.11 clearly show that economic evaluation favor gas absorber membranes over gas separation membranes. This is because gas absorption membranes produce a pure product stream at low cost, both in terms of energy and capital.

*Carbon Dioxide Removal from Various Natural Gas Streams*
These streams would be naturally occurring fuel gases, enhanced with recovery gases, landfill gases and biogases. The key components of these gases are methane

**Table 3.11.** Power Generation Efficiency and Avoided $CO_2$ Cost for Various Power Generation Cycles Using MEA (Riemer, 1993)

| Power Generation Cycle | Efficiency | | | Avoided $CO_2$ Cost $/tonne | |
|---|---|---|---|---|---|
| | No $CO_2$ Removal | With $CO_2$ Removal (Conventional) | With $CO_2$ Removal (Memb. Abs.) | Conv. Plant | Memb. Gas Abs. |
| PCFP | 39.9 | 29.1 | 29.7 | 49.2 | 45.1 |
| GFCC | 52.0 | 45.4 | 46.6 | 38.8 | 31.4 |
| IGCC | 41.7 | 31.3 | 32.3 | 47.2 | 41.3 |

and carbon dioxide. The carbon dioxide removal process is demanding because methane is a valuable component.

The removal of acid gases from natural gas requires highly selective membranes with high flux. In this application there are also stringent requirements for the removal of $H_2S$ and reduction of water to prevent hydrate formation. Selective polymeric membranes offer an edge over other technologies because all the undesired components ($CO_2$, $H_2S$, and $H_2O$) have much higher gas permeabilities than $CH_4$.

The performance of a cellulose acetate membrane system was compared with various amine scrubbers for about one million cubic meter per day natural gas stream containing 11% $CO_2$. The product stream was to contain 2% $CO_2$ and 4 lb water/MMSCF of gas. The data summarized in Figure 3.14 show that the membrane system is competitive. Further, the membrane system offered an added feature of turndown to half to one third the capacity, which was needed in this application (Cook and Losin, 1995). Membrane systems with performance properties superior to those given in this example are now available, offering still better economics over the conventional scrubbers.

*Synthesis Gases*

In syn gas production, the resulting gases may contain 5–16% carbon dioxide, which needs to be removed from $H_2$/CO mixture. The $CO_2$ content of the feed gas depends on the syn gas process used (steam reforming, partial oxidation), feed stock (naptha, natural gas, coal) and degree of carbon dioxide recycle. The combustible components in the synthesis gas are a mixture of CO and hydrogen. In a power plant without the $CO_2$ removal option, this gas mixture is burnt to

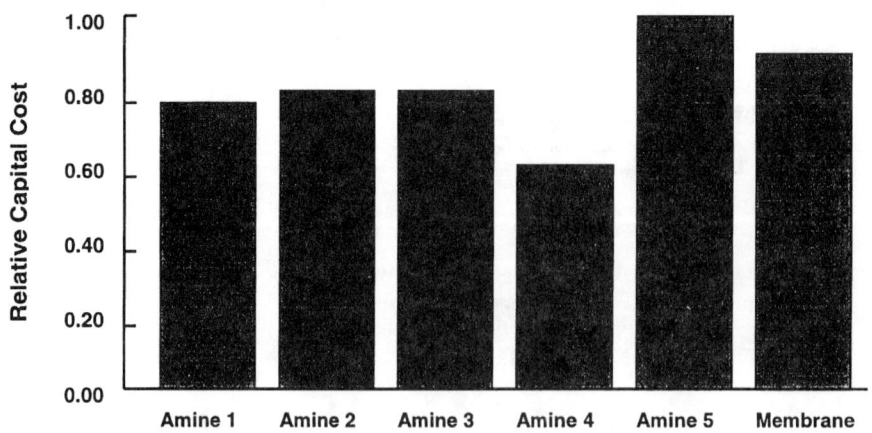

**Figure 3.14**. Economics of $CO_2$ Removal by Membranes (Cook and Losin, 1995).

generate electricity. For plants with the $CO_2$ removal option, an efficient process scheme may use membranes to divide the gas mixture into hydrogen-rich and CO-rich gas streams which can be burnt separately. The hydrogen-rich stream may contain a lower percent of $CO_2$ and need not be treated for $CO_2$ recovery. For the CO-rich stream, $CO_2$ can be recovered by membranes or absorption as described above, or $CO_2$ can be recycled for a CO2-rich combustion process.

Commercially available polymeric gas separation membranes are usable for separating syn gas into two streams. The $H_2/CO$ separation properties of the various membranes are given in Table 3.12.

In the IGCC power plant, $CO_2$ removal cost was estimated to be $16/tonne with an efficiency penalty of 6% for $CO_2$-rich combustion when compared to IGCC without $CO_2$ removal. However, $CO_2$-rich combustion requires development of a new gas turbine suitable for the $CO_2/O_2$ atmosphere.

In coal gasification or natural gas reforming processes, one process route involves conversion of synthesis gas to a mixture of $H_2$ and $CO_2$ via the shift reaction. Commercial membranes are available that are either $CO_2$ selective or $H_2$ selective. These are listed in Table 3.13. Due to low selectivities, these separations are not very efficient. It is estimated that (Castro, 1991) the separation factor must exceed 50 in order to achieve an efficient separation. For IGCC, $CO_2$ removal by absorption process (e.g., Selexol or Purisol) have been reported to be more efficient (Riemer, 1993). In $CO_2$ removal from flue gases by analogy, one may assume a membrane absorber economics to be favorable over a conventional absorber.

**Table 3.12.** Hydrogen/Carbon Monoxide Separation Membranes (Riemer, 1993)

| Company | Membrane | Selectivity |
|---|---|---|
| Air Products (Permea) | Polysulfone | 60 |
| UOP (Separex) | Cellulose Acetate | 30–50 |
| Ube | Polyimide | 60–100 |
| GKKS | Polyetherimide | 100 |

**Table 3.13** Carbon Dioxide/Hydrogen Separation Membranes (Riemer, 1993)

| Polymer | Fast Gas | Selectivity |
|---|---|---|
| Polydimethyl Siloxane (GKSS) | $CO_2$ | 5 |
| Polyimide (Ube) | $H_2$ | 10 |

## 3.2.2. Solvent Vapor Recovery from Gas Streams

Vent gas streams containing solvent vapors or other volatile organic compounds (VOCs) are produced throughout the chemical processing, petrochemical and pharmaceutical industries. Safe disposal of the gas is of great concern to plant operators. Often several vent streams are pooled and incinerated or flared by an end-of-pipe treatment unit. This procedure eliminates VOC emissions but significant amounts of feedstock or product are lost. Membrane-based vapor separation technology, developed during the past ten years, has emerged as an alternative treatment technology, which can also recover valuable VOCs. Approximately 60 membrane VOC recovery plants have been installed worldwide.

### 3.2.2.1. Technology Description

*Membranes and Modules*
Two types of membrane are used in gas separation systems. The most widely known type is made from rigid, tough glassy polymers such as polysulfone. Such membranes permeate nitrogen and air and retain the VOCs. The second type is made of softer rubbery polymers such as silicone rubber. These membranes are hydrophobic and thus preferentially sorb and permeate the VOCs, retaining air and nitrogen.

Although both glassy and rubbery membranes can be used to separate VOCs from air, rubbery membranes are preferred for a number of reasons. First, VOCs are usually the minor components of the vent gas, so when VOC-permeable rubbery membranes are used, very little gas has to permeate the membrane to remove the bulk of the VOCs. Thus, relatively small membrane areas are needed. If air-permeable glassy membranes are used, the major component has to permeate the membrane. Also, the permeability of rubbery membranes is usually 1000 or more times higher than the permeability of glassy membranes, which reduces the membrane area required.

Rubbery VOC-permeable membranes can also provide very high removals of VOCs from the feed gas. This is important in VOC-control projects requiring 99–99.9% VOC removal. The ability of VOC-permeable membranes to achieve this degree of removal is illustrated in Figure 3.15. In this example calculation, a pressurized feed gas containing 1% VOCs is passed to a membrane module. The rubbery membrane is assumed to have a VOC/air selectivity of 20. In one example (top), the membrane area is chosen such that the VOC content of the gas leaving the module is reduced by 90%, to 0.1% VOCs. The VOCs are concentrated into a permeate gas with ~17% of the original feed gas volume and enriched to 5.4% VOCs. If a larger membrane area is used (middle), the concentration of VOCs in the gas leaving the module can be reduced another order of

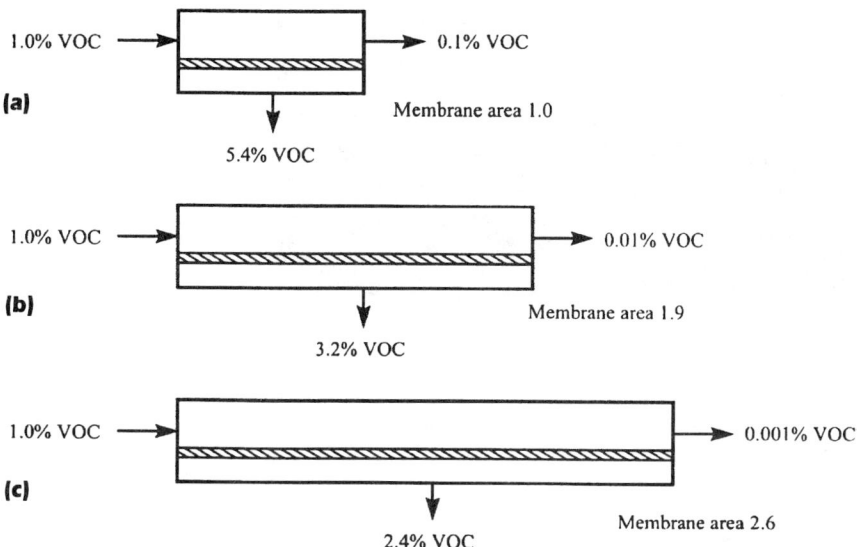

**Figure 3.15.** As the degree of volatile organic compound (VOC) removal increases from 90 to 99.9%, the required membrane area increases and the enrichment of VOCs in the permeate decreases. This trade-off determines the design of membrane separation systems. Calculations are based on a feed-to-permeate pressure ratio of 40 and a VOC-to-air selectivity of 20.

magnitude to 0.01% VOC, representing 99% VOC removal. However, the VOC concentration in the permeate is now reduced to 3.2%. Similarly, an even larger membrane module (bottom) could achieve 99.9% VOC removal. However, more than 40% of the feed gas would then permeate the membrane module, and the VOC concentration in the permeate stream would be only 2.4 times higher than in the original feed gas. As these calculations show, a trade-off between the degree of removal of VOCs from one stream and the degree of enrichment in the other is inevitable.

The most commonly used VOC-permeable rubbery polymer is silicone rubber (polydimethylsiloxane). The polymer is formed into a composite membrane as shown in Figure 3.16. The thin rubbery selective layer is coated onto a highly permeable porous support polymer which provides mechanical strength.

### 3.2.2.2. Engineering, Economic, Environmental and Energy Considerations

Current commercial vapor separation membrane systems use between 10 and 500 m² of membrane. Several methods are used to incorporate these large areas of membrane into membrane modules. Vapor separation systems use flat sheets of membrane that can be packaged as a layered structure (plate-and-frame

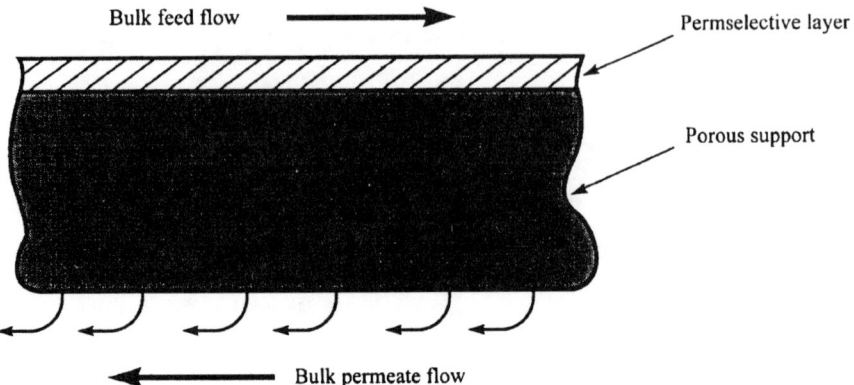

**Figure 3.16.** Thin, relatively soft, silicone rubber, the most common VOC-selective membrane material, is coated onto a highly permeable porous support polymer. The porous material provides the mechanical strength required to support the thin selective layer.

module design) or rolled with appropriate spacers around a central permeate collection pipe (spiral-wound module design).

*Applications of Membrane Systems*
More than 60 membrane vapor separation systems have been installed worldwide. In Europe and Japan, the combination of high energy costs and stringent environmental regulations has stimulated installation of systems to recover hydrocarbon vapors from petroleum transfer operations. In the United States, most of the systems have been installed in chemical processing plants to recover CFCs, vinyl chloride, propylene and other high-value materials. Table 3.14 summarizes existing and emerging applications for membrane vapor technology. To illustrate the types of systems used in these processes, two specific applications of the technology are described below.

*Polyolefin Polymerization Vents*
A significant new application of membrane vapor separation technology is recovery of ethylene and propylene from polyolefin manufacturing purge-gas vents. In a typical olefin polymerization process, monomer plus catalyst, various comonomers, solvents and stabilizers are contacted at high pressure in a polymerization reactor. After polymerization, the small polymer particles are sent to large resin bins, through which nitrogen is circulated to purge absorbed monomer and process solvents from the resin. The composition of the off-gas from this step varies widely and depends on the particular proprietary polymerization process

## 3. Membrane Technologies

**Table 3.14.** Existing and Emerging Applications for Membrane Vapor Recovery Technology

| Existing | Emerging |
|---|---|
| Gasoline vapor recovery at storage facilities | Recovery of feedstock from oxidation reactor vents |
| Recovery of vinyl chloride monomer from PVC reactor vents | Recovery of olefins from reactor purge gas streams |
| Recovery of CFCs from process vents and transfer operations | Recovery of gasoline vapor from gas station storage tanks |
| Recovery of olefins from resin purge bin off-gas in polyolefin production | |

used, the product polymer (polyethylene or polypropylene), and the plant size. Generally, the value of the recoverable monomer is very high, so these vent streams represent a considerable resource recovery opportunity. Until the development of membrane technology no appropriate recovery technology existed. The streams are not good candidates for absorption processes, condensation can be used, but very low temperatures are required to achieve significant removal of the very volatile ethylene or propylene.

Several membrane system designs can be used for this separation. Figure 3.17 shows a process in which the permeating monomer-rich gas is processed twice to achieve a sufficient separation. This is called a two-stage process. The membrane system fractionates the vent gas into two streams: a monomer/solvent-rich stream that is recycled to the polymerization reactor, and a nitrogen stream that is recycled to the resin purge bin.

Typical olefin polymerization plant vent gases contain 500–1000 lb/h of recoverable monomer and solvent, and an additional 1000 to 2000 lb/h of recoverable nitrogen. The combined value of these two streams is more than $1 million/year, so recovery and reuse is well worthwhile. The first plant of this type became operational in December 1996. Since then, four more units have been ordered.

### Distillation Vents

Inert gas vents are common in distillation operations because inerts often enter the column with the feed stream. During distillation these components move to the top of the column and must be vented for efficient operation. The vent gas is saturated with the light product, and, although the gas flow may be small, the loss

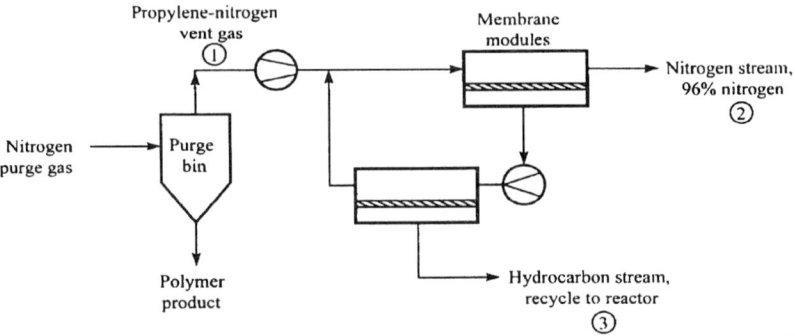

| Stream | Purge-Bin Vent ① | Nitrogen ② | Hydrocarbon ③ |
|---|---|---|---|
| Flow (scfm)[lb/h] | 700[3,520] | 616 [2,940] | 84 [580] |
| Composition (%) | | | |
| $C_2$ | 0.3 | 0.1 | 1.5 |
| $C_3$ | 13 | 3.5 | 83 |
| Nitrogen | 87 | 96.4 | 15.5 |

**Figure 3.17.** A two-stage membrane separation process can be used to separate organics from nitrogen in a polypropylene resin purge gas. The table indicates gas flow and composition at different points in the process.

of product can be significant. An example is shown in Figure 3.18, in which the light product from distillation is $CF_2Cl_2$ (CFC-12). The inert vent stream is only 21 scfm but contains an average of 120 kg/h of CFC-12. A two-step membrane recycle design is used. First, the low pressure vent gas is compressed and passed to a condenser where it is cooled; a portion of the organic vapor condenses and is removed as a liquid. In the second step, the noncondensed portion of the feed gas, which still contains a significant amount of organic vapor, passes across the surface of an organic-vapor-permeable membrane. The membrane separates the gas into two streams: a permeate enriched in organic vapor and a residue containing the noncondensable gases. The organic-vapor-enriched permeate is recycled back to the compressor inlet. A photograph of such a system is also shown in Figure 3.18. The complete vapor recovery system is quite small, with a footprint of 7 ft × 8 ft and is installed at the top of the column.

*Competitive Technology*
The position of membrane technology in the VOC recovery and control market is illustrated in Figure 3.19. The three commonly used VOC-recovery processes are condensation, adsorption on activated carbon, and membranes. Condensation can be a very simple and low-cost process, but at moderate temperatures the VOC content of the vent gas leaving the condenser (saturated with vapor at the

## 3. Membrane Technologies

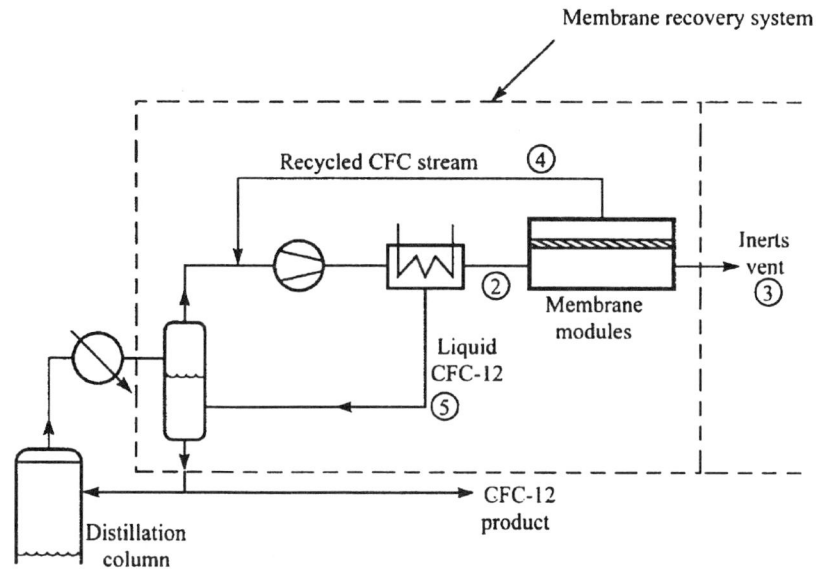

| COMPONENT | STREAM COMPOSITION | | | | |
|---|---|---|---|---|---|
| | ① | ② | ③ | ④ | ⑤ |
| CFC-12 | 63 | 65 | 4.7 | 65 | >99 |
| NITROGEN | 37 | 35 | 95.3 | 35 | |
| FLOW (KG/H) | 140 | 341 | 20.5 | 201 | 120 |

**Figure 3.18.** Flow schematic and photograph of the membrane system used to recover CFC-12 from a nitrogen distillation column vent. The unit recovers approximately 120 kg/h of CFC-12.

temperature of the condenser) is still too high to meet EPA regulations. Low condenser temperatures provide better VOC recovery but increase the cost significantly. Moreover, the water vapor present in most process streams freezes at low temperatures, reducing reliability. For these reasons, condensation systems are generally limited to concentrated gas streams from which good fractional removal of VOCs is possible.

Adsorption, a process in which VOCs are removed by passing the gas stream through a bed of activated carbon or other adsorbent material, is widely used. Periodically, the bed is regenerated by heat or vacuum, and the VOCs are recovered as a condensate. The size and cost of a carbon adsorption system are proportional to the mass of VOCs in the gas stream but are relatively independent of the volume of the gas stream. For this reason, carbon adsorption is the favored low-cost process for large, dilute streams.

On the other hand, the cost of membrane systems is relatively independent of the concentration of the gas stream to be treated; it is proportional to the volume. Thus, membrane systems are generally most cost competitive for relatively concentrated gas streams containing 1000 ppm (preferably, > 5000 ppm) VOCs.

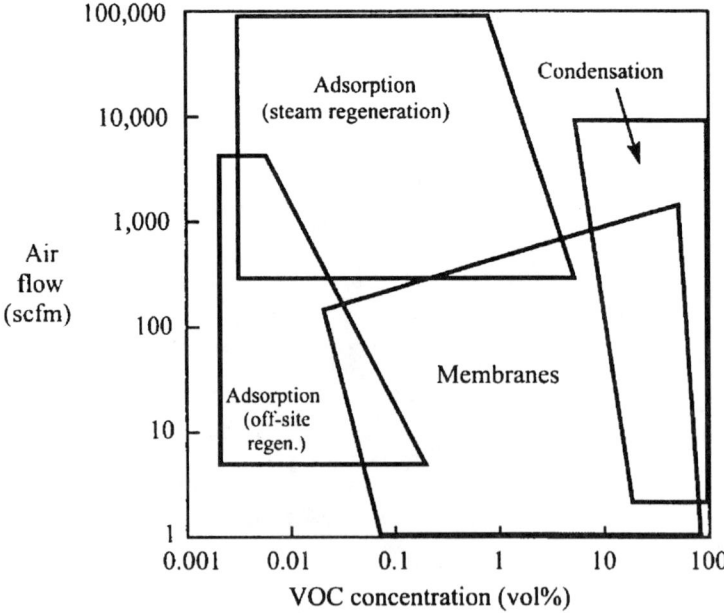

**Figure 3.19.** Application areas of some of the major VOC treatment technologies. Only recovery technologies are included.

# 3. Membrane Technologies

## 3.2.3. Metal Ion Recovery from Aqueous Waste Streams

### Introduction

Membrane processes have found wide application for the treatment of aqueous-based systems involving material recovery and reuse (Ho and Sirkar, 1992; Noble and Stern, 1995; Scott, 1995; Mulder, 1996; Sourirajan and Matsuura, 1985; Rautenbach and Albrecht, 1989). Conventional membrane processes include reverse osmosis (RO), nanofiltration (NF), ultrafiltration (UF), microfiltration (MF), and electrodialysis (ED). All of these conventional membrane processes (with the exception of ED) are pressure driven with the degree of separation depending on the nature, size, and charge of the component to be separated. For instance, RO can reject various heavy metals at >98% level, while MF can filter out precipitated particles of heavy metals. The reported (Sourirajan and Matsuura, 1985) RO rejections of metal salt (sulfate as anion) solutions such as $Cd^{2+}$, $Cu^{2+}$, and $Ni^{2+}$ are greater than 99.9% with cellulose acetate membranes. Charged NF processes can be used to separate heavy metals (low concentration) from high concentrations of NaCl where selective separations may be desirable. Based on the limitations of the conventional membrane processes, several advanced membrane applications have been reported including polymer enhanced UF (PEUF), micellular enhanced UF (MEUF), functionalized MF membranes, and membrane contactors. In addition to pressure driven processes, metals recovery from acidic solutions have also been practiced using ED, diffusion dialysis, and Donan dialysis. By using these membrane-based technologies in hybrid systems, many different selective separations involving metal recovery and water reuse can be obtained.

### Background

All pressure-driven membrane processes (RO/NF/UF/MF) are characterized with the same parameters: (water) flux $(J_w)$, rejection $(R)$, and water recovery $(r)$. The measured rejection is typically defined as: $R = 1 - C_i''/ C_i'$, where single prime refers to feed concentration and double prime refers to permeate concentration. Typical operating feed pressure ranges (in MPa) are 2.7–14 for RO, 0.4–2.0 for NF, 0.2–0.7 for UF, and 0.03–0.3 for MF.

Many models have been proposed to describe flux and rejection through pressure driven membranes (Ho and Sirkar, 1992; Mulder, 1996; Sourirajan and Matsuura, 1985). Of course, in addition to establishing water and metal salt transport through membranes, one would also need to optimize variables to minimize concentration polarization and membrane fouling. RO membrane systems can often be accurately described by the solution-diffusion (SD) mechanism. The flux obtained is directly proportional to the net transmembrane pressure applied

(applied pressure – osmotic pressure difference). Thus, even for dilute metal solutions, when the feed contains high nonpollutant dissolved solids (such as NaCl), the osmotic pressure effect could be substantial. For complex waste mixtures, where all of the individual component concentrations may not be known, the rule of thumb is 10 psi (0.07 Mpa) for 1000 mg/L total dissolved solids can be used.

NF membranes are usually charged (carboxylic groups, sulfonic groups, etc.), and, as a result, ion repulsion (Donnan exclusion) is a major factor in determining salt rejection. More highly charged ions such as $SO_4^{-2}$ (counter-ions to heavy metals) are rejected considerably more than monovalent ions, and thus the rejections of metal sulfate salts will be considerably greater than the rejection of metal chloride salts. For oxyanion metals, such as Cr(VI), the rejection is a function of pH and thus one would expect the rejection of $CrO_4^{-2} > HCrO^{4-}$. One of the main advantages of NF membranes is selective separation of divalent metal ions and moderate to high flux at low pressures.

For separations involving the addition of chelating agents (polymeric or monomeric), the principle of separation is a complexation of metal ions to a chelating agent and the resulting separation of the metal–chelate complex by membranes. In MEUF and PEUF, minimizing the amount of free (uncomplexed) metals allows for maximum separation. With functionalized MF and membrane contactors, the rate of uptake as well as the complexation must be considered.

In ED, the electrical potential difference is the main driving force for separation. The extent of separation of metal ions is a strong function of the permselectivity of the ion exchange membranes used. Required membrane area is directly proportional to the desired flow rate of the product water, difference in concentration between feed and product water, number of cells (a series of ion exchange membranes between and anode and a cathode) in the stack, and is inversely proportional to the current density and current utilization. Of course it should be noted that in contrast to pressure driven membrane processes, for ED, water transport across the membranes must be negligible.

*Membrane Selection and Applications*

Selecting a membrane process for metal recovery is very dependent on the nature of the component to be recovered. Figure 3.20 shows a breakdown of some of the possible membrane processes for use in in-line recycling and metal containing waste treatment and some basic processing conditions telling when to use a particular membrane-based process. As shown in Figure 3.20, the separation can be very dependent on the nature of the stream and the types of membrane processes. For instance, RO and NF can separate dissolved components while UF and MF either need precipitation or ligand enhancement to separate metals. Further, with both ligand enhanced and functionalized membranes the polymer or surfactant (micelles) must have affinity to bind with the membrane.

## 3. Membrane Technologies

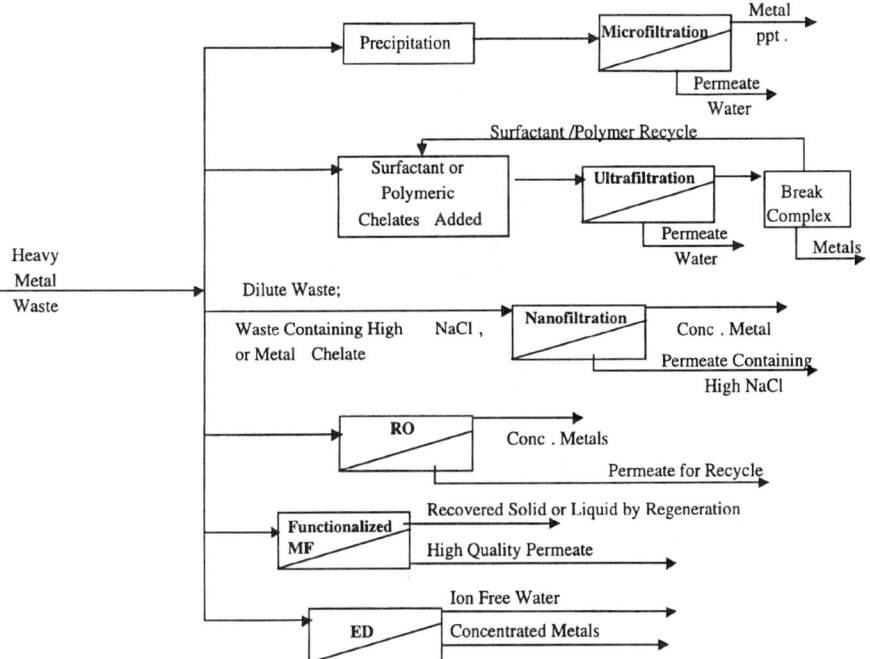

**Figure 3.20.** Various membrane-based metal recovery schemes for metal containing wastes.

Traditionally, membrane applications have been limited by fouling, selectivity, concentration factor, etc., and so the applications in the metals recovery area were restricted. Commercial membranes used for direct salt separations are classified in three areas: (a) seawater desalination membranes, (b) brackish water membranes, and (c) NF ("loose" RO) membranes. Very seldom, membranes are developed for specific metal separations. Rather, the commercial desalination membranes are used with possible adjustment of the membrane chemistry and modules. Figure 3.21 shows NaCl rejection characteristics and water flux values of various commercial membranes. One can safely assume the rejection of heavy metal salts will be higher than NaCl rejction. Advances in both membrane materials, module design, and in-line processing has increased the range of applications in which membranes apply.

The most established membrane technology in metals recovery is RO. A typical RO application (provide by Osmonics/Desal Co.) involving nickel recovery (Kloos, personal communication, 1997) and water reuse is shown in Figure 3.22. The process is conducted in two passes because the membrane being used needs to be highly acid resistant. As shown, this process treats the metal waste

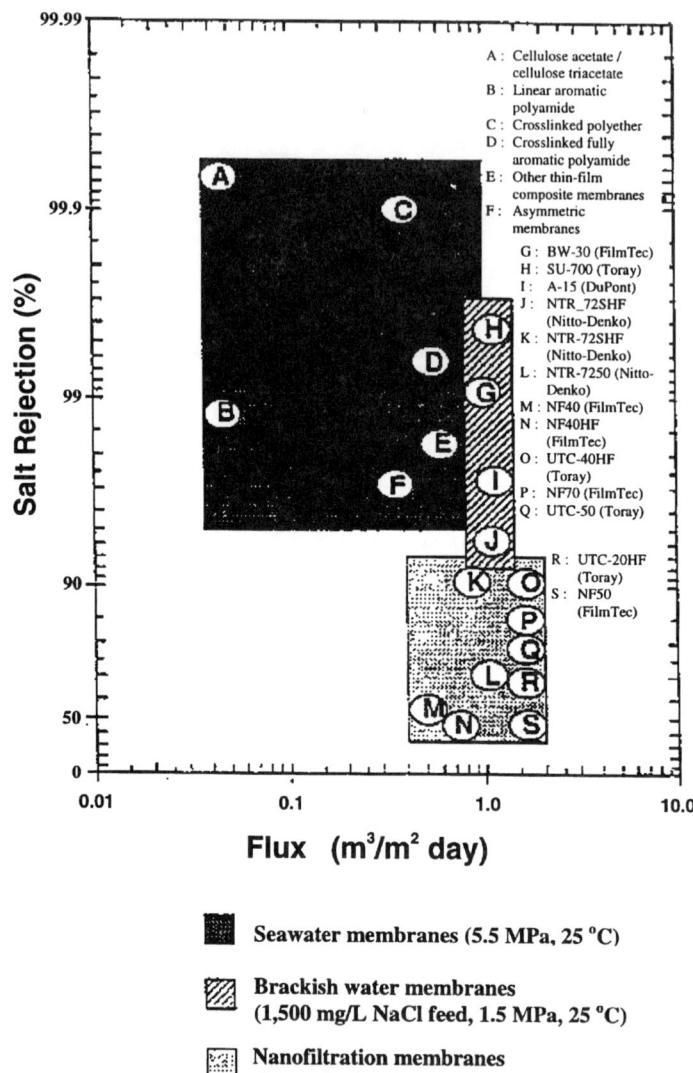

**Figure 3.21.** Water flux and NaCl rejections of commercial RO and NF membranes (Bhattacharyya et al. 1997).

with nearly 90% water recovery while yielding an end permeate with less than 2 ppm metal (after two passes). Many other examples of RO for heavy metal recycling/recovery have been reported in the literature, especially in the areas of plating. Some of the treated metals include nickel, cadmium, chromium, and molybdenum (Kloos, personal communication, 1997; Nirmal et al., 1992; Rob-

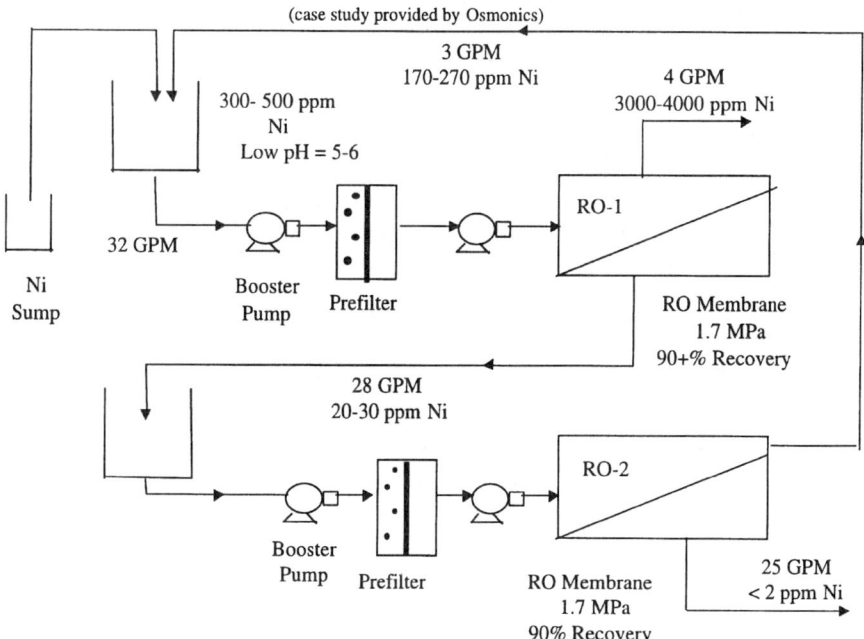

**Figure 3.22.** Nickel acetate recovery scheme for a RO membrane system (Kloos, personal communication, 1997).

inson, 1983; Schoemann et al., 1992). In all of these cases RO is an excellent solution because it provides a complete recovery scheme (most components are rejected).

Simple calculations can be performed to calculate flux on RO membranes. For instance, with a cellulose acetate Membratek Membrane (Schoemann et al. 1992) the observed pure water flux is 0.56 $m^3/m^2/d$. Given that the feed solution was 2100 mg/L Ni and water recovery maintained at 80%, it can be found that the membrane flux should be 0.5 $m^3/m^2/d$ based on simple calculation using the osmotic pressure effect. However, the observed flux is only 0.31 $m^3/m^2/d$, which demonstrates that concentration polarization and/or fouling is taking place. Thus, in actual industrial waste streams, it may be necessary to run bench scale experiments to establish fouling and concentration polarization behavior as a function of operating conditions.

If heavy metals exist in a precipitated form, then MF or UF (also rejects some dissolved species) can be used for recovery. Industrial cases where MF has been used are battery recycling plants and the electronics industry, rejecting metals such as cadmium, mercury, chromium, copper, lead, etc. (Broom et al., 1994; Dineen, 1994; Enoch et al., 1994; Karrs et al., 1992; Nunno and Freeman,

1987). MF, however, is limited by the solubility of the metal in question. For instance, typically lime is used for precipitation. Even in precipitation cases, the metal hydroxide still has some solubility (often greater than 5 mg/L) and will never be rejected by MF while the metal is in solution.

ED is an electrically driven process. An example of how ED could be used with plating rinse solutions has been reported by Smith and Foreman (Smith and Foreman, 1997). They have looked at ED of a process stream to remove tin and zinc (present as citrate salts in the plating bath) at concentrations of 300 mg/L each and pH 6-7 (anion and cation exchange membranes, Tosflex DF-34 and Nafion 350, respectively). A recovery scheme with electrowinning and electrostripping was also studied to yield a concentrated metal stream. The authors showed a reduction in concentration to less than 15 mg/L of each metal with possible recoveries of 70% tin and 100% zinc. However, since zinc is highly electronegative it is impossible to deposit it at 100% current efficiency. The current required for the process was around 300 Coulombs (Smith and Foreman, 1997). ED has also been used in the case of iron, chromium, and nickel, copper, and zinc (Ramachandraiah et al. 1996; Wisniewski and Wisniewska, 1997; Wisniewski and Suder, 1995) showing 92–99% removal of iron and nickel and 65–68% removal of chromium. Hansen et al. (1997) have also studied eledrodialytic salt remediation (heavy metals) by ion exchange membranes.

NF has also been studied in heavy metal recovery/recycling. An interesting example of NF in a plating solution was presented by Cadotte et al. (1988). The authors used Filmtec XP-20 membrane to try to selectively recover Cu–EDTA complex from by-product sulfate salts. This sort of separation would give the bath a longer working life. The membrane required for this operation must be able to operate at pH 12. Although the NF membrane was found to reject 99% of the copper complex, it also rejected 85% of the sodium sulfate. New NF membrane materials with lower sulfate rejections would be required for a membrane process to be used in this application. Other processes which have used or are looking into NF type processes have been reported (Alami-Younssi et al., 1994; Linde and Jonsson, 1995; Prabhaker et al., 1996).

### 3.2.3.1. Technology Description

*Modules and Membranes*
Although traditional membrane modules and materials work very well in several applications, limitations exist which lead to the need for different systems. For instance, where high fouling may occur, such as a landfill leachate, or where a high recovery is needed (as in a solids recovery to sludge scheme), a spiral-wound module may have limitations at high water recovery. In the case where a highly acidic metal-containing stream was to be processed, cellulose acetate ($5 < pH < 7$)

## 3. Membrane Technologies

could not be the material of choice and other membrane materials would be needed. Four brief case studies will be presented below which demonstrate the use of advanced modules and materials in extreme conditions.

A landfill leachate often contains a combination of organics and metals. For this reason, pretreatment would be difficult and expensive. Further, since the desired effect is high water recovery, traditional membrane modules cannot operate at sufficient pressures to achieve the separation goal. However, the Rochem DT-module (Figure 3.23) has wide feed channel spacing and can operate at very high pressures (up to 200 bar) which make it ideal for this type of application (Rautenbach and Linn, 1996; U.S. EPA, 1995; Vail and Sanford, 1992). One site where the DT-module has been used is in Ihlenberg, Germany (Rautenbach and Linn, 1996). This waste is a mixture of nitrogen compounds, dissolved solids, and heavy metals including lead, chromium, zinc, cadmium, arsenic, and mercury. Water recovery rates of >97% have been shown at this site using a combination of high pressure RO and NF (Rautenbach and Linn, 1996). However, Rochem membrane systems have higher capital costs as compared to spiral-wound RO membranes. For this reason, most applications of the DT-module have been limited to highly contaminated leachate sites.

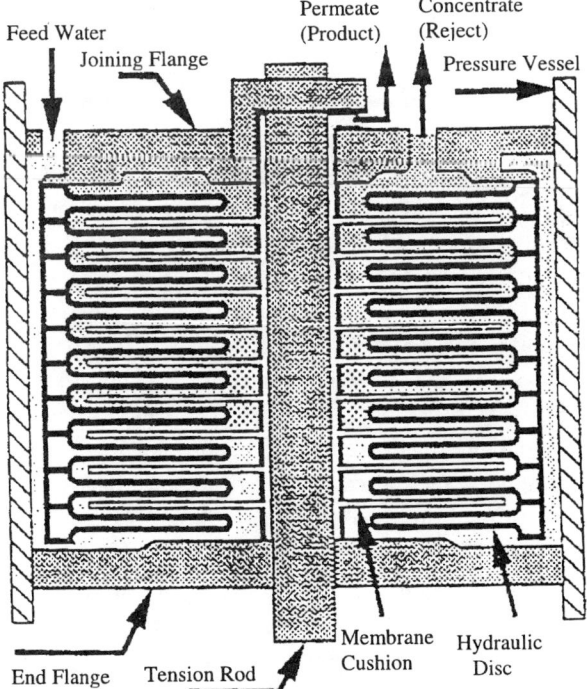

**Figure 3.23.** Schematic of a Rochem DT (disc tube)-module (US EPA, 1995).

Besides water recovery, another important area of module technology is reduction of fouling. In the area of recovery of metal precipitate from sludge, this would be a major problem since high concentrations of solids would be present. One way to reduce fouling is by using membrane motion to create shear at the membrane surface. Spin-Tek and Membrex (Du Pont) membrane systems have used a feed flow through a system similar to a centrifuge and "spinning" the membrane around in a circular motion while the permeate comes through (Hetstkin et al., 1997). New Logic International has designed a membrane module known as Vibratory Shear Enhanced Processing system (V*SEP) which vibrates a filter pack (16-inch outer diameter) at 60 Hz with an oscillation amplitude of 1 inch. This creates a shear at the membrane surface 10 times higher than traditional filtration devices (Culkin and Armando, 1992). One application of the V*SEP unit showed that it was able to convert a feed flow of 0.35 GPM with 0.7% solids to 0.34 GPM permeate containing 500–600 ppm solids and a concentrate of 0.01 GPM containing 60% solids at a GE manufacturing plant (Hetstkin et al., 1997; Shapiro et al., 1995). This conversion to near "paste" would not be possible with traditional MF systems and allows for expensive evaporation processes to be reduced. Again, however, there is a cost penalty to pay. Although operating costs tend to be lower for V*SEP systems (Culkin and Armando, 1992), there is a high installed cost compared to well established spiral-wound technology.

Finally, advanced membrane materials are also opening doorways in more innovative applications. Some of the advanced materials used are high temperature polymeric RO and NF membranes (Snow et al., 1996), ceramic membranes (Hetstkin et al., 1997), and alumina NF membranes (Alami-Younssi et al., 1994). One example of a ceramic membrane being used is the removal of lead particles from a machining coolant at a manufacturing plant (Hetstkin et al., 1997). Because of the high amounts of organics present and extreme temperature conditions, polymeric membranes could not be used. So, precipitation of lead with phosphate was performed with a separation using a MF membrane of approximately 500 Å pore size. The membrane was found to reject nearly 96% of the lead particles making it possible to keep reusing the machining coolant without it ever becoming hazardous waste (Hetstkin et al., 1997).

Another interesting example of an advanced membrane material is similar to that of Figure 3.20 in that it is a two-pass RO, but it is combined with NF in a slightly different way. The process (Osmonics/Desal) stream is an acidic waste stream from a copper rod refinery. Because of the highly acidic pH (0.9–2), traditional polymeric membranes are not used. The initial feed, at pH 1.2, has 2% acid and 1200 mg/L $CuSO_4$. The goal of the separation was to produce a high quality permeate low in copper sulfate, and another stream highly concentrated in both acid and copper. In these processes, the RO membranes rejected both metal ions and the acid, and thus concentrated the acid (after one pass) from 2 to 10% and copper from 1200 to 8100 mg/L. The retentate, concentrated by another RO unit, is then recycled back to the first unit with a recovered permeate of <3 mg/L $CuSO_4$ and 50 ppm acid at pH 2.8. The NF membrane successfully further concentrated the stream (first RO retentate) to 29,400 mg/L $CuSO_4$ with 100% acid pass. At a total flow rate of 409 $m^3/d$, this process has a payback period of less than one year (Osmonics, 1997).

*Ligand Enhanced Membrane Processing*
Another type of advanced membrane processing system is ligand enhanced processing. For instance, it has been outlined earlier that RO can remove dissolved metals while UF can not. But, if the dissolved metal is complexed with a macromolecule, UF will be able to reject the macromolecule thus also rejecting the metal. This is precisely the principle behind PEUF. Some of the metals that have been separated using this process are mercury (Uludag et al., 1997), chromium (Sriratana et al., 1996), copper (Mundkur et al., 1993), and magnesium (Tabatabai et al., 1995a,b). Along the same principle of PEUF is MEUF which uses surfactants to create micellelar complexes in which the metal is bound to and rejected by the membrane. Some of the metals that have been studied with MEUF are cadmium, lead, copper, nickel, and zinc (Ahmadi et al., 1994; Fillipi et al., 1997; Goff et al., 1997; Huang et al., 1994). A detailed description of ligands used in UF as well as NF (Gaubert et al., 1997a,b) are shown in Table 3.15. One of the particular advantages of using this process is that the membrane is not the only thing dictating selectivity. The selectivity of the ligand can also influence the separation. Further, the use of UF allows membranes made of ceramics and metals (more resistant to extreme conditions) to be used for filtering metals. The major disadvantage of this area is that an additional compound must be added and a complex broken after the desired separation occurs. This same principle of complexation has been demonstrated with EDTA-Cu complexes in ED because the charge of the species changes the separation characteristics (Cherif et al., 1993).

*Functionalized MF*
Another advanced membrane separation that will be discussed is functionalized MF membranes. Traditionally, ion-exchange materials are used in bead form where pores are small and diffusion plays a prominent role. MF membranes could be used as an exchanger, but in general, the surface areas are too low (<50 $m^2/g$) as opposed to ion-exchange beads (600–1000 $m^2/g$) and so capacities are limited.

**Table 3.15.** Ligand Enhanced Processing with Various Membranes

| Membrane Technology | Membrane Type | P MPa | Pure Water Flux m/d | Metals | Conc. mM | Rej. % | Ligand |
|---|---|---|---|---|---|---|---|
| RO Schoeman et al., 1992 | Membratek CA | 4.0 | 0.55 | Ni | 33 | 97 | None |
| NF Linde, 1995 | PCI Membrane AFC-40 (charged) | 2.0 | 4.2 | NaCl CaCl$_2$ MgCl$_2$ | 330 330 330 | 0 5 55 | None |
| NF Linde, 1995 | PCI Membrane AFC-40 (charged) | 2.0 | 4.2 | MgSO$_4$ | 50 | 85 | None |
| NF Gaubert et al., 1997a | Filmtec NF-70 | 0.5 | 0.84 | Sr | 0.28 | 60 | Poly(acrylicaride) (2000 MW) |
| NF Gaubert et al., 1997b | Filmtec NF-70 | 1.5 | 1.8 | Cs Na | 0.075 87 | 95 85 | Resorcinorine (440 MW) |
| UF Fillipi, et al., 1997 | Regenerated Cellulose 6000 MW Cut-Off | Na | Na | Cu Ca | 0.7 0.293 | 99.99 37.2 | Alkyl—diketone (244 MW) |
| UF Huang et al., 1994 | Osmonics HG-03 2000 MW Cut-Off | 0.28 | 1.56 | Ni | 1.2 | 100 | deoxycholic acid (393 MW) |
| UF Mundkur and Watters, 1993 | Osmonics HG-03 2000 MW Cut-Off | 0.28 | 1.56 | Cd | 1.2 | 95 | Sodium-dodecyl sulfate (288 MW) |
| UF Mundkur and Watters, 1993 | Romicon HF 1.1-45-XM50 50,000 MW Cut-Off | 0.17 | NA | Cu | 1.7 | 45 | Carboxy-methyl cellulose ($7 \times 10^6$ MW) |

## 3. Membrane Technologies

| Membrane Technology | Membrane Type | P MPa | Pure Water Flux m/d | Metals | Conc. m$M$ | Rej. % | Ligand |
|---|---|---|---|---|---|---|---|
| UF Mundkur and Watters, 1993 | Romicon HF 1.1-45-XM50 50,000 MW Cut-Off | 0.17 | NA | Cu | 1.7 | 30 | Polystyrene-sulfonate (6 × 10$^5$ MW) |
| UF Sriratana et al., 1996 | Spectrum CA 10,000 MW Cut-Off | 0.41 | NA | Cr | 20 | 99.8 | Poly(diallyl-dimethyl-ammonium chloride) (2.4 × 10$^5$ MW) |
| UF Uludab et al., 1997 | Osmonics HG-01 5,000 MW Cut-Off | 0.1 | 0.12 | Hg | 100 | 97 | polyethyleneimine (50 × 10$^5$ MW) |

NA = Not Available

However, if long chain polymers containing multiple functional groups are attached into the pores, the MF membrane can be used for exchange with capacities at or considerably above ion-exchange. This is the approach reported by Bhattacharyya et al. (1997a,b). In this case, cellulosic (or silica) membranes are derivatized with aldehyde (or epoxide) functionalities and poly glutamic or poly aspartic acid are grafted via Schiff base reaction. The resulting membrane has been used for metal ion entrapment (in acidic pH conditions) with capacities as high as 11 meq/g (Pb results) (Bhattacharyya et al., 1997a). Further, using a 40-mg/L initial concentration Pb stream it has been demonstrated that rapid sorption takes place with a concentration factor (liquid to solid) of 400. Many other exchange/chelation groups have been attached (radiation induced graft attaching) including sodium styrene sulfonate\acrylic acid (Sugiyama et al., 1993), amidoxime, phosphoric acid, and iminodiacetate functionalities (Tsuneda et al., 1991) with attachment efficiencies from 1.2 to 7.1 mmol/g. Another area of functionalized membrane work being looked into is immobilized metal-sorbing vesicles which uses surface modified membranes to activate avidin–biotin chemistry for metal sorption with cadmium and lead (Shamsai and Monbouquette, 1997; Walsh and Monbouquette, 1993; van Zanten et al., 1995). Their work has demonstrated that for multilayered sorption vesicles, longer spacer chains gave sharper breakthrough curves. This is important because the tail of a breakthrough curve is usually unavailable for metal sorption. All of these functionalized processes have the advantage that sorption is not limited by diffusion through small pores so exchange is rapid. This technology lends to the possibility of post treatment of RO concentrate for further concentration.

*Membrane Contactors*

One final area of note is the use of membrane contactors for metal separation. This process is similar to liquid\liquid extraction although a membrane (usually hollow fiber) is used to create contact between the two liquids and selectively remove the desired metal component. This is the approach taken by Yang et al. (1996), and by Ho and Sirkar (1992). They found by extracting cations and anions simultaneously, the rate of extraction is superior to extracting one at a time. This is easily accomplished in a membrane system by running an anion complexing component in the tube side of half of the hollow fibers and a cation complexer in the other half. Using a metal ion system with Cu and Zn (cationic) and Cr (anionic), they found a drop in concentration of Cu from 229 to 8.5 mg/L and Cr from 51 to 8.9 mg/L while Zn only was reduced from 365 to 344 mg/L.

*Choosing a Membrane System*

Choosing the proper separation system for metals recovery and waste treatment is often the most challenging part of the separation. As outlined above and shown in Figure 3.20, there are many different options for separation. Table 3.16 shows some examples of when a particular separation technique should be used. In general, RO provides complete separation of metals from solutions but has limited operating conditions (pH and temperature limitations) and limited concentration factors. ED is the only process discussed where the minor component (the metal) is transferred through the membrane and so the cost is directly dependent on the concentration. Some of the more advanced techniques like PEUF, MEUF, and functionalized membranes can be used with other conventional membrane techniques to achieve very high separation factors.

### 3.2.3.2. Engineering, Economic, Environmental, and Energy Considerations

*Membrane Economics*

Membrane economics can be a function of both nonpollutant dissolved solids and contaminated metal concentrations. Typical costs of conventional spiral-wound membrane modules are $101/(m^3/d)$ for seawater membranes and $42/(m^3/d)$ for brackish water membranes (Ho and Sirkar, 1992; Bhattacharyya et al., 1997). Streams such as seawater and municipal water are well characterized and so cost estimations and models have been derived (Ho and Sirkar, 1992;

**Table 3.16.** Selecting a Membrane-Based Process

| Process | MW Cut Off (approx.) | Advantages | Drawbacks | Applications |
|---|---|---|---|---|
| RO | None | • High rejection of most components | • High pressure<br>• Significant pretreatment required<br>• Only polymeric membrane systems | • Plating waste<br>• Landfill leachate<br>• Desalting<br>• Ultrapure water |
| NF | >200 | • Selective removal of divalent from monovalent anions<br>• Lower pressure than RO | • Suitable for dilute solutions | • Selective separations of heavy metals from NaCl<br>• Selective separation of salts from dyes |
| UF | 1000–300,000 | • Selective separation of macromolecules from salts<br>• Ceramic and alumina membranes which give superior membrane durability | Moderate fouling<br>• Dissolved metals too "small" to reject | • Removal of macromolecules from metals<br>• Polymer and micellelar enhanced processing |
| MF | 0.05–5 $\mu$m | • Wide variety of materials<br>• Low operating pressures | • Severe fouling<br>• Only particle rejection | • Filtration of precipitated metals |
| ED | Not applicable | • Removes metals as minor component, so cost dependent on concentration membranes generally more acid resistant than RO | • High energy costs | • Metals from highly acidic streams<br>• Plating wastes |
| Func. MF PEUF, MEUF | Not applicable | • Selective separations controlled by ligand characteristics | • Emerging technology | • Selective separations with high concentration factors |

Manwell and McGowan, 1994; Pickering and Wiesner, 1993). A brief listing of some of the costs for treatment of process streams containing metals is given in Table 3.17. However, even in this table, processing conditions have a huge impact on economics. For instance, the nickel acetate (Figure 3.22) RO stream is treated by two-pass RO because of the extreme pH conditions. If the pH of this stream was within normal operating range (5 < pH < 7) it is possible that one pass RO could be used and costs could be significantly reduced. Further, other dissolved solids existing in the waste stream may also play a significant role in membrane design economics since they will still contribute to osmotic pressure effects (RO and NF applications). Along these same lines, membrane life must also be considered. Typically, a membrane in a seawater/brackish water applications can have a lifetime as long as 5–10 years but with process streams the membrane life is typically only 1–3 years.

### 3.2.3.3. Vendors and Contacts

It has been shown that many membrane technologies in metals recovery and treatment areas are both technologically and economically feasible. Further, some basic design principles and cost information has been given. However, to implement many of these technologies, working with some of the researchers in the field may be beneficial. Table 3.18 is a listing of selected researchers in the area of membrane-based metal separations. Table 3.19 is a brief listing of some of the membrane module and system manufacturers and contacts (where available). A complete listing of major membrane companies and researchers can be found in the membrane directory (Membrane Directory, 1996).

## 3.2.5. Pervaporation/Aqueous Streams

### 3.2.5.1. Technology Description

Pervaporation is a membrane process used to separate liquid mixtures. The feed liquid contacts one side of a membrane which selectively permeates one of the feed components. Pervaporation is just beginning to be applied to removal of volatile organic compounds (VOCs) from aqueous waste streams, producing dischargeable water and a small-volume VOC concentrate for disposal or recovery. The membranes used for this application allow the VOCs to pass through rapidly while holding back the water, which permeates at a much slower rate, resulting in an effective separation. A schematic drawing of the pervaporation process is shown in Figure 3.24. The aqueous waste stream is passed across the surface of the membrane at a temperature between 40 and 80°C. The VOC preferentially permeates the membrane as a vapor. The permeate vapor is cooled and condensed, providing the pressure gradient that drives the process. The

**Table 3.17.** Economic Evaluation of Some Membrane-Based Metals Removal Processes

| Site | Waste | Membrane Tech. | Major Metals | Conc. (mg/L) | Flow (m³/d) | Comments | Membrane Type | Capital Cost ($/gal/d) | Op. Cost ($/1000 gal) |
|---|---|---|---|---|---|---|---|---|---|
| West Thurrock Treat. Plant, London, England; Broom et al., 1994 | Mixed plating effluent for river discharge | MF | Cd Cr Cu Hg Ni Pb Zn | 0.6–8.7 3.5–30 3.3–35.0 1.3–2.3 3.1–11.6 1.7–3.8 7.8–31.6 | 200 | Highly acidic waste stream | Dynamic membrane from suspen. solids on cloth support | 3.44 | 1.17 |
| GE Pilot Waste Water Treat. Plant; Hestekin et al., 1997 | Suspended solids | V*SEP (MF) | Suspen solids | 0.7% | 1.9 | Waste had to be very highly conc. | V*SEP | 6.25 | 5.3 |
| Not Available; Kloos, pers. commun., 1997 | Ni acetate process solution | RO | Ni | 300–500 | 174 | Two-pass RO | Osmonics RO membrane | 6.00 | 1.00–3.00 |
| Generic; Tabatabai et al., 1995 | Softening of polluted waste-water | PEUF | Ca, Mg | 603 | 3800 | Poly styrene sulfonate complex. | 10,000 MW Cut-Off Continental Water System | 0.5 | 3.51 |

**Table 3.18.** Selected Researchers in Membrane Technologies (Metal Separations)

| Separation Technology | Affiliation | Contact |
|---|---|---|
| PEUF, MEUF | University of Oklahoma Norman, OK 73019 | J. Scamehorn |
| Functionalized MF Membranes | University of Kentucky Lexington, KY 40506-0046 | D. Bhattacharyya |
| Advanced Wastewater Treatment (Rochem DT-modules) | Westinghouse Savannah River Aiken, SC 29808 | J. Siler |
| Hollow Fiber Membrane Contactor | New Jersey Inst. of Technology Newark, NJ 07102 | K. Sirkar |
| Advanced Wastewater Treatment (V*SEP membrane modules) | GE Corporate R&D Schenectady, NY 12301 | B. Mo Kim |
| ED of Heavy Metals | Los Alamos National Lab Los Alamos, NM 87545 | T. Foreman |
| Metal Sorbing Vesicles | University of California Los Angeles, CA 90024-1592 | H. Monbouquette |

**Table 3.19.** Selected Membrane Vendors

| Membrane Company | Separation Technologies | Contact Name (when available) |
|---|---|---|
| Osmonics/Desal Minnetonka, MN | RO, NF, UF, MF | S. Kloos |
| Dow/Filmtec Minneapolis, MN | RO, NF | |
| Rochem Torrence, CA | RO, NF, UF Disc Tube Modules and High Pressure Systems | D. Lamonica |
| Spintek Hunington Beach, CA | Spinning Centrifugal Membranes UF/RO | |
| New Logic International Emeryvill, CA | V*SEP UF/MF App. | |
| Ionics Watertown, MA | ED, RO | |
| Nitto Denko America (Hydranautics) Santa Barbara, CA | RO, MF | |
| UOP Fluid Systems San Diego, CA | RO, NF, UF | |
| Membrane Products Rehovot, Israel | NF, UF | |

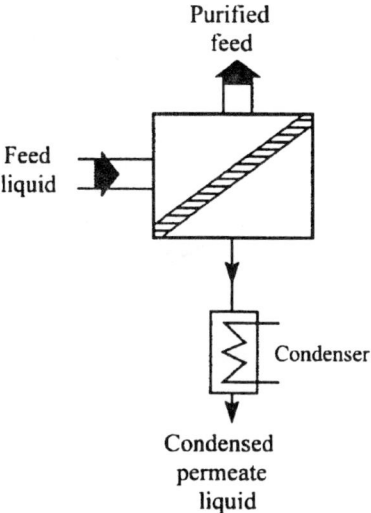

**Figure 3.24.** Schematic of the basic pervaporation process.

small-volume condensate is concentrated in VOC and frequently phase-separates to provide a relatively pure VOC fraction. The purified water exits on the feed side of the membrane.

The separation achieved is determined by the individual rates of permeation through the membrane and the relative volatility of the components of the feed mixture. Typical separation factors, that is, permeate-to-feed concentration ratios, for dilute VOC solutions are shown in Table 3.20. The best candidate VOCs for removal by pervaporation are volatile hydrophobic compounds such as

**Table 3.20.** Typical Pervaporation Separation Factors (VOC Removal from Water)

| Separation Factor for VOC over Water | Volatile Organic Compound (VOC) |
|---|---|
| 200–1000 | Benzene, toluene, ethyl benzene, xylenes, TCE, chloroform, vinyl chloride, ethylene dichloride, methylene chloride, perchlorofluorocarbons, hexane |
| 20–200 | Ethyl acetate, propanols, butanols, MEK, acetone, aniline, amyl alcohol |
| 5–20 | Methanol, ethanol, phenol, acetaldehyde |
| 1–5 | Acetic acid, ethylene glycol, DMF, DMAC |

chlorinated solvents and the BTEX (benzene, toluene, ethylbenzene, and xylenes) aromatics, but good separations are achieved even with moderately hydrophobic VOCs such as methyl ethyl ketone (MEK) and butanol. Very hydrophilic or nonvolatile compounds such as acetic acid or ethylene glycol are not good candidates for recovery by pervaporation.

### 3.2.5.2. Engineering, Economic, Environmental, and Energy Considerations

*Pervaporation System Design*

The design of a pervaporation system depends on the particular stream being treated. For flow rates in the 100 to 5000 gal/day range, a cyclic batch system design as illustrated in Figure 3.25 is generally preferred. In such a system, feed solution accumulates in a surge tank. Some of the solution is then transferred to the feed tank and circulated at high velocity through the pervaporation modules until the VOC concentration reaches the desired level. The treated water is then removed from the feed tank, and the tank is loaded with a new batch of untreated solution. Operation of the system is automated by a programmable logic control unit. The system runs either through a preset number of cycles or until no feed solution remains to be treated. The results of operating this type of system on toluene-contaminated water containing about 500 ppm toluene is also shown in Figure 3.25.

The dependence of the rate of removal of VOCs from the feed solution on the type of VOC is illustrated by the data in Figure 3.26. The efficiency of pervaporation is greatest for the more hydrophobic VOCs. Hydrophilic VOCs such as ethanol, methanol, acetic acid, and glycol are removed poorly; isopropanol removal is better, but still marginal. However, butanol, acetone, and ethyl acetate are removed well, and with VOCs such as toluene, benzene, methylene chloride, and trichloroethylene, membrane separation factors of 1000 or more are routinely obtained. This means that the permeate is more than 1000 times more concentrated than the feed solution.

Operation of batch pervaporation systems of the type shown in Figure 3.25 can be adjusted to accommodate the characteristics of the feed solution and the degree of the VOC removal required. The key variables are the feed water temperature and the batch cycle time. At higher feed water temperatures, the permeation rate through the membrane is higher, hence the rate of removal is increased. In practice, the better performance at higher temperatures has to be balanced against a decrease in the selectivity of the membrane and the cost of the energy used to heat the feed. Typical feed water temperatures are in the range 40 to 80°C. The cycle time of the programmable logic controller can also be adjusted to

## 3. Membrane Technologies

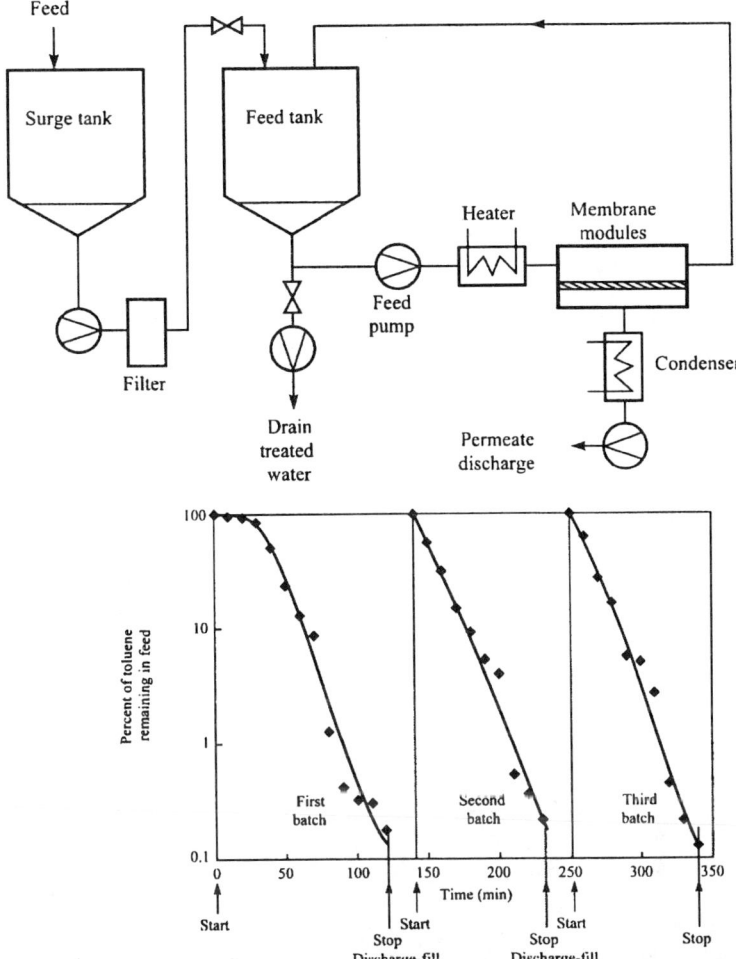

**Figure 3.25.** Flow diagram and typical performance for a 50-gal cyclic batch pervaporation system. The treatment time for the first 50-gal batch was set at 120 minutes because the unit was cold; thereafter, the cycle time was set at 90 minutes. The system achieved 99.8% removal of toluene from the feed water.

achieve any desired degree of VOC removal or the same degree of removal for different VOCs.

A photograph of a small (100–300 gal/day) system installed to treat wastewater contaminated with methylene chloride used in glass cleaning is shown in Figure 3.27. The local POTW issued a discharge permit that required treatment to less than 34 ppb of methylene chloride in sanitary water discharge. Prior to installing the pervaporation system, the plant was trucking the

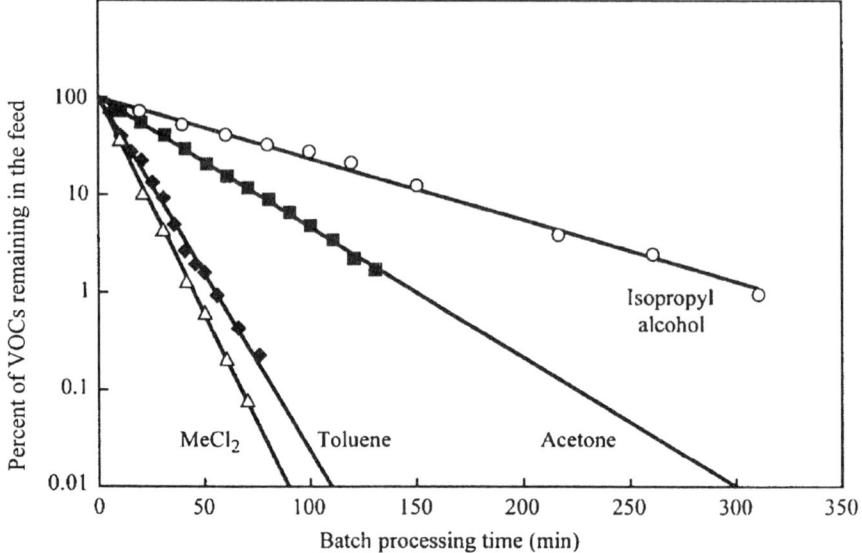

**Figure 3.26.** Removal of various VOCs from 50-gallon batches of test solution. The rate of removal from the feed solution is greatest for more hydrophobic and volatile VOCs.

**Figure 3.27.** Photograph of a 100–300 gal/day pervaporation system installed to treat wastewater contaminated with methylene chloride. This system has been operating at the Applied Biosystems Division of Perkin Elmer in Foster City, California since 1995.

wastewater to an offsite hazardous waste recycler at a cost of $3.83/gal. The pervaporation unit has now been operating for two years and has saved the company an estimated $261,000/year in avoided hauling costs and treatment.

*Competitive Position of Pervaporation*

Small-volume wastewater streams are difficult and expensive to treat by most conventional VOC-control technologies. The principal processes available to the wastewater producers and the competitive position of pervaporation are shown in Figure 3.28. If the stream is very dilute (less than 10–100 ppm VOC) chemical oxidation, UV destruction, or air stripping followed by carbon adsorption to capture the VOC may be appropriate. The cost of these processes scales directly with the VOC concentration; above 100 ppm VOC they would normally be too expensive. In the range 100 ppm to 50,000 ppm (5%) VOC, competitive processes are steam stripping and pervaporation. Steam stripping, a well established technology, is most efficient for streams greater than 10–20 gal/min, but for smaller streams, it is too capital cost intensive to be economical. For such streams the new technology of pervaporation is preferred. If the water contains more than 5 to 10% VOC, the value of the organic may make recovery by distillation feasible. Trucking to an offsite disposal site is always possible with minimal capital cost, but is generally so expensive that it is not used for streams larger than 0.3 gal/min (500 gal/day).

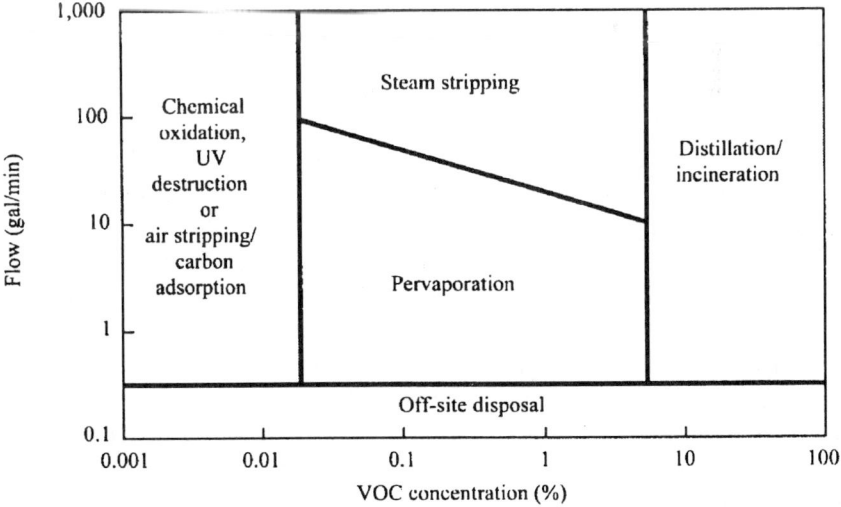

**Figure 3.28.** Pervaporation is best applied to wastewaters containing 100 ppm to 50,000 ppm (5%) VOC with flow rates less than about 20 gal/min. The process is particularly suitable for the removal of hydrophobic VOCs.

## 3.3. Selected Emerging Separative Reactors Using Membranes

### 3.3.1. Overview: Membranes in Reactors

Membrane reactor systems combine a chemical reaction and membrane separation into a single unit operation. This conjoined process produces synergistic process enhancements in the form of a higher product purity, yield, or selectivity, while reducing the number of process steps. The economics of these systems are as yet unclear, but it is the hope that the improved efficiency will result in lower operating costs. Lower capital costs are possible, when the number of unit operations is reduced.

The complexity of a membrane reactor system is greater than conventional fixed or fluidized bed reactor. The added complexity is warranted when the conventional reactor produces unwanted results, such as by-products of environmental concern or low yields per pass, and thus costly systems. Opportunities for reactor replacement must first be benchmarked against the prevailing technology. If an existing process produces yields in excess of 90% per pass, then optimization, and not radical reactor redesign, is needed. If, however, the existing process produces yields below 10–20% per pass, then an alternative reactor design, such as a membrane reactor, may well be required. As with any process substitution, rigorous economic analyses coupled with environmental considerations based on by-product formation will determine the optimal reactor design. Membrane reactors are one within a suite of many new and exciting reactor designs.

Two generic types of membrane reactors exist based on the functional requirement of the membrane: product separation or selective reactant addition. In the first mode, the product is selectively separated as it is formed. In the latter mode, the membrane serves as a means to selectively control the addition of one reactant (such as oxygen) to the reaction zone. The field has been extensively described in several recent review articles (Hsieh 1989, 1996; Noble 1992; Zaman and Chakma, 1994).

The first type of membrane reactor facilitates *in situ* product separation. A permselective membrane removes one product from the reaction zone, which allows the reaction to proceed farther to completion. For reactions that have low thermodynamic yields, simultaneous product separation is an added reaction driving force. This process is explored in greater detail in Section 3.3.3. The focus of this discussion will center on inorganic membranes which in general can withstand the rigors of the harsh reaction conditions. The available inorganic permselective membranes only separate hydrogen at elevated reaction temperatures. This limits the scope of affected reactions and points to the need for novel materials development in this area.

Low temperature reactions have been investigated with permselective polymeric membrane reactors to separate products and shift an unfavorable reaction equilibrium (Song and Lee, 1993; Song et al., 1991). These membranes have upper temperature limits near 473°K.

Another possible reactor configuration includes the *in situ* separation of a highly reactive intermediate to prevent subsequent undesired side reactions (Agarwalla and Lund, 1992; Bernstein and Lund, 1993). Unfortunately, there are no known materials that demonstrate the desired permselectivity (i.e., olefins from alkanes) at elevated reaction temperatures (above 673°K).

The second generic type of membrane reactor configuration selectively adds one reactant to the reaction zone. The primary focus of this research area is partial oxidation reactions, which will be covered in Section 3.3.2. Other reactions which may have applications for this mode of operation include alkylation reactions.

Applicable reactions for consideration in membrane reactors have one of the following attributes: (a) thermodynamic limitations on product yield, (b) parallel reactions with the desired product formed in the lowest order reaction, (c) series reactions with the desired product formed in the first step and which degrade in a series step, or (d) a series-parallel reaction pathway with specific operating windows based upon the reaction kinetics.

*Thermodynamically limited reactions*: example $A + B = C + D$, such as cyclohexane dehydrogenation ($C_6H_{12} = C_6H_6 + 3H_2$).

Removal of the product as it forms (such as C, D, or $H_2$) drives the reaction toward completion. Thermodynamics predicts the equilibrium conversion for a closed reaction system as a function of temperature, pressure, and composition. A membrane reactor is an open system and affects each of these parameters by removing mass from the reaction zone. Along with the local changes in composition (or partial pressure), the removal of mass changes the local total pressure and may affect the local temperature if sufficient mass is removed.

*Parallel reactions*: example $A + B = C + D$ ($r_1 = k_1 A^{z1} B$) and $A + B = E + F$ ($r_2 = k_2 A^{z2} B$), where $z1 < z2$.

The controlled addition of component A (when $z1 < z2$) into the reaction zone through a membrane favors the production of the desired product C. In a conventional fixed-bed reactor, equimolar feed rates at the top of the reactor favor product E, which is formed in the higher ordered reaction step. The reaction would otherwise need to be run with a dilute A feed stream to achieve a high product selectivity. In a membrane reactor, the reactant A is fed along the length of the reaction zone to lower the local partial pressure/composition of A and favor the production of C.

***Series reactions:*** example A = B + C and B = D + E, such as a selective dehydrogenation $C_6H_{12} = C_6H_{10} + H_2$ and $C_6H_{10} = C_6H_8 + H_2$ or a partial oxidation reaction.

Removal of the desired product B as it forms prevents unwanted series decomposition reactions with the valuable product. Most partial oxidation reactions suffer from unwanted deep oxidation of the desired product. The ability to selectively remove a high-value product from this process would result in considerable yield enhancements.

***Series-parallel reactions:*** example A + B = C + D and C + B = E + F such as ethylene oxide production ($C_2H_6 + O_2 = C_2H_4O + H_2O$, $C_2H_6 + O_2 = 2CO_2 + 3H_2O$, $C_2H_4O + O_2 = 2CO_2 + 2H_2O$).

This complex reaction pathway does not have a simple explanation to determine which reactor design produces the highest per pass yield. Complex operating windows exist, where the membrane reactor outperforms the fixed-bed reactor and vice versa. The apparent reaction kinetics (form of the rate expression, rate constants, and reaction orders) form a multidimensional array which define the operating windows (Tonkovich et al., 1996a,b). As an example, for a simple pseudo-homogeneous reaction network corresponding to A + ½B = C + D and C + B = E + F ($r_1 = k_1 p_a p_b^{0.5}$ and $r_2 = k_2 p_c p_b$), the feed ratio of A : B and the rate constant ratio $k_1 : k_2$ play an important role in reactor selection and operation. Other variables, including the relative reaction rate orders and form of the rate expression (simple kinetics vs. Langmuir-Hinshelwood kinetics) also make dramatic changes to the operating windows. Figure 3.29a shows the windows where a membrane reactor outperforms the fixed-bed reactor. Figure 3.29b shows the corresponding yields of the desired product C.

A number of issues are critical to determine the suitability of a combined reaction and separation process in a membrane reactor. These include

- membrane materials for selective separations at the required reaction temperature,
- rate compatibility between product formation and separation,
- materials robustness to withstand thermal stresses formed during reaction,
- membrane functionality (catalytic or inert to not promote unwanted reactions), and
- reactor sealing.

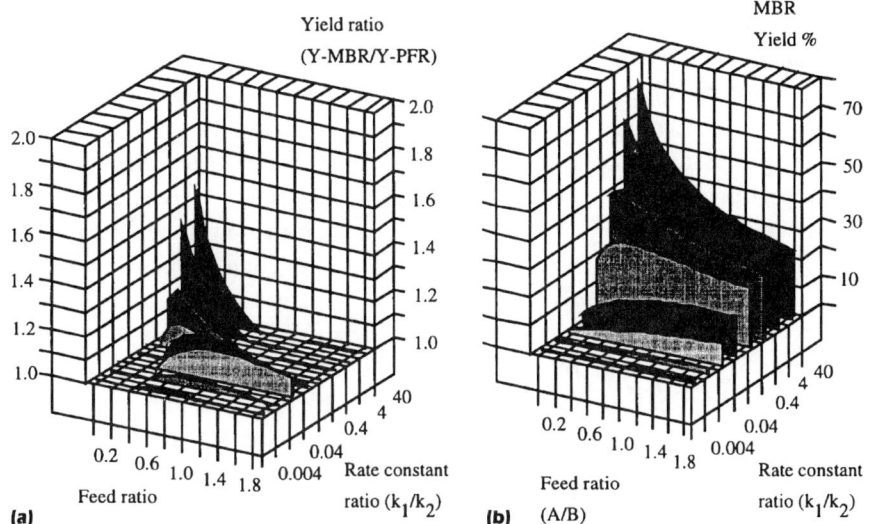

**Figure 3.29.** (a) Regions where a membrane reactor outperforms a fixed bed reactor for pseudo-homogeneous kinetics. (b) Membrane reactor yield that corresponds to part (a) regions.

The most fundamental incompatibility between merging reaction and separation results from temperature differences. The critical component to each individual unit operation which provides the process selectivity is either the catalyst for reactors or membrane for separations. Each of these are traditionally optimized independently based on the best materials available. A conjoined process, such as a membrane reactor, requires a global optimization rather than combined local optimization. As an example, many industrial synthesis reactions of interest operate at relatively high temperatures (100°C to 800°C) to achieve fast processing rates and small reactor volumes. Conversely, most selective separations occur through the use of low temperature polymeric-based membranes, which cannot withstand the rigors of reaction operating conditions. Inorganic membranes afford the promise for operating under similar conditions.

Membrane selection is a function of process temperature and the corresponding permeance of reactants or products through the membrane material. The ratio of permeances with other components found in the reaction zone is also an important consideration when selecting membrane materials.

Process rates are a critical issue to consider while investigating the applicability of a membrane reactor. Separation rates that are considerably slower than product formation rates will have little impact in shifting reaction equilibria. Slow mass transfer between the point of product formation on a catalyst surface and product separation at the membrane surface may also limit the effectiveness of this system. The distance of the membrane from the catalyst is an important parameter. If the system is dominated by slow kinetics, then the rate issue is less clear. Fast product removal will add an extra reaction driving force, but the magnitude of this effect depends on the intrinsic reaction kinetics.

The selection of robust materials is critical for deploying integrated reaction and separation processes. The ability to transport heat from or to the reaction zone is a functional requirement for membranes. Further, they must be able to withstand local hot or cold spots that are formed as a function of the local reaction exothermicity or endothermicity. Thermal cycling may also exacerbate this problem, as external heat transfer equipment respond to local changes in temperature.

Many of the reactions of interest for membrane reactor systems are catalytic. Careful consideration of the catalyst and its deployment is important in the design and operation of a membrane reactor. The membrane may act as the catalyst or the catalyst support. The membrane may also act as a physical barrier between the catalyst and the multiple points of reactant addition. In the latter case, it is critical that the membrane is catalytically inert and does not promote unwanted reactions.

Reactor sealing is a serious consideration for process implementation. Special care needs to be addressed to join often exotic ceramic membrane materials with the normal metallic reactor materials of construction. One option to mitigate this issue is to develop graded seals to join ceramic to metals. In other cases, if the membrane material can withstand large thermal gradients, critical seals can be made in cooler sections of the reactor.

An increase in conversion, under given reaction conditions (i.e. temperature and pressure), can also reduce operating conditions, thus lowering operating expenditures to achieve desired conversions. Increased environmental pressures require the development of clean chemical production processes to reduce waste generation. The selective removal or addition of reactants in a chemical process not only drives the reaction equilibrium forward, but also lowers the environmental impact by reducing the production of by-products. Another advantage to reactive membrane systems is that thermal runaway and explosive gas mixtures can be mitigated in properly designed systems. When the transport properties, transmembrane fluxes, pressure gradients, and temperatures are well understood in a membrane process, the membrane will serve as a transport barrier between flowing streams. This process is ideally suited for fast, heterogeneous, and exothermic reactions.

Membrane reactor systems are also classified as a function of membrane type:

- ***Organic,*** primarily polymeric materials or mixed organic/inorganic materials may be used such as polyphosphazenes

- **Porous inorganic** membranes such as ceramics (including zeolites), glass, or sintered stainless steel
- **Nonporous (dense) inorganic** membranes which include palladium, Pd-alloys, and ion conducting materials (such as perovskites or zirconia).

Research activity from the early 1960s until the mid 1980s was primarily focused on the development of polymeric membranes. The limitations of polymeric materials to withstand harsh conditions and high temperature applications limit their application to the biotechnological field, where relatively mild conditions are necessary due to the species involved in these reaction systems (microorganisms, enzymes, etc.). The focus changed in the mid 1980s with the development of ceramic and metal membranes. Current research and literature has been targeting the production of very-thin, defect-free membranes with both high chemical tolerances and thermal stability. The characteristics of ceramic materials makes them ideally suited for these applications. However, nonoptimized transmembrane fluxes and poor solute selectivity have restricted their use in commercial membrane processes.

Inorganic membranes can be classified into two categories, porous and dense. Porous membranes are permeable to all reactive species, but at different rates. Membranes with exceptionally small pores may behave as a molecular sieve, if one solute is larger than the pore opening. Separation properties such as permselectivity and permeability depend on pore size and the pore distribution. nonporous membranes, also known as dense or solid electrolyte membranes, are impervious to gases and liquids, but will transport one or more ionic species. Transport is a result of a voltage differential or the chemical potential gradient of a migrating ion. These membranes are highly permselective to one component.

Solid electrolyte membranes have the unique ability to selectively transport certain ionic species. Their specific ionic conductivity makes it possible to accurately measure the concentration of the migrating ion species, resulting in greater control over the reactive process. Common nonporous membranes fall into three categories, simple or complex oxides, simple or complex halides and oxide soluions.

Palladium-based membranes have been studied in the continuous removal of hydrogen in dehydrogenation reactions. Hydrogen is the only component selectively transported through this membrane under ideal conditions (i.e., defect-free films). Dehydrogenation reactions are typically equilibrium limited and the continuous removal of hydrogen from this reaction system drives the reaction forward to provide a substantial increase in the reaction conversion. Further research is required on nonporous membranes to increase transport fluxes, decrease defect density, develop supported ultra-thin films, and prevent metal sintering.

Porous membranes have been commercially available for many years primarily for use as a filtration device in the food and biotechnology industries. Recently they have been studied for use in reaction systems. Traditionally, these membranes are composed of alumina (both gamma and alpha phases), porous Vycor glass, sintered stainless steel, silver–zirconia, and polymeric materials. These membranes usually exhibit low selectivity, but have much higher permeation rates than nonporous membranes. A porous membrane reactor can also be categorized as a separation device that is used in conjunction with a reactor or used as a catalyst support in combination with separation. In the latter case, the catalyst is impregnated within the pores of the membrane, where the reaction takes place.

Further research is needed to develop thin, selective membranes suitable to withstand harsh reaction conditions with permeation rates capable of satisfying requirements of commercial production processes. An ideal membrane is sufficiently thin to allow transmembrane transport but thick enough to withstand applied pressure gradients and harsh operating conditions. One example is the development of thin defect-free "molecular sieve" membranes. Zeolite-based membranes are one approach to this technology. The zeolite material is applied in a very thin layer over a porous alumina or stainless steel support and can serve as a nanofiltration separator or used in combination with its acidic catalytic activity. Zeolites, although somewhat brittle by nature are ideally suited to perform molecular scale separation, while maintaining their stability in relatively harsh environments and high temperatures. Zeolites can shrink and crack like all ceramics when they are dried during production. The challenge for researchers is to produce these materials into a thickness small enough to accommodate permeance rates required for production processes, while maintaining a defect free pore size distribution.

Other research fronts to develop small-pore membranes include the development of mesoporous membranes. Mesoporous materials are created using a sol–gel process where an organic surfactant tail on an inorganic molecule (Si, Ti, Zr) self assembles into a honeycomb structure. The organic is later removed (burned out) to leave a highly regular inorganic porous network. The average pore diameter ranges from 20 Å to 100 Å. These pores are housed within a larger macroporous support membrane which provides the mechanical strength for the system.

Additional challenges remain before the commercial application of membrane reactors can be realized, these include

- economically attractive manufacture of ultra-thin, defect-free selective membrane layers over large surface areas of porous supports,
- leak-free reaction systems with embedded high temperature seals,
- elimination or reduction of sweep gases which dilute product streams, and
- enhanced membrane and catalyst functionality, including reduced susceptibility to poisoning and fouling.

## 3.3.2. Hydrocarbon-Selective Oxidation

Hydrocarbon partial oxidation reactions have applications in membrane reactors to improve per pass yields and selectivities. Partial oxidation reactions are inherently unselective and often make by-products of little or no value. Oxygen-rich feeds result in low product selectivities, and high hydrocarbon conversions. Oxygen-lean feeds produce high product selectivities, but low per pass hydrocarbon conversions. Low product yields result from both operational modes in traditional fixed-bed reactor designs.

In general, reaction yields limited by slow kinetics are improved through catalysis. Reaction yields limited by unfavorable thermodynamic driving forces are improved through nontraditional reactor designs that alter local temperature, pressure, or concentration. Partial oxidation reactions are limited by both kinetics and thermodynamics. The rates of the undesired reactions are typically much faster than the desired reaction rate, and there is a strong thermodynamic driving force for product overoxidation. Yield improvements can be achieved by addressing either limitation.

Per pass yields for these series-parallel reactions could be improved through the development of catalysts that operate far from equilibrium and at very short reaction times to inhibit product overoxidation. Yields can also be improved through the use of nontraditional reactor designs that mitigate the thermodynamic driving force to overoxidize the desired product.

One nontraditional reactor design might simultaneously separate the more reactive hydrocarbon product from the reaction conditions to minimize product overoxidation. While this approach is preferable, hydrocarbon separations, including alkanes from olefins, alcohols, and the like, are not yet feasible at the required synthesis temperatures. The development of low-temperature catalysts or high-temperature separation schemes would facilitate the direct coupling of reaction and product separation in a membrane reactor.

Another nontraditional reactor design improves partial oxidation yields by capitalizing upon the oxygen concentration effect. The thermodynamic driving force for combustion increases as the hydrocarbon to oxygen ratio decreases. High oxygen partial pressures promote overoxidation, while low oxygen partial pressures favor the desired intermediate product. However, operating with a low oxygen partial pressure limits the hydrocarbon conversion per pass.

Inorganic membrane reactors, in general, combine chemical reactions with separation to disturb thermodynamic driving forces that limit product yields in traditional reactor designs. A distributed-feed membrane reactor adds one reactant (oxygen in partial oxidation reactions) radially along the length of a permeable reactor wall. Overall hydrocarbon to oxygen feed rates near stoichiometric levels can be added, while simultaneously minimizing local oxygen partial pressures within the reactor. The former permits higher hydrocarbon conversions, while the latter maintains high product selectivity.

### 3.3.2.1. Technology Description

A membrane reactor is not a panacea for all reactions with poor yields per pass, but it does have tremendous potential for several reaction types, including partial oxidation and alkylation reactions.

The membrane reactor uses the membrane to control how the reactants contact each other. Either a permselective or nonpermselective membrane, which has a high trans-membrane pressure drop, is used to distribute the feed. This mode of operation has applications for series-parallel reactions where a high partial pressure of one reactant promotes undesired reaction pathways. For example, in partial oxidation reactions, the oxygen partial pressure determines the product selectivity. Oxygen fed in stoichiometric proportions with the hydrocarbon promotes combustion. Distributing the oxygen along the length of the reactor lowers the *in situ* oxygen partial pressure and thus promotes the desired reaction. Reactor configurations using dense membranes to transport atomic oxygen have been described (Cales and Baumard, 1982; Eng and Stoukides, 1991; Fujimoto et al., 1991; Gryaznov et al., 1986; Itoh et al., 1993; Nigara and Cales, 1986; Nozaki et al., 1992; Omata et al., 1989; Stoukides and Vayenas, 1981).

The use of a porous membrane approximates the effect of a dense oxygen permeable membrane, and does so under more robust reaction conditions than currently experimentally feasible with the latter (Coronas et al., 1994a,b; Lafarga et al., 1994; Tonkovich et al., 1995, 1996a,b). The effect of splitting the oxygen feed can be investigated with either membrane. However, the effect of atomic oxygen reactions with the hydrocarbon, which are postulated to occur at the surface of the dense catalytic membrane prior to oxygen recombination, are not measurable with the porous membrane. The results in both dense and porous membrane systems become equivalent in the limit of fast atomic oxygen recombination in the boundary layer surrounding the dense membrane. Subsequent catalytic and/or homogeneous reactions with molecular oxygen then occur in both membrane systems.

Porous membranes increase the experimental range of investigation (i.e., require lower operating temperatures and can achieve any desired oxygen flux) and could potentially be used to determine which partial oxidation reactions are applicable for membrane reactors. This, in turn, will serve to set criteria (operating temperature, pressure, durability, and oxygen fluxes) for the development of new oxygen permselective membranes. Modeling efforts have begun to identify the applicable operating regions for this mode of operation (Harold et al., 1993).

Another membrane reactor configuration promotes reaction within the membrane pores. Reactants are fed from opposing sides of the membrane and combine on a catalyst impregnated within the membrane pores (Sloot et al., 1990, 1991; Veldsink et al., 1992; Zaspalis et al., 1991a–d).

*Example—Oxidative Coupling Reaction*
When reaction products can be separated from one and other using conventional means, such as distilling, chemical equilibrium concentration can be continuously shifted to obtain virtually 100% conversion of reactants to products. The oxidative coupling reaction is not equilibrium limited, but it is a low conversion process when the deep oxidation of the hydrocarbon products to $CO_2$ and $H_2O$ is minimized. The oxidative coupling reaction also requires sufficiently elevated temperatures (>700°C) such that mixing catalyst and adsorbent is unsatisfactory due to the chemical structure integrity of these compounds under these harsh conditions. Unfortunately, the high costs associated with recycle and separation of these low conversion processes makes their economic and environmental impact too great for implementation into an industrial process. Yields in excess of 30% per pass are required for an economically attractive process (Srivastava et al., 1992).

Methane is the most abundant component of natural gas. Considerable interest has been focused on the direct conversion of this most abundant compound into more valuable components, specifically basic feedstocks and liquid fuels (methanol) and building blocks (ethane and ethylene). The ethane and ethylene yields are typically ~20% for the most selective catalysts in a conventional reactor system, therefore this system is not commercially feasible. A ceramic membrane reactor has been investigated for this reaction (Omata et al., 1989) to limit the contact of $C_2$ products with oxygen and to prevent nitrogen buildup in the recycled methane stream. In principle, modes of contact which maintain a low oxygen concentration in the reaction process, should favor the desired reaction and thus improve the selectivity at a given conversion level. The reaction takes place between the methane and the oxygen that diffused through the membrane. Omata et al. (1989) used PbO/MgO coating on a porous alumina tube to effect methane oxidative coupling. The reactor was successful in attaining a high selectivity for the $C_2$ products, but the yields did not reach the desired level for commercial implementation. The rate of the permeating oxygen is the most important factor for the reactor being developed (Lafarga, 1994). Reaction temperatures for this application were 1100–1200°C. Controlling the flux of oxygen into the reaction mechanism is the key factor in regulating the desired conversion to make the process viable.

Researchers have given their attention lately to the development of oxide-ion conducting membrane for limiting control of oxygen into reaction processes. One class of materials that exhibits high oxide ion conductivity is based on the perovskite structure. By incorporating vacancies into the structure, an important requirement for perovskite ion conductors, it is possible to attain high oxygen ion conductivity (Kendall et al., 1995; ten Elshof et al., 1995).

Another material under investigation into the oxidative coupling of methane is the use of zeolite-based ceramic membrane reactors. Recent advances in the application of zeolite films on the surface of alumina membrane create porous membranes permeable to oxygen at temperatures required for these processes. The porosity and thickness of the zeolite layer has been a key factor that control of the permeance of the oxygen. The control of the migrating oxygen into the reaction system is regulated through the flux of the oxygen through the membrane. The zeolite (ZSM-5) has also been reported to serve as a catalytic mechanism for the oxidative coupling

reaction. Further research is required into the investigation of these materials as a membrane material, however, the ultimate goal might well be a single step process from methane to liquid hydrocarbon feedstocks.

*Example—Synthesis Gas Production*
Production of synthesis gas in an inorganic membrane reactor has received considerable attention in recent years. Although research has investigated both porous (Santos et al., 1995; Tonkovich et al., 1996b) and dense membranes, the most exciting results are found with the ionic conducting dense membranes (Balachandran et al., 1997; Pei et al., 1995; Sato et al., 1995). Yields to carbon monoxide and hydrogen in excess of 90% per pass have been reported with membrane lifetimes of thousands of hours. Development of this technology has progressed into a commercial phase with industrial partners.

### 3.3.2.2. Engineering, Economic, Environmental, and Energy

*Engineering*
The primary engineering consideration in the deployment of membrane reactors for selective hydrocarbon oxidation applications is scale-up. Proof-of-principle demonstrations have shown enhanced product selectivity and yield for a number of test reactions. These demonstrations, however, were typically conducted at a bench-scale with one or a few membranes and relatively low processing rates. The issues that must be considered during scale-up are as follows:

- Manifolding and connections with multiple membranes
- Heat removal from these highly exothermic reactions
- Membrane failure mechanisms which minimize catastrophic process failure
- Fouling from real world feedstocks
- Regeneration methods for membrane and catalyst fouling

Multiple membranes, either tubular or flat plate, will be required to handle the large processing rates required in the chemical processing industry. Sealing single tubes is an experimental issue for all bench-scale tests and becomes a serious consideration when designing full-scale processes. The most probable approach to sealing multiple catalytic membranes will focus on scale-up parameters developed for membrane separation applications. Multiple tubes (or sheets) will be manifolded together in modules. Multiple modules will be required to increase the process capacity.

Heat removal is an important scale-up parameter. The exothermic partial oxidation reactions can readily form hot-spots which can lead to thermal runaway and explosions if the system is not properly designed. Further, the membranes must be able to withstand a thermal gradient from the hot reactor centerline to the heat removal device at the end of the membrane module. An alternate design would consider mixing membrane tubes and cooling tubes within the membrane module to assist in rapid heat removal.

Design for membrane failure is an important consideration. The failure of a single or several membrane tubes (or sheets) must not create a safety situation. The design of module shut-down for repair must be automatic when the integrity of any membrane within a module becomes an issue. As materials development efforts create robust materials, this issue will be lessened but will never completely be eliminated. The drive to increase both product throughput and selectivity, pushes the membrane development path toward increasingly thin films on mechanical supports. There will always be trade-offs among flux, selectivity, and membrane integrity. A zero-defect density throughout the lifetime of a membrane will be a difficult goal to achieve.

Fouling is an issue to consider during scale-up. Real world feedstocks often contain particulates and unwanted materials. As an example, methane upgrading processes will most likely be scaled-up using natural gas as the feedstock. Natural gas is roughly 90% methane, with an assortment of other materials, including ethane, ethylene, propane, and sulfur bearing compounds. Most laboratory studies have focused on clean methane feeds.

Regeneration of both catalysts and membranes is also an important consideration. Ideally, regeneration processing can clean both surfaces at the same time and with the same procedure. The most likely foulant during hydrocarbon processing is carbon deposition. Both the catalyst and membrane must withstand either oxygen or steam to regenerate the surface. Depending upon the catalyst employed, a second regeneration step may be required such as a reduction step to remove an oxide layer off a noble metal catalyst. This processing step must not affect the membrane performance.

*Economic*

A rigorous description of membrane reactor process economics for oxidation reactions is needed. In part, this analysis is a strong function of the membrane performance; many membranes are not yet mature for these applications. The strong economic incentive for these processes is that they permit tremendous increases in per pass yields. As an example, the test reaction of ethane oxidative dehydrogenation to ethylene showed that per pass yields for an unoptimized system could be increased from 12% to 52% based entirely on the membrane reactor design (Tonkovich et al., 1996a). The high reported yields of synthesis gas from methane also appear to have the potential for favorable economics based upon the interest of the commercial partners.

*Environmental*

The intent of a conjoined reaction and separation process, such as a membrane reactor, is to achieve a synergistic environmental effect. Minimization of by-product formation is a goal, but in large part has focused on reducing the formation of carbon dioxide versus toxic by-products. While the reduction of greenhouse gas formation is an important consideration, especially in light of world-wide reduction goals, hydrocarbon oxidation processes only add a small contribution to the overall greenhouse gas effect. Considering the top 50 commodity chemicals, the maximum reduction in carbon dioxide is slightly in excess of 13 billion pounds per year. This represents 5.9 million metric tons of $CO_2$. While this number appears to be large, when compared to the overall carbon dioxide emissions per year in the United States (in excess of 52 billion metric tons of carbon dioxide—DOE/EIA-0573), this reduction represents roughly 0.1% of the total.

*Energy*

The maximum potential energy savings is roughly 227 trillion BTU (or 0.23 quads) if every commodity chemical produced by a selective oxidation route achieved its maximum efficiency (Tonkovich 1994, Tonkovich and Gerber 1995). Individual processes are described in Table 3.21

### 3.3.2.3. Contacts and Suggested Vendors

Development of membrane reactors for hydrocarbon selective oxidation reactions is in commercial development for synthesis gas production from methane by PraxAir, Argonne National Laboratory, British Petroleum, and Amoco based on dense ion conducting membranes.

## 3. Membrane Technologies

**Table 3.21.** Drivers for Process Improvements in the Top 50 Commodity Chemicals (Produced by Oxidation Reactions)

| Commodity Chemical | Production volume (bil lb/yr) | Production Process | Process Temp. (°C) | Max. Energy Savings (tril. BTU) | Max. Feedstock Savings (mil $) | Max. $CO_2$ reduction (bil lb/yr) |
|---|---|---|---|---|---|---|
| Nitric acid | 16.08 | Ammonia oxidation | 820–950 | 1.3 | 35.75 | 0 |
| Ethylene dichloride | 15.94 | Oxy-chlorination | 370–500 | 10.8 | 73.1 | 0.821 |
| Terephthalic acid | 5.64 | Air oxidation of $p$-xylene | 120–175 | 7.8 | 68.4 | 0.629 |
| Ethylene oxide | 5.56 | Ethylene epoxidation | 260–280 | 28.9 | 312.8 | 3.512 |
| Phenol | 3.71 | Cumene peroxidation | 100–140 | 12.4 | 29.2 | 0.483 |
| Acetic acid | 3.6 | Acetalydehyde or $n$-butane oxidation | 130–250 | 1.8 | 1.8 | 0.027 |
| Butadiene | 3.18 | Butane oxidative dehydrogenation | 340–600 | 80.6 | 412.9 | 4.99 |
| Acrylonitrile | 2.83 | Propylene ammoxidation | | 23.5 | 145.5 | 1.629 |
| Propylene oxide | 2.7 | Propylene peroxidation | 80–150 | 15.5 | 48.2 | 0.683 |
| Vinyl acetate | 2.66 | Catalytic oxidation | 175–200 | 15.7 | 26.9 | 0.302 |
| Acetone | 2.39 | Cumene peroxidation or oxidative dehydrogenation | 100–140 or 400–600 | 8.1 | 0 | 0 |
| Adipic Acid | 1.75 | Cyclohexane oxidation | 125–160 | 20.4 | 20.2 | 0.352 |
| Total | | | | 226.8 | 1174.80 | 13.43 |

## 3.3.3. Dehydrogenation Reactions

For reactions that have low thermodynamic yields, such as dehydrogenation reactions, simultaneous product separation is an added reaction driving force (Ali et al., 1994; Armor 1989; Champagnie et al., 1990, 1991, 1992; Edlund and Pledger, 1993; Falconer et al., 1993; Gallaher et al., 1993; Gryaznov 1992; Hsieh 1991; Ilias and Govind, 1989; Ioannides and Gavalas, 1993; Keizer et al., 1991; Matsuda et al., 1993; Mohan and Govind, 1988a,b; Shu et al., 1991; Sun and Khang, 1988; Tiscareno-Lechuga et al., 1993 Tsotsis et al., 1992, 1993; Uemiya et al., 1990, 1991; Zhao et al., 1990; Ziaka et al., 1993).

### 3.3.3.1. Technology Description

A majority of separative reaction techniques involve the dehydrogenation, in which the removal of $H_2$ displaces the equilibrium toward the products. In dehydrogenation reactions, a hydrocarbon molecule is converted into a more unsaturated hydrocarbon by breaking C–H bonds and forming C=C bonds. The equilibrium shift toward the products limits the need for elevated temperatures along with reduced recycle and downstream separation requirements. The majority of investigators claim exceeding equilibrium conversions and better selectivities achieved with membrane reactors. By incorporating a membrane to remove a product, a different equilibrium state is created since the system is no longer a closed one. Another important implication of conversion enhancement through the use of inorganic membrane reactors is that reactions may be carried out at lower temperatures to achieve the same conversions. Work in the literature includes several types of membrane reactors that have been developed to overcome equilibrium limitations. These include (Ali and Rippin, 1995):

- Pd and Pd alloys in the form of tubes, foils or thin films deposited on porous ceramic supports,
- inert porous ceramic or glass membranes with a packed catalyst bed,
- catalytically active porous ceramic membranes with or without a packed catalyst bed,
- Pd alloy membranes as catalysts and $H_2$ permeators without a packed catalyst bed.

One possible application for the use of membrane reactors in dehydrogenation reactions, is the direct dehydrogenation of ethylbenzene to styrene. About 85% of the commercial production of styrene is through various types of dehydrogenation reactions. Current production methods involve the reaction being carried out in the gas phase with steam over a catalyst consisting primarily of iron. The reaction is thermodynamically limited, which forces the operating conditions toward high temperatures and low pressures. This causes reduction in the styrene production because of the increased degree of side reactions occurring over the harsh conditions.

Laboratory studies have shown that the use of membrane reactors for these processes results in conversions as high as 70%, an approximate 15% higher than current methods. This is accompanied by increases in selectivity of 2–5% toward styrene products to the 90–95% range.

Another commercially viable application for membrane reactors is in the dissociation of hydrogen sulfide. Porous alumina and glass membranes have been found to be effective for decomposing $H_2S$ at 800°C by impregnating membrane tubes with $MoS_2$ or $WS_2$ catalysts. More than twice the equilibrium conversion is reached, compared to conventional packed beds. Viscous flow through the membrane is believed to be the limiting factor for increasing the conversion (Abe, 1987). The chemical environment for $H_2S$ is quite corrosive, especially at high temperatures. Membrane materials that can withstand these environments currently posesa problem for researchers. However, the development of thin film zeolite membranes over either sintered stainless steel or alumina and composite–metal membranes presents exciting opportunities.

### 3.3.3.2. Engineering, Economic, Environmental, and Energy

*Engineering*
The engineering considerations for deployment of membrane reactors for dehydrogenation reactions focus on robust membrane materials and process scale-up. The literature is full of examples where the bench-scale proof-of-principle experiments show product yield enhancements that result from hydrogen removal. With few exceptions, the membranes are based on palladium and its binary or ternary alloys. Most examples have relied upon thick self-supported tubes which can operate without leaks. Several examples cite the results of research to develop ultra-thin palladium films on porous supports (which provide structural integrity). This latter approach affords a great reduction in the amount of the expensive metal required for the process. It however, adds a number of technical challenges to maintain a permselective barrier which rejects all solutes other than hydrogen. Defects and pin holes are a serious issue which needs to be considered.

Scale-up of these processes has similar issues with the membrane reactors for selective hydrocarbon oxidation:

- Manifolding and connections with multiple membranes
- Heat addition to these highly endothermic reactions
- Membrane failure mechanisms
- Fouling from real world feedstocks
- Regeneration methods for membrane and catalyst fouling

*Economic*

A study completed by researchers at Chevron compared the economics of dehydrogenation reactions in membrane reactors with conventional fixed-bed reactor technology (Mohr, 1996). Three reactions were evaluated: styrene from ethylbenzene, propylene from propane, and ethylene from ethane. The authors found a complex relationship with each reaction and discuss the trade-offs between minimizing partial pressure for conversion effects and maximizing partial pressure for membrane driving force. There are regions where the economics appear to be attractive.

Based on increased hydrogen fluxes and reduced material costs, composite-metal membranes being developed by Bend Research, Inc., are anticipated to have processing costs of approximately one tenth of existing decomposition technologies. Composite-metal "plate-and-frame" modules, containing up to 10 ft$^2$ of membrane area are being developed and put into service. They have demonstrated that the current generation of membranes delivers high hydrogen flux and is not adversely effected by $H_2S$, $H_2O$, $CO$, $CO_2$, $CH_4$, or $N_2$. They claim these membranes have been operated for as much as 400 days at 400°C without any flux decrease. The projected cost of on-site hydrogen manufacture with this type of system is less than $0.50/100 SCFH. This includes the system and energy requirements. This means that hydrogen recovery from metallurgical operations can be accomplished at less than half the cost of purchasing hydrogen

Another study was completed which compared the economics of methane reforming with simultaneous hydrogen removal through a membrane with a conventional fixed bed reactor (Cox et al., 1995a,b). This report found that at a scale commensurate with the smallest of conventional reformers (1 million SCF $H_2$/day), the capital cost of the membrane reactor was roughly 50% of the conventional technology.

In general, the economics become unclear if hydrogen recovery for use in conventional processing is a desired goal. The membrane reactor produces low pressure hydrogen versus the high pressure hydrogen produced in a conventional reforming reactor. The cost of compressing hydrogen must be added in the process costs. For applications that seek only to maximize the hydrocarbon yield or produce low pressure hydrogen for fuel cell or other applications, the economics appear to be more attractive for this process.

*Environmental*

The environmental impact of membrane reactors for dehydrogenation reactions is unclear. The primary focus of research has been on improving the hydrocarbon yield per pass while producing a pure hydrogen stream or combusting hydrogen to provide the endothermic reaction energy. The formation of other hydrocar-

## 3. Membrane Technologies

bon side products should be lessened as the selectivity improves, but this impact has not been quantified.

*Energy*

The maximum potential energy savings is roughly 26 trillion BTU (or 0.026 quads) if every commodity chemical produced by a dehydrogenation route achieved its maximum efficiency (Tonkovich 1994; Tonkovich and Gerber 1995). Individual processes are described in Table 3.22.

### 3.3.4. Methane Reforming Reactions

Natural gas is an abundant resource in various parts of the world. Methane is the major component of natural gas, often comprising over 90 mol% of the hydrocarbon fraction of the gas. Methane itself is primarily used as a fuel, while the nonmethane components can be separated and used as feedstocks for the production of chemicals or liquid fuels. Various direct and indirect routes for the production of useful chemicals from methane have been proposed (Ross et al., 1996). The representative direct method is oxidative coupling of methane to ethylene (Amenomiya et al., 1990). Methods for reforming of methane to carbon monoxide and hydrogen (syngas) are normally grouped in the indirect method category.

**Table 3.22.** Drivers for Process Improvements in the Top 50 Commodity Chemicals (Produced by Dehydrogenation Reactions)

| Commodity Chemical | Production volume (bil lb/yr) | Production Process | Process Temp. (°C) | Max. Energy Savings (tril. BTU) | Max Feedstock Savings (mil $) | Max. $CO_2$ reduction (bil lb/yr) |
|---|---|---|---|---|---|---|
| Styrene | 8.94 | Ethylbenzene dehydrogenation | 600–700 | 20 | 187 | 2.302 |
| Formaldehyde | 6.98 | Methanol dehydrogenation | 450–650 | 6.1 | 50 | 1.08 |
| Isobutylene | 1.29 | Isomerization and dehydrogenation | | 0.3 | 7 | 0.082 |
| Total | | | | 26.4 | 244 | 3.464 |

*Synthesis gas from methane reforming
  ammonia
  methanol
  MTBE

The proven technology for converting methane to syngas is steam reforming.

$$CH_4 + H_2O = CO + 3H_2 \qquad (3.12)$$

More hydrogen can be produced by the water-gas shift reaction as

$$CO + H_2O \rightarrow CO_2 + H_2 \qquad (3.13)$$

Carbon dioxide reforming of methane has recently attracted renewed interest (Wang et al., 1996)

$$CO_2 + CH_4 \rightarrow 2CO + 2H_2 \qquad (3.14)$$

The reforming reactions (3.12) and (3.14) are endothermic. For example, the steam reforming of methane requires a large energy input: $\Delta H^0 = 206$ kJ/mol. Furthermore, the reaction equilibrium constants are low at lower temperatures. Sufficient conversion can be achieved only at high temperatures.

Another approach to reforming of methane is through partial oxidative reaction with oxygen:

$$CH_4 + 0.5O_2 \rightarrow CO + 2H_2 \qquad (3.15)$$

This is an exothermic reaction ($\Delta H^0 = -36$ kJ/mol) with no conversion problem. However, selectivity is a critical issue as the competing reaction is formation of carbon dioxide:

$$CH_4 + 2O_2 \rightarrow CO_2 + 2H_2O \qquad (3.16)$$

### 3.3.4.1. Technology Description

Membrane reactors can be categorized into the group with product removal to enhance the conversion and the group with control feed of reactants to optimize selectivity of the reaction (Hsieh, 1996). Both groups of the membrane reactors have been studied for methane reforming to syngas. For steam reforming of methane or water-gas shift reaction, the membranes used are hydrogen permselective. Membrane reactors for partial oxidative reaction of methane are based on oxygen permselective ceramic membranes. The membrane reactor processes will make an impact on environment protection through reduction in carbon dioxide formation and saving in energy consumption associated with improvement of the process efficiency and elimination of the separation processes.

*Steam Reforming and Water–Gas Shifting Reactions*
Figure 3.30 shows schematic of the membrane reactor for steam reforming (or water–gas shift reaction). The reactor is made of a hydrogen permselective inor-

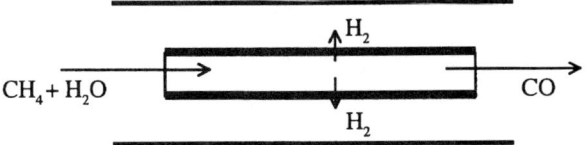

**Figure 3.30.** Schematic representation of membrane reactor for steam-reforming of methane

ganic membrane tube. A catalyst for the methane reforming (e.g. 15% NiO on calcium aluminate) is packed in the membrane tube. Membranes studied include mesoporous $\gamma$-alumina (Tsotsis et al., 1992) and dense palladium (Basile et al., 1996; Uemiya et al., 1991) membranes. The permselectivity of the mesoporous membranes is determined by the Knudsen factor (square root of the ratio of gas molecular weights). Therefore they do not give very high selectivity for hydrogen over methane, carbon monoxide, and water. Palladium-based membranes theoretically are permeable only to hydrogen, but not to methane, carbon oxides and water.

In reaction, methane and water (or water and carbon monoxide) are fed into the catalyst-packed membrane tube. The shell side is evacuated or purged with sweeping gas to provide the driving force for hydrogen permeation. Since the membrane tube is hydrogen permselective, hydrogen formed in the reactor will be simultaneously removed from the reactor. The removal of one of the products will enhance the conversion of methane, resulting in a higher product yield under the same reaction conditions. Furthermore, hydrogen is simultaneously removed on the reactor so additional separation task is minimized.

Tsotsis et al. (1992) have demonstrated experimentally that the membrane reactor (4-nm pore commercial $\gamma$-alumina) gives methane conversion of about 20% larger than the equilibrium value in 450–650°C for steam reforming of methane. Uemiya et al. (1991) performed water–gas shift reaction on electroless-plated Pd membrane reactors. Complete conversion was obtained at 400°C. Sogge and Strom (1997) performed an economic analysis on the membrane reactor technology for production of syngas by steam reforming. They found that the investments and the natural gas consumption of the membrane reactor technology are respectively about 10–20% and 5–10% lower than those of a conventional process. They concluded that the membrane reactor process steam reforming can be favorably applied for hydrogen production.

*Methane Reforming by Partial Oxidative Reaction*
Membrane reactors can be effectively employed for production of syngas through partial oxidation of methane [Reaction (3.15)]. Figure 3.31 shows sche-

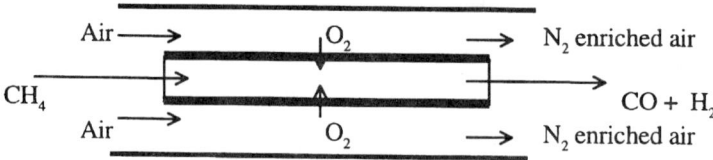

**Figure 3.31.** Schematic representation of membrane reactor for production of syngas by partial oxidative reaction of methane.

matic of the membrane reactor for methane reforming to syngas by partial oxidative reaction. The reactor in this case is made of an oxygen permselective ceramic membranes. The reactor is packed with a catalyst (such as Rh-alumina) for partial oxidation of methane. Methane is fed into the reactor tube side and oxygen (or air if membrane is oxygen semipermeable) is fed into the shell side. The effluent from the reactor tube side is syngas.

The dependence of the formation rate for syngas [Reaction (3.15)] on oxygen partial pressure is different from that for carbon dioxide [Reaction (3.16)]. In general, higher oxygen partial pressure gives higher activity and lower selectivity for formation of the desirable product (like syngas). In the conventional reactor, methane and oxygen are co-fed into the reactor, and the oxygen partial pressure, starting from the maximum, decreases along the reactor axial direction. The oxygen partial pressure along the reactor axial direction can not be optimized in the conventional reactor considering both activity and selectivity. However, the oxygen concentration profile can be optimized along the axial direction in the membrane reactor to gives a significantly improved average selectivity while maintaining reasonably higher reaction activity for formation of the desirable products (Kao et al., 1997). Thus the membrane reactor will give a substantially higher product yield than the conventional reactor.

If the ceramic membrane is only permeable to oxygen (oxygen semipermeable membrane), air instead of pure oxygen can be used as the oxidant for these reactions. In this case, the membrane reactors offer additional advantages such as elimination of oxygen production plant for the partial oxidation reaction process. Table 3.23 compares oxygen permeation fluxes of several ceramic and polymeric membranes. As shown, the oxygen permeation flux through mixed-conducting type perovskite-type ceramic $La_{1-x}Sr_xCo_{1-y}Fe_yO_3$ (LSCF) membrane at high temperatures (>800°C) is higher than the polymeric and porous Vycor glass membranes, and comparable to the 4-nm pore commercial $\gamma$-alumina membrane. The intrinsic oxygen permeability of the LSCF membranes is much larger than the polymeric or porous ceramic membranes. The $O_2/N_2$ separation factor of these dense ceramic membranes is theoretically infinite. Therefore, these dense mixed conducting ceramic membranes are the most

**Table 3.23.** Oxygen Permeation Flux, Intrinsic Permeability, and Selectivity of Various Membranes at 850°C (All data for inorganic membranes are obtained in Lin's laboratory)

| Membrane Type | Membrane (top-layer) Thickness, $L$ | Measured Oxygen[a] Permeation Flux $J_{O_2}$ | | Ideal Separ. Factor, $J_{O_2}/J_{N_2}$ | Oxygen permeability, $J_{O_2} \cdot L$ (mol/cm·s) |
|---|---|---|---|---|---|
| | | mol/cm²·s | cc/cm²·min | | |
| Cellulose Acetate[b] | 0.5 μm | $9.8 \times 10^{-9}$ | 0.01 | 3.0 | $4.9 \times 10^{-13}$ |
| Porous $\gamma$-$Al_2O_3$ | 5 μm | $1.3 \times 10^{-6}$ | 1.8 | 1.0 | $6.5 \times 10^{-10}$ |
| Vycor glass | 1.1 mm | $2.5 \times 10^{-8}$ | 0.04 | 1.1 | $2.7 \times 10^{-9}$ |
| 8%$Y_2O_3$–$ZrO_2$ | 10 μm | $2.0 \times 10^{-9}$ | 0.003 | ∞ | $2.0 \times 10^{-12}$ |
| $Cu_{0.75}Y_{0.5}Bi_{1.25}O_{3-\delta}$ | 3 mm | $1.0 \times 10^{-8}$ | 0.02 | ∞ | $3.0 \times 10^{-9}$ |
| $La_{0.2}Sr_{0.2}Co_{0.6}Fe_{0.3}O_3$ | 3 mm | $2.0 \times 10^{-7}$ | 0.3 | ∞ | $6.0 \times 10^{-8}$ |

[a] $P_{O_2(high)} = 0.21$ atm, $P_{O_2(low)} \approx 0.0001$ atm.
[b] At 25 °C

promising membrane materials for use in the membrane reactor for partial oxidative reaction of methane.

Oxygen transport through the dense mixed-conducting ceramic membranes is a result of oxygen ionic and electronic conduction using mechanisms that involve defects such as lattice vacancies and electron (electron–hole) (Lin et al., 1994). Perovskite of $ABO_3$ structure, with trivalent transition metal ions in B position and lanthanum ion or another rare earth in A position, are $p$-type electronic conductors. Doping appreciable amount of acceptor on either A sites (e.g., Sr as dopant) or B site (e.g., Fe as dopant) is necessary to produce the highly mobile oxygen vacancies, giving rise to high oxygen ionic conductivity. Teraoka et al. (1985) were the first to report extremely high oxygen permeation flux through defect perovskite-type LSCF membranes which are mixed-conductors. High oxygen permeance through the mixed-conducting ceramic membranes were later confirmed in several other laboratories.

Tasi et al. (1997) reported production of syngas by Reaction (3.17) on a disk-shaped perovskite-type ($La_{0.2}Ba_{0.8}Fe_{0.8}Co_{0.2}O_3$) ceramic membrane reactor. The work was aimed at demonstrating a concept of simultaneous reaction and separation of oxygen from air. Improvement of the syngas product yield was not obtained because of the specific reactor configuration used. Successful effort on converting methane to syngas and air separation on ceramic membrane reactor was demonstrated by a team of Argonne National Laboratories and Amoco

(Balachandran et al., 1997). The experiments were conducted on tubular membrane (see Figure 3.31) made of mixed-conducting SrFeCo$_{0.5}$O$_x$ whose structure is however still not well understood yet (Gugilla and Manthriam, 1997). The membrane tubes studied were 30 cm long, 6.5 mm in outer diameter and 0.25–1.2 mm in wall thickness. With Rh/alumina catalyst packed in the membrane tube, they obtained over 98% methane conversion at CO and H$_2$ selectivity over 90% at 850°C. With the air as the oxygen source, the membrane reactor allows permeation of oxygen at extremely high flux of 2 cc (STP)/min·cm$^2$ (900°C). The membrane reactor was operated successfully for longer than 1000 hours.

### 3.3.4.2. Engineering, Economic, Environmental, and Energy

*Engineering*

The dense ceramic membrane technology for methane reforming has attracted attention of major chemical and petroleum companies in the U.S. Companies currently involved in the research and development of the mixed-conducting ceramic membranes include Air Products, Praxair, BP, and Amoco. In the past 5 years, these companies spent over $15 million, plus equal amount of funds matched by the federal governments, on research of the membrane reactor process based on the mixed-conducting ceramic membrane. In 1997, the DOE has selected a team led by Air Products for an 8-year, $84 million cost-shared project to develop precommercial mixed-conducting ceramic membrane reactor process for methane reforming to syngas. Other members of this team include Arco, Argonne National Labs, Babcock and Wilcox, Ceramatec, Chevron, Eltron Research, Horsk Hydro, Pacific Northwest National Labs, and two universities in Pennsylvenia. Another team including BP Chemical, Praxair, Amoco, Sasol and Statoil has also put a concerted effort to commercialize the dense ceramic membrane technology for syngas production but the financial details of the alliance between the five companies have not been released.

The feasibility of the dense membrane technology for efficient reforming of methane has been demonstrated in the small laboratory scale. Several challenges are facing us in the course of developing a large scale ceramic membrane process for methane reforming. Cost-effective method needs to be developed for fabrication of large quantity dense ceramic membrane tubes, preferably in the asymmetric structure (thin dense film coated on porous support). Material chemical stability under reducing atmosphere remains to be further improved. Effective high temperature sealing technology is not available yet, and will be a key to the success of the large-scale membrane process. Heat-removal from the membrane reactor is difficult although coupling the exothermic partial oxidative methane reforming reaction with the endothermic steam reforming reaction appears to be a good possibility. The membrane in the membrane reactor serves more as an oxygen distributor to obtain optimum oxygen partial pressure along the reactor axial position. Radial distribution of the oxygen concentration could affect the overall performance for large-scale membrane process. Many engineering problems will be also encountered relating high temperature ($>800^\circ C$) mass and energy transferring devices and equipment in membrane reactor process. Despite these challenges, there is no doubt that this membrane process for syngas production will become a major commercial success in membrane area if a concerted and persistent effort is devoted toward the research and development of the membrane technology.

### 3.3.5 Industry Implementation Considerations

**Economics and Productivity Issues**
One must consider the overall chemical process and economics when assessing any new reactor technology. In this section some practical issues are discussed before focusing on how these issues relate to the membrane reactor.

The typical chemical process comprises the chemical reactor, separations system, fluid handling equipment, heat exchange equipment, instrumentation, and process control system. The chemical reactor is considered to be the heart of the process because in it reactants are converted to products, both desired and undesired. Five primary issues must be addressed during the development, design, and eventual commercialization of chemical reactors.

*i. Safety*
A chemical reactor should be inherently safe. Meeting the requisite safety criteria may be more challenging with membrane reactors because of inherent design and operational features. The membrane serves as a transport barrier between to flowing streams. Leaks in the device which result in the formation of mixtures of undesirable compositions in either or both flowing streams, which may be reactive, must be avoided. Another key aspect in the design of the membrane reactor is to tailor the membrane morphology to achieve a set of transmembrane fluxes of components for the given operating conditions, such as transmembrane pressure gradients, temperature, etc. For example, in the applications described in this report in which one stream contains a hydrocarbon and the other oxygen, it is imperative that the transport features of the membrane be fully characterized in order to avoid a flammable mixture. Table 3.24 provides a breakdown of chemical plant accidents (Marrs et al., 1989). It is interesting to note that in spite of the hazards associated with oxidation processes, only 2% of the accidents were attrib-

**Table 3.24.** Breakdown of Chemical Plant Accidents[a]

| Process Type | Percent of Incidents |
|---|---|
| Polymerization | 48 |
| Nitration | 11 |
| Sulfonation | 10 |
| Hydrolysis | 7 |
| Salt formation | 6 |
| Halogenation | 6 |
| Friedel-Crafts | 4 |
| Amination | 3 |
| Diazonation | 3 |
| Oxidation | 2 |
| Esterification | 1 |

[a] Taken from Marrs et al. (1989).

uted to oxidation. A likely factor is the focus on the potential hazards associated with flammable mixtures. This focus must be intensified in the development of the membrane reactor.

*ii. Productivity*

Ultimately, a reactor must produce product at a specified production rate. A reactor that cannot safely and reliably meet the nameplate capacity is not viable. The productivity of a reactor is the key metric in this category. Figure 3.32 captures in schematic form findings of Westerterp (1992) regarding reactor productivity. A line is provided along which the range of feasible reactor productivities encountered in chemical processes are indicated. The units of reactor productivity $(P_R)$ are in kilomoles of product per cubic meter of reactor volume per hour. Reactor productivity $P_R$, which is also called the space–time yield, is defined by the formula

$$P_R = \frac{SXF_0}{V_R} \qquad (3.17)$$

where $S$ is the selectivity in moles of desired product formed per mole of specified reactant reacted, $X$ is the conversion of the specified reactant, $F_0$ is the molar feed

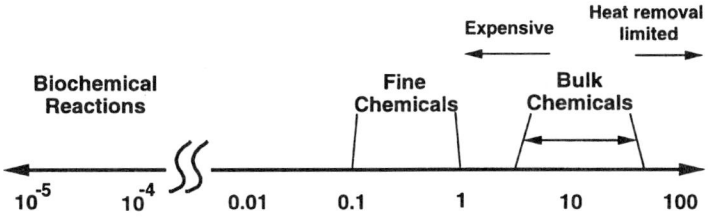

**Figure 3.32.** Classification or reaction types according to reactor productivity (after Westerterp, 1992).

rate of the reactant, and $V_R$ is the reactor volume. The range spans $10^{-5}$ to 100 kmol/m³ hr. Westerterp has identified several ranges of interest along this scale. The range between about 5 and 50 kmol/m³ hr includes most bulk continuous chemical and oil industry reactors.

Both physical and chemical processes determine the reactor productivity. Generally speaking, slow reactions, which are those that have productivities less than about 0.1 kmole/m³ hr, are kinetically limited. The lower limit is an economic boundary below which the reactor becomes too large and hence too costly. The range between 0.1 and 1 kmol/m³ hr includes most batch reactors in which fine chemicals are produced. Biochemical reactors have much lower productivity, in the range of $5 \times 10^{-4}$ kmol/m³ hr. Modifications to the reactor design will have limited effect on the reactor performance. On the other hand, fast reactions, which have productivities exceeding about 1 kmole/m³ hr, are typically limited by physical transport processes. In these cases the reactor geometry is critical. The upper limit of the fast range, 50 kmol/m³ hr, is determined by the feasible rate at which energy can be withdrawn during an exothermic reaction. Obviously, this heat removal limit does not have any relevance for an adiabatic reaction, that is unless one considers that the volume of a downstream heat exchanger is part of the reactor.

The inherent feature of the membrane reactor is the existence of a diffusional barrier between two streams. As described in other sections of this monograph, this barrier serves as a medium through which reactants and/or products are transported. While there are exceptions, the barrier creates a mass transfer resistance which reduces the rate of the overall reaction(s). The reduction in rate is traded for an improvement in overall performance, such as overall conversion or selectivity. For example, the diffusional transport of hydrogen through a hydrogen-permselective membrane is the rate limiting step in the dehydrogenation application. The continuous removal of hydrogen provides for a wider window of achievable conversion. The membrane reactor exploits a selective transport process to side-step reaction equilibrium limitations. The sacrifice of reactor productivity for either conversion or selectivity must be economically justified. The membrane area per unit volume becomes a critical design parameter. The higher the area per unit volume the lower the material cost.

*iii. Selectivity*

Most commercial chemical reaction systems involve more than one reaction. A major challenge is to maximize the selectivity to the desired product in order to minimize raw material consumption and waste generation at a given production rate.

There are increased environmental pressures to reduce waste generated in chemical processes. New chemical routes to products must be developed in which fewer byproducts are generated. These new chemical routes may require unconventional and even exotic reactor designs and operating strategies. Designing a reactor to achieve a specified selectivity level must be done on a process-by-process basis. That is, interactions within the reactor between the kinetics of the reactions, heat transport, mass transport, and momentum transport are quite specific to the particular reaction system. A notable exception is when the reactions attain a state of reaction equilibrium. In such a case only the thermodynamics and not the kinetics/transport are important.

*iv. Energy Management*

Energy management is a crucial aspect of chemical reactor design. In particular, the ability to efficiently remove heat during an exothermic reaction will make or break the development of a new reactor type. Irreversible exothermic reactions can runaway without adequate heat removal. This can lead to a serious safety problem, damaged equipment or catalyst, and reduced desired product yield. Reversible exothermic reactions will encounter a thermodynamic conversion limitation without adequate heat removal. On the other hand, endothermic reactions such as alkane dehydrogenations require considerable heat input.

Membrane reactor applications described in this report fall into both categories. The need to establish adequate membrane area and heat exchange area complicates the design and scale-up.

*v. Cost*

Despite its importance, the chemical reactor has a cost that is typically only a small fraction of the permanent investment. More specifically, the reactor cost is about 10–20% of the bare equipment cost. Thus, in absolute dollars, the conventional reactor is usually less than 10% of the total project investment.

On the other hand, the influence of the reactor on the final product price can be significant. The reason for this influence is that the composition of the reactor effluent, or the product distribution, is a direct measure of the overall reactor performance. The product distribution is the main variable which impacts the downstream separations and materials handling requirements. These process units occupy a much more significant fraction of the total investment. It should be noted that the project investment itself is frequently the most important determinant of the viability of a process. The fraction of the price of a chemical attributed to the investment varies from process to process, but the range 50–75% is not unreasonable. Thus, based on influence, the reactor is crucial.

The reactor is typically not a major fraction of the total investment of a chemical process. Despite this, an increased emphasis on capital and operating cost minimization in the chemical industry must be considered during the development of membrane reactors. The multifunctional features of the membrane reactor offers the potential for process simplification and capital and operating expense.

The cost of manufacturing a chemical can be reduced in several ways. An effective way is to develop a new process chemistry or technology that gives a higher desired product yield while maintaining the same production level. Yield losses can occur at any point in the process (purge streams, etc.), although the critical point is the reactor itself. Higher desired product yields mean a reduction in both the investment and operating costs associated with the separation system. More selective reactors can then not only reduce the complexity and cost of the separation system, but meet more stringent environmental regulations. Unfortunately, since chemical reactors are customized, this suggests that more selective reactors will be more costly than their less selective predecessors. Given the relatively low cost of conventional, less selective reactors, however, additional investment needed for the "greener" reactors may be warranted.

Another way to reduce cost is to use the same process technology but take advantage of economies of scale. The chemical industry has traditionally produced chemicals at large complexes. The conversion of raw materials to final products may involve several chemical steps with associated separations. Typically large-capacity processing facilities provide the most cost effective means of production because of savings associated with management of materials and energy and the sharing of resources. Thus, a higher capacity means reduced incremental costs associated with annualized capital, labor, and other factors. (Investment scales to the 0.6 power while labor scales to the 0.2 power.) This option is feasible as long as there is a demand to meet the increased supply, however.

In practice, chemical reactors in conventional chemical processes follow the guiding principles of economies of scale. That is, a single reactor is usually designed and operated to meet production needs. Exceptions to this rule apply to reaction systems in which regeneration or catalyst replacement must be carried out on a reasonably frequent basis. Then at least two reactors are placed in parallel to maintain continuous production. Another exception is when additional reactors are added in time as the desired production rate increases. In such situations, however, a single, large reactor eventually replaces the several smaller ones. A third exception applies to reactors which are not easily scaled-up. Examples include electrochemical reactors and photochemical reactors. In these examples, limitations in electrode size and irradiation efficiency, respectively, undermine the use of a single, large reactor. Thus, a battery of reactors configured in parallel are needed to meet production needs. Membrane reactors are likely to fall into this category.

## References

Abe, F. (1987) European Patent Application 228,885 to NGK Insulators.

Agarwalla, S., and Lund, C. (1992) Use of a membrane reactor to improve selectivity to intermediate products in consecutive catalytic reactions, *J. Mem. Sci.*, 70, 129.

Agrawal, R., and Xu, J. (1996) *AIChE J.*, 42(8), 2141–2154.

Ahmadi, S., Tseng, L. K., Batchelor, B., and Koseoglu, S. S. (1994) Micellar-enhanced ultrafiltration of heavy metals using lecithin. *Sep. Sci. Tech.*, 29, 2435–2450.

Alami-Younssi, S., Larbot, A., Persin, M., Sarrazin, J., and Cot, L. (1994) Gamma alumina nanofiltration membrane. Application to the rejection of metallic cations. *J. Memb. Sci.*, 91, 87–95.

Ali, J., Newson, E., and Rippin, D. (1994) Exceeding equilibrium conversion with a catalytic membrane reactor for the dehydrogenation of methylcyclohexane. *Chem. Eng. Sci.*, 49(13), 2129.

Ali, J. K., and Rippin, D. W. T. (1995) Comparing mono- and bimetallic noble-metal catalysts in a catalytic membrane reactor for methylcyclohexane dehydrogenation. *Ind. Chem. Eng. Res.*, 34, 722–729.

Amenomiya, Y., Birss, V.I., Goledzinowski, M., Galuszka, J., and Sanger, A.R. (1990) *Catal. Rev.-Sci. Eng.*, 32, 163.

Anand, M., *Novel Selective Surface Flow (SSF™) Membranes for the Recovery of Hydrogen from Waste Gas Streams, Phase I: Exploratory Development, Final Report*, DOE/AL/94461-1, Air Products and Chemicals, Inc., Allentown, Pennsylvania, August 1995.

Antonson, C. R., Gardner, R. J., King, C. F., and Ko, D. Y., (1977) *Ind. Eng. Chem. Process Des. Dev.* 16 (4), 473–479.

Armor, J. (1989), *Appl. Cat.*, 49(1), 1.

Balachandran, U., Dusek, J.T., Maiya, P.S., Ma, B. (1997) Ceramic membrane reactor for converting methane to syngas, *Catalysis Today*, 36(3), 265–272.

Basile, A., Drioli, E., Santella, F., Violante,V., Capannelli, G., and Vitulli, G. (1996) *Gas Sep. Purif.*, 10, 53.

Bernstein, L., and Lund, C. (1993) Membrane reactors for catalytic series and series-parallel reactions, *J. Mem. Sci.*, 77, 155.

Bhattacharyya, D., Williams, M. E., Ray, R. J., McCray, S. B. (1992) Reverse osmosis. In W. S. Winston Ho and K. K. Sirkar (Eds.), *Membrane Handbook,* Chaps. 21, 22, 23, 24 and 25, Chapman and Hall, New York.

Bhattacharyya, D., Mangum, W.C., and Williams, M.E. (1997) Reverse Osmosis. In *Kirk–Othmer Encyclopedia of Chemical Technology*, 4th ed., Vol. 21, pp. 303–335, John Wiley, New York.

Bhattacharyya, D., Hestekin, J. A., Brushaber, P., Bachas, L. G., and Sikdar S. K. (1997a) Novel poly glutamic acid functionalized microfiltration membranes for sorption of heavy metals at high capacity. *J. Memb. Sci.*

Bhattacharyya, D., Hestekin, J. A., Ritchie, S., Bachas, L. G., and Sikdar S. K. (1997b) Membrane techniques for removal and recovery of metals from waste streams, *Proc. 1997 Topical Conf. on Separation Sci. and Tech.,* Annual AIChE Meeting, Nov., pp. 675–680.

Bikson, B., and Giglia, S. (1992) Double Etched Hollow Fiber Bundle and Fluid Separation Apparatus, US Patent 5,160,042.

Broom, G. P., Squires, R. C., Simpson, M.P.J., and Martin, I. (1994) The treatment of heavy metal effluents by crossflow microfiltration. *J. Memb. Sci.*, 87, 219–230.

Cadolle, J.E. (1981) US Patent 4,277,344.

Cadotte, J., Forester, R., Kim, M., Petersen, R., and Stocker, T. (1988) Nanofiltration membranes broaden the use of membrane separation technology. *Desalination,* 70, 77–88.

Cales, B., and Baumard, J. (1982) Production of hydrogen by direct thermal decomposition of water with the aid of a semipermeable membrane, *High Temp.-High Pres.*, 14, 681.

Callahan, R. A., "Multistage Semi-permeable Membrane Process and Applications for Gas Separations," US Patent 5,482,539.

Caskey T. I., Trimmer, J. L., and Jorgensen, J. L. (1990, 1991), "Hollow Fiber Membrane Fluid Separation Adapted for Boreside Feed Which Contains Multiple Concentric Stages," US Patent 5,013,437 (1991), 4,929,259 (1990).

Cen, Y., Lichtenthaler, R. N. (1995). Vapor permeation. In R.D. Noble and S.A. Stern (Eds.), *Membrane Separations Technology: Principles and Applications,* Chap. 3, Elsevier, New York.

Champagnie, A., Tsotsis, T., Minet, R., and Webster, I. (1990) A high temperature catalytic membrane reactor for ethane dehydrogenation, *Chem. Eng. Sci.*, 45(8), 2423.

Champagnie, A., Tsotsis, T., Minet, R., Ziaka, Z., and Wagner, E. (1991), The use of high temperature membrane reactors for the enhancement of selectivity and yield of catalytic reactions, *Key Eng. Matls.,* 61–62, 599.

Champagnie, A., Tsotsis, T., Minet, R., and Wagner, E. (1992) The study of ethane dehydrogenation in a catalytic membrane reactor, *J. Catal.,* 134, 713.

Cherif, A.T., Elmidaoui, A., and Gavach, C. (1993) Separation of $Ag^+$, $Zn^{2+}$ and $Cu^{2+}$ ions by electrodialysis with a monovalent cation specific membrane and EDTA. *J. Memb. Sci.,* 76, 39–49.

Cheryan, M. (1986) *Ultrafiltration Handbook,* Technomic Publishing Co., Lancaster, PA.

Coplan, M. J. (1987) Module for Multistage Gas Separation, US Patent 4,676,808.

Coronas, J., Menendez, M., and Santamaria, J. (1994a) Methane oxidative coupling using porous ceramic membrane reactors—II. Reaction studies, *Chem. Eng. Sci.*, **49**(12), 2015.

Coronas, J., Menendez, M., and Santamaria, J. (1994b) Use of a ceramic membrane reactor for the oxidative dehydrogenation of ethane to ethylene and higher hydrocarbons, *Proceedings of the 7th International Symposium on Synthetic Membranes in Science and Industry.*

Cox, J. L., Tonkovich, A. Y., Elliott, D. C., and Baker, E. G. (1995a) *Indirect Liquefaction of Biomass: A Fresh Approach*, Preprints, Div. of Fuel Chem., **40**(3), 719.

Cox, J. L., Tonkovich, A. Y., Elliott, D. C., andBaker, E. G. (1995b) *Hydrogen from biomass: a fresh approach,* Proceedings of the Second Biomass Conference of the Americas, p. 657.

Culkin, B., and Armando, A. D. (1992) New separation system extends the use of membranes, *Filt. & Sep.,* September/October.

Davis, R.H. (1992) Microfiltration. In W.S. Winston Ho and K.K. Sirkar (Eds.), *Membrane Handbook,* Chaps. 31 and 33, Chapman and Hall, New York.

Davis, R.H., and Grant, D.C. (1992) Theory for deadend microfiltration. In W.S. Winston Ho and K.K. Sirkar (Eds.) *Membrane Handbook,* Chap. 32, Chapman and Hall, New York.

Dineen, M. (1994) Plating waste processing includes microfiltration. *Water/Eng. Manag.,* 141(11), 22–23.

DOE/EIA-0573 (94), Emissions of Greenhouse Gases in the United States 1987–1994.

Doshi, J. J. (1987) Enhanced Gas Separation Process, US Patent 4,690,695.

Edlund, D., and Pledger, W. (1993) Thermolysis of hydrogen sulfide in a metal-membrane reactor, *J. Mem. Sci.*, 77, 255.

Edward, D.W. (1989) Modula, Shell-less, Air Permeator, US Patent 4,871,379.

Ekiner, O. M., et al. (1992) US Patent 5,085,676.

Elhaus, R.-W. and Mäkel, E. (1991) Apparatus for Pulling Extrusions from a Press, US Patent 4,871,379.

Eng, D., and Stoukides, M. (1991) Catalytic and electrocatalytic methane oxidation with solid oxide membranes, *Catal. Rev.-Sci. Eng.*, **33**(3,4), 375.

Enoch, G. D., van den Broeke, W. F., and Spiering, W. (1994) Removal of heavy metals and suspended solids from wastewater from wet lime(stone)-gypsum flue gas desulphurization plants by means of hydrophobic and hydrophilic crossflow microfiltration membranes. *J. Memb. Sci.*, 87, 191–198.

Fain, D. E. (1996), Inorganic membranes: Where are we? What can we expect, *Proceedings of the Fourteenth Annual Membrane Technology/Separation Planning Conference,* Boston, Massachusetts, October 28–30.

Fain, D. E., and Roettger, G. E. (1996) Ceramic membranes for high temperature hydrogen separation, *Proceedings of the Conference on Fossil Energy Materials,* Knoxville, Tennessee, May 14–16.

Fain, D. E., and Roettger, G. E. (1997) Ceramic membranes for high temperature hydrogen separation, *Proceedings of the Conference on Fossil Energy Materials*, Knoxville, Tennessee, May 20–22.

Falconer, J., Noble, R., and Sperry, D. (1993) Catalytic membrane reactors, In S. Stern and R. Noble (Eds.), *Membrane Separations Technology: Principles and Applications*, Elsevier, New York.

Fillipi, B.R., Scamehorn, J.F., Taylor, R.W., and Christian, S. D. (1997) Selective removal of copper from an aqueous solution using ligand-modified micellar-enhanced ultrafiltration using an alkyl—diketone ligand. *Sep. Sci. Tech.*, 32, 2401–2424.

Fleming, H. L., and Slater, C. S. (1992) Pervaporation. In W.S. Winston Ho and K.K. Sirkar (Eds.), *Membrane Handbook*, Chaps. 7–10, Chapman and Hall, New York.

Fujimoto, K., Asami, K., Omata, K., and Hashimoto, S. (1991) Selective oxidative coupling of methane with a membrane reactor. In A.Holmen (Ed.), *Natural Gas Conversion*, Elsevier, Amsterdam, p. 525.

Gallaher, G., Gerdes, T., and Liu, P. (1993) Experimental evaluation of dehydrogenations using catalytic membrane processes, *Sep. Sci. Technol.*, **28**(1–3), 309.

Ganzi, G. (1988) Electrodeionization for high purity water production, *AIChE Symp. Ser.*, 84, No. 261, 73.

Gaubert, E., Barnier, H., Maurel, A., Foos, J., Guy, A., Bardot, C., and Lemaire, M. (1997a) Selective strontium removal from a dodium nitrate aqueous medium by nanofiltration–complexation. *Sep. Sci. Tech.*, 32, 585–597.

Gaubert, E., Barnier, H., Nicod, L., Favre-Reguillon, A., Goos, J., Guy, A., Bardot, C., and Lemaire, M. (1997b) Selective cesium removal from a sodium nitrate aqueous medium by nanofiltration–complexation. *Sep. Sci. Tech.*, 32, 2309–2320.

Goel, V., Accomazzo, M. A., DiLco, A. J., Meier, P, Pitt, A., Pluskal, M. (1992) Deadend microfiltration. In W.S. Winston Ho and K.K. Sirkar (Eds.) *Membrane Handbook*, Chap. 34, Chapman and Hall, New York.

Gryaznov, V., Vedernikov, V., and Gul'yanova, S. (1986) Participation of oxygen, having diffused through a silver membrane catalyst, in heterogeneous oxidation processes, *Kin. Kat.*, **27**(1), 129.

Gryaznov, M. (1992) Platinum metals as components of catalyst–membrane systems, *Plat. Met. Rev.*, **36**(2), 70.

Guggilla, S. and Manthriam, A. (1997) *J. Electrochem. Soc.*, 144, L120.

Hansen, H. K., Ottosen, L. M., Laursen, S., and Villumsen V. (1997) Electrochemical analysis of ion-exchange membranes with respect to a possible use in electrodialytic decontamination of soil polluted with heavy metals. *Sep. Sci. Tech.*, 32, 1997, 2425–2444.

Harold, M., Zaspalis, V., Keizer, K., and A. Burggraaf (1993) Intermediate product yield enhancement with a catalytic inorganic membrane—I. Analytical model for the case of isothermal differential operation, *Chem. Eng. Sci.*, **48**(15), 2705–2725.

Hestekin, J. A., Bhattacharyya, D., Sikdar, S. K., and Kim, B. M. (1997) Applications of membranes for treatment of hazardous wastewaters. In *Encyclopedia of Environmental Analysis and Remediation*, John Wiley, New York.

Ho, W. S. Winston, Li, N. N. (1992). Emulsion liquid membranes. In W. S. Winston Ho and K.K. Sirkar (Eds.), *Membrane Handbook,* Chaps. 36–37, Chapman and Hall, New York.

Ho, W. S. Winston, and Sirkar, K.K. (Eds.) (1992) *Membrane Handbook,* Chapman and Hall, New York.

Hsieh, H. (1989) Inorganic membrane reactors—A review, *AIChE Symposium series 268,* **85,** 53.

Hsieh, H. (1991) Inorganic membrane reactors, *Catal. Rev.-Sci. Eng.,* **33**(1,2), 1.

Hsieh, H. (1996) *Inorganic Membranes for Reaction and Separation,* Elsevier, New York.

Huang, R.Y.M. (Ed.) (1991) *Pervaporation Membrane Separation Processes,* Elsevier, New York.

Huang, Y. C., Batchelor, B., and Koseoglu, S. S. (1994) Crossflow surfactant-based ultrafiltration of heavy metals from waste streams. *Sep. Sci. Tech.,* 29, 1979–1998.

Ilias, S., and Govind, R. (1989) Development of high temperature membranes for membrane reactor: an overview, *AIChE Symp. Ser.* 268, **85,** 18.

Ioannides, T., and Gavalas, G. (1993) Catalytic isobutane dehydrogenation in a dense silica membrane reactor, *J. Mem. Sci.,* 77, 207.

Itoh, N., Sanchez, M., Xu, W., Haraya, K., and Hongo, M. (1993) Application of a membrane reactor system to thermal decomposition of $CO_2$, *J. Mem. Sci.,* 77, 245.

Johnson, H. K. (1988) Device and Method for Separating Individual Fluids from a Mixture of Fluids, US Patent 4,752,305.

Karrs, S. R., Hapach, G. S. and McConahy, J. L. (1992) Recovery techniques reduce hazardous waste in continuous strip plating. *Water Env. Tech.,* November, 21–28.

Kessler, S., and Klein, E. (1992) Dialysis. In W. S. Winston Ho and K. K. Sirkar (Eds.), *Membrane Handbook,* Chaps. 11–15, Chapman and Hall, New York.

Kesting, R. E. et al., US Patents 4,871,494; 4,880,441 (1989).

Kesting, R. E., and Fritzsche, A. K. (1993) *Polymeric Gas Separation Membranes,* pp. 328–331, John Wiley, New York.

Kao, Y.K., Luo, L., and Lin, Y. S. (1997) *Ind. Eng. Chem. Res.,* 36, 3583.

Keizer, K., Zaspalis, V., and Burggraaf, T. (1991) Passive and catalytically active membranes for affecting chemical reactions In P. Vincenzini (Ed.), *Ceramics Today—Tomorrow's Ceramics,* p. 2511, Elsevier, New York.

Kendall, K.R., Navas, C., Thomas J. K., and zur Loye, H-C. (1995), *Solid State Ionics,* 82, 215–223

Kobuchi, Y., Motomura, H., Noma, Y., and Hanada, F. (1987) Application of ion exchange membranes to recover acids by diffusion dialysis, *J. Membrane Sci.,* 27, 173–179.

Koresh, J. E., and Soffer, A. (1985) Carbon Molecular sieve membranes. General properties and the permeability of $CH_4/H_2$," *Symposium on Separation Science and Technology for Energy Applications,* Knoxville, Tennessee, October 20, CONF-851011, pp. 973–982.

Koros, W. J., and Fleming, G. K. (1993) Membrane-based gas separation, *J. Membrane Sci.,* 83, 1–80.

Kulkarni, S. S., Funk, E. W., and Li, N. N. (1992) Ultrafiltration. In W. S. Winston Ho and K.K. Sirkar (Eds.), *Membrane Handbook,* Chaps. 26-30, Chapman and Hall, New York.

Lafarga, D., Santamaria, J., and Menendez, M. (1994) Methane oxidative coupling using porous ceramic membrane reactors—I. Reactor development, *Chem. Eng. Sci.,* **49**(12), 2205-2013.

Le Goff, P., Benadda, B., Comel, C., and Gourdon, R. (1997) Ultrafiltration for the removal of cadmium in waste streams from industrial waste incineration. *Sep. Sci. Tech.,* 32, 1615-1628.

Linde, K., and Jonsson, A. (1995) Nanofiltration of salt solutions and landfill leachate, *Desalination,* 103, 223-232.

Lin, Y. S., Wang, W., and Han, J. (1994) *AIChE J.,* **40**, 786-798

Loeb, S., and Sourirajan, S. (1963) *ACS Adv. Chem. Ser. 38,* 117.

Manwell, J. F., and McGowan, J. G. (1994) Recent renewable energy driven desalination system research and development in North America. *Desalination,* 94, 229-241.

Marrs, G. P., Lees, F. P. Barton, J. and Scilly, N. (1989) *Chem. Eng. Res. Devel.,* 67, 381.

Matsuda, T., Koike, I., Kubo, N., and Kikuchi, E. (1993), Dehydrogenation of isobutane to isobutene in a palladium membrane reactor, *App. Cat. A, Gen.,* **96**, 3.

*Membrane* Directory (1996), *J. Memb. Sci., 122,* No. 1-2.

Mir, L., Michaels, S. L., Goel, V., and Kaiser, R. (1992) Crossflow microfiltration: applications, design and cost. In W. S. Winston Ho and K. K. Sirkar (Eds.), *Membrane Handbook,* Chap. 35, Chapman and Hall, New York.

Mohan, K., and Govind, R. (1988a), Studies on a membrane reactor, *Sep. Sci. Technol.,* **23**(12,13), 1715.

Mohan, K., and Govind, R. (1988b), Analysis of equilibrium shift in isothermal reactors with a permselective wall, *AIChE J.,* **34**(9), 1493.

Mohr, D. (1996) Membrane reactors for dehydrogenation reactions: conceptual design issues.

Mulder, M. (1996) *Basic Principles of Membrane Technology,* Kluwer Academic Publisher, Boston.

Mundkur, S. D. and Watters, J. C. (1993) Polyelectrolyte-enhanced ultrafiltration of copper from a waste stream. *Sep. Sci. Tech.,* 28, 1157-1168.

Nicholas, R. W. (1990) Hollow Fiber Separation Module and Method for the Use Thereof, US Patent 4,959,152.

Nielsen, W. K., Madsen, R. F., and Olsen, O. J. (1980) *Desalination,* 32, 309.

Nigara, Y., and Cales, B. (1986) Production of carbon monoxide by direct thermal splitting of carbon dioxide at high temperatures, *Bull. Chem. Soc. Jpn.,* **59**, 1997.

Nirmal, J. D., Pandya, V. P., Desai, N. V., and Rangarajan, R. (1992) Cellulose triacetate membrane for applications in plating, fertilizer, and textile dye industry wastes. *Sep. Sci. Tech.,* 27(15), 2083-2098.

Noble, R. (1992) Overview—Catalytic membrane reactors, *Div. Pet. Chem ACS,* **37**(1), 11.

Noble, R. D. and Stern, S.A. (Eds.) (1995) *Membrane Separations Technology: Principles and Applications,* Elsevier, New York.

Nozaki, T., Yamazaki, O., Omata, K., and Fujimoto, K. (1992) Selective oxidative coupling of methane with membrane reactor, *Chem. Eng. Sci.*, 47(9–11), 2945.

Nunno, T. J. and Freeman H. M. (1987) Hazardous waste management: case studies of minimizing plating bath wastes in the electronics products industry. *JAPCA,* 37(6), 723–729.

Omata, K., Hashimoto, S., Tominaga, H., and Fujimoto, K. (1989) Oxidative coupling of methane using a membrane reactor, *Appl. Cat.*, 52, L1.

Osmonics Web Site (1997) www.osmonics.com, Nanofiltration-126 Acid Waste, December (reported with permission).

Pei, S., Kleefisch, M. S., Kobylinski, T. P., and Faber, J. (1995) Failure mechanisms of ceramic membrane reactors in partial oxidation of methane to synthesis gas, *Cat. Lett.*, 30(1–4), 201–212.

Perrin, J. (1989) Hollow Fiber Multimembrane Cells and Permeators, US Patent 4,880,440.

Pickering, K.D., and Wiesner, M. R. (1993) Cost model for low-pressure membrane filtration. *J. Env. Eng.,* 119, 772–797.

Poddar, T. K., Majumdar, S., and Sirkar, K. K. (1996a) Removal of VOCs from air by membrane-based absorption and stripping, *J. Membrane Sci.,* 120, 221–237.

Poddar, T. K., Majumdar, S., and Sirkar, K. K., (1996b) Membrane-based absorption of VOCs from a gas stream, *AIChE J.*, 120, 3267–3282.

Prabhakar, S., Balasubramaniyan, C. Hanra, M. S. Misra, B. M. Roy, S. B. Meghal, A. M. and Mukherjee, T. K. (1996) Performance evaluation of reverse osmosis (RO) and nanofiltration (NF) membranes for the decontamination of ammonium diuranate effluents. *Sep. Sci. Tech.,* 31, 533–544.

Prasad, R., Sirkar, K. K. (1992) Membrane-based solvent extraction, in W. S. Winston Ho and K. K. Sirkar (Eds.), *Membrane Handbook,* Chap. 41, Chapman and Hall, New York.

Puri, P. S., (1988) US Patent 4,756,932.

Puri, P. S., and Kalthod, D. G. (1993) Multiple Stage Countercurrent Hollow Fiber Membrane Module, US Patnet 5,176,725.

Puri, P. S., et al. (1996) US Patent 5,514,413.

Puri, P.S., (1996a) Gas separation membranes—current status, *Membrane J.,* 6(3), 117–126.

Puri, P. S. (1996b) Paper presented at the seminar on the Ecological Applications of Innovative Membrane Technology in the Chemical Industry, Cetraro (Italy).

Ramachandraiah, G., Thampy, S. K., Narayanan, P. K., Chauhnan, D. K., Nageswara Rao, N., and Indusekhar, V. K. (1996) Separation and concentration of metals present in industrial effluent and sludge samples by using electrodialysis, coulometry, and photocatalysis. *Sep. Sci. Tech.*, 31, 523–532.

Rao, M. B., and Sircar, S. (1993). Nanoporous carbon membranes for separation of gas mixtures by selective surface flow, *J. Membrane Sci.,* 85, 253.

Rautenbach, R., and Albrecht, R. (1989) *Membrane Separation Processes*, John Wiley, New York.

Rautenbach, R., and Linn, T. (1996) High-pressure reverse osmosis and nanofiltration, a zero discharge process combination for the treatment of waste water with severe fouling/scaling potential. *Desalination,* 105, 63–70.

Reck, R. A. (1995) Paper presented at the 2nd International Symposium on $CO_2$ Fixation and Efficient Utilization of Energy, Tokyo.

Reed, B. W., Semmens, M. J., Cussler, E. L. (1995) Membrane contactors. In R.D. Noble and S.A. Stern (Eds.). *Membrane Separations Technology: Principles and Applications,* Chap. 10, Elsevier, New York.

Richels, R. (1996) Paper presented at the 3rd International Conference on Carbon Dioxide Removal, Cambridge, MA, USA.

Riemer, P. (1993) "The Capture of Carbon Dioxide from Fossil Fuel fired Power Stations," IEA Greenhouse Gas R&D Program, Cheltenham.

Robinson, G. T. (1983) Recovery pays off for Chicago job shop plater. *Prod. Finish.,* June.

Ross, J.R.H., van Keulen, A.N.J., Hegarty, M.E.S., and Seshan, K. (1996) *Catal. Today,* 30, 193.

Santos, A., Coronas, J., Menendez, M., and Santamaria, J. (1995) Catalytic partial oxidation of methane to synthesis gas in a ceramic membrane reactor, *Cat. Lett.,* 30(1–4), 189–199.

Sato, K., Nakamura, J., Uchijima, T., and Hayakawa, T. (1995), Partial oxidation of $CH_4$ to synthesis gas using an Rh–Vertical–Bar–YSV–Vertical–Bar–Ag electrochemical membrane reactor, *J. Chem. Soc. Faraday Trans.,* 91(11), 1655–1661.

Schaffer, L. H., and Mintz, M. S. (1980) Electrodialysis. In K. S. Spiegler and A.D.K. Laird (Eds.) *Principles of Desalination,* 2nd ed, Part A, Chap. 6, Academic Press, New York.

Schoeman, J. J., van Standen, J. F., Saayman, H. M., and Vorster, W. A. (1992) Evaluation of reverse osmosis for electroplating effluent treatment. *Water Sci. Tech,* 25(10), 79–93.

Scott, K. (1995) *Handbook of Industrial Membranes,* Elsevier, New York.

Sengupta, A., Sodaro, R. A., and Reed, B. W. (1995) Oxygen removal from water using process-scale Extra-Flow® membrane contactors and systems. Paper presented at the 7th Annual Meeting of the North American Membrane Society, Portland, OR, May 23.

Shamsai, B. M., and H. G. Monbouquette (1997) $Cd^{2+}$ extraction from dilute solution by polymerized metal-sorbing vesicles immobilized in porous nylon membranes, *Proc. 1997 Topical Conf. on Separation Sci. and Tech., Annual AIChE Meeting,* pp. 681–683.

Shapiro, A. P., Thornton, R. F. Kim, B. M., and John, F. E. (1995) Case study of waste water minimization at a General Electric manufacturing plant, *Env. Prog.,* 14(3), 176.

Shu, J., Grandjean, B., Van Neste, A., and Kaliaguine, S. (1991), Catalytic palladium-based membrane reactors: A review, *Can. J. Chem. Eng.,* 69, 1036.

Sirkar, K. K. (1992) Other membrane processes. In W. S. Winston Ho and K. K. Sirkar (Eds.), *Membrane Handbook,* Chap. 46, Chapman and Hall, New York.

Sirkar, K. K. (1995) Membrane separations: newer concepts and applications for the food industry. In R. K. Singh and S.S.H. Rizvi (Eds.), *Bioseparations Processes in Foods,* Chap. 10, Marcel Dekker, New York.

Sirkar, K. K. (1997) Membrane separation technologies: Current developments, *Chem. Eng. Communications,* 157, 145–184.

Sloot, H., Versteeg, G., and van Swaaij, W. (1990) A permselective membrane reactor for chemical processes normally requiring strict stoichiometric feed rates of reactants, *Chem. Eng. Sci.*, **45**(8), 2415.

Sloot, H., Versteeg, G., Smolders, C., and van Swaaij, W. (1991), A non-permselective membrane reactor for the selective catalytic reduction of $NO_x$ with ammonia, *Key Eng. Mat.*, **61**, 62, 261.

Smith, W. H., and Foreman, T. (1997) Electrowinning/electrostripping and electrodialysis processes for the recovery and recycle of metals from plating rinse solutions. *Sep. Sci. Tech.,* 32, 669–679.

Snow, M. J., de Winter, D., Buckingham, R., Campell, J., and Wagner, J., (1996) New techniques for extreme conditions: high temperature reverse osmosis and nanofiltration. *Desalination,* 105, 57–61.

Sogge, J., and Strom, T. (1997). *Stud. Surf. Sci. Catal.*, 107, 561

Song, I., Lee, W., and Kim, J. (1991), Application of heteropoly acid catalyst in an inert polymer membrane catalytic reactor in ethanol dehydration, *Cat. Lett.*, **9**, 339.

Song, I., and Lee, W. (1993) Methyl $t$-butyl ether decomposition in an inert membrane reactor composed of 12-tungstophosphoric acid catalyst and polyphenylene oxide membrane, *App. Cat. A: Gen.*, **96**, 53.

Sourirajan, S., and Matsura, T. (1985) *Reverse Osmosis/Ultrafiltration Process Principles,* pp. 773–834. National Research Council of Canada

Sriratana, S., Scamehorn, J. F., Chavadej, S., Saiwan, C., Haller, K. J., Christian, S. D., and Tucker, E. E. (1996) Use of polyelectrolyte-enhanced ultrafiltration to remove chromate from water. *Sep. Sci. Tech.*, 31, 2493–2504.

Srivastava, R. D., Zhou, P., Stiegel, G. J., Rao, V., and Cinquegrane, G., (1992) Direct conversion of methane to liquid fuels and chemicals, *Specialist Periodical Report, Catalysis* **9**, 183–228.

Stern, S. A., (1994) Polymers for gas separations: the next decade (Review), *J. Membrane Sci.,* 94, 1–65.

Stoukides, M., and Vayenas, C. (1981) *J. Catalysis,* **70**, 137.

Strathmann, H. (1992) Electrodialysis. In W. S. Winston Ho and K. K. Sirkar (Eds.), *Membrane Handbook,* Chaps. 16–20, Chapman and Hall, New York.

Sugiyama, S., Tsuneda, S., Saito, K., Furusaki, S., Sugo, T., and Makuuchi, K. (1993) Attachment of sulfonic acid groups to various shapes of polyethylene, polypropylene and polytetrafluoroethylene by radiation-induced graft polymerization *React. Polym.,* 21, 187–191.

Sun, Y., and Khang, S. (1988) Catalytic membrane for simultaneous chemical reaction and separation applied to a dehydrogenation reaction, *Ind. Eng. Chem. Res.*, **27**, 1136.

Tabatabai, A., Scamehorn, J. F., and Christian, S. D. (1995a) Economic feasibility study of polyelectrolyte-enhanced ultrafiltration (PEUF) for water softening, *J. Membrane Sci.,* 100, 193–207.

Tabatabai, A., Scamehorn, J. F., and Christian, S. D. (1995b) Water softening using polyelectrolyte-enhanced ultrafiltration. *Sep. Sci. Tech.,* 30, 211–224.

Tasi, C.-Y., Dixon, A. G., Moser W. R., and Ma, Y. H. (1997) *AIChE J.*, 43, 2741.

ten Elshof, J. E., Bouwmeester, H.J.M., and Verweij, H. (1995) Oxidative coupling of methane in a mixed-conducting perovskite membrane reactor, *Applied Catalysis A, General,* 130, 195–212

Teraoka, Y., Zhang, H., Furukawa, S., and Yamazoe, N. (1985) *Chem. Lett.,* 1743.

Tiscareno-Lechuga, F., Hill, C., and Anderson, M. (1993) Experimental studies of the non-oxidative dehydrogenation of ethylbenzene using a membrane reactor, *App. Cat., A Gen,* 96, 33.

Tonkovich, A.L.Y., Secker, R., Reed, E., Roberts, G., and Cox, J. (1995), Inorganic membrane reactors: a design for bi-molecular reactant addition, *Sep. Sci. Technol.,* 30(7–9), 1609.

Tonkovich, A.L.Y., Zilka, J. L., Jimenez, D. M., Roberts, G. L., and Cox, J. L. (1996a) Experimental investigations of inorganic membrane reactors: a distributed feed approach for partial oxidation reactions, *Chem. Eng. Sci.,* 51(5), 789–806.

Tonkovich, A.L.Y., Jimenez, D. M., Zilka, J. L., and Roberts, G. L. (1996b) Inorganic membrane reactors for the oxidative coupling of methane, *Chem. Eng. Sci.,* 51(11), 3051–3056.

Tonkovich, A.L.Y., and Gerber, M. A. (1995) "The Top 50 Commodity Chemicals: Impact of Catalytic Process Limitations on Energy, Environment, and Economics," PNL-10684, report prepared for the U. S. Department of Energy, EE-Office of Industrial Technology.

Tonkovich, A.L.Y. (1994) "Impact of Catalysis on the Production of the Top 50 U.S. Commodity Chemicals," PNL-9432, report prepared for the U.S. Department of Energy, EE-Office of Industrial Technology.

Tsotsis, T., Champagnie, A., Vasileiadis, S., Ziaka, Z., and Minet, R. (1993), The enhancement of reaction yield through the use of high temperature membrane reactors, *Sep. Sci. Technol.,* 28(1–3), 397.

Tsotsis, T., Champagnie, A., Vasileiadis, S., Ziaka, Z., and Minet, R. (1992), Packed bed catalytic membrane reactors, *Chem. Eng. Sci.,* 47(9–11), 2903.

Tsuneda, S., Saito, K., Furusaki, S., Sugo, T., and Okamoto, J. (1991) Metal collection using chelating hollow fiber membrane. *J. Membrane Sci.,* 58, 221–234.

Uemiya, S., Sato, N., Ando, H., Matsuda, T., and Kikuchi, E. (1990) Promotion of methane steam reforming by use of palladium membrane, *Sekiyu Gakkaishi,* 33(6), 418.

Uemiya, S., Sato, N., Ando, H., Matsuda, T., and Kikuchi, E. (1991) Steam reforming of methane in a hydrogen-permeable membrane reactor, *App. Cat.,* 67, 223.

Uemiya, S., Sato, N. Ando, H., and Kikuchi, E. (1991) *Ind. Eng. Chem. Res.,* 30, 585.

Uludag, Y., Ozbelge, H. O., and Yilmaz, L. (1997) Removal of mercury from aqueous solutions via polymer-enhanced ultrafiltration. *J. Memb. Sci.,* 129, 93–99.

United States Environmental Protection Agency SITE Technology Capsule Report (Project Officer D. Grosse of US EPA), "Rochem Separation Systems, Inc. Disc Tube Module Technology." September, 1995.

Vail, P., and Sanford, P. (1992) A membrane process for recycle and reuse of wastewater streams, *Ind. Water Treatment,* July/August, 34–39.

van Zanten, J. H., Chang, D.S.W. Stanish, I., and Monbouquette H. G. (1995) Selective extraction of $Pb^{2+}$ by metal-sorbing vesicles bearing ionophores of a new class. *J. Memb. Sci.,* 99, 49–56.

Veldsink, J., van Damme, R., Versteeg, G., and van Swaaij, W. (1992) A catalytically active membrane reactor for fast, exothermic, heterogeneously catalysed reactions, *Chem. Eng. Sci.,* 47(9–11), 2939.

Walsh, A. J., and Monbouquette, H. G. (1993) Extraction of $Cd^{2+}$ and $Pb^{2+}$ from dilute aqueous solution using metal-sorbing vesicles in a hollow-fiber cartridge. *J. Memb. Sci.,* 84, 107–121.

Wang, S., Lu, G. W., and Millar, G. J. (1996), *Energy and Fuels,* 10, 896.

Westerterp, R. (1992) *Chem. Eng. Sci.,* 47, 2195–2206.

Williams, M. E., Bhattacharyya, D., Ray, R. J., McCray, S. B. (1992) Selected applications. In W. S. Winston Ho and K. K. Sirkar (Eds.), *Membrane Handbook,* Chap. 24, Chapman and Hall, New York.

Wisniewski, J. and Suder, S. (1995) Water recovery from etching effluents for the purpose of rinsing stainless steel. *Desalination,* 101, 245–253.

Wisniewski, J., and Wisniewski, G. (1997) Acids and iron salts removal from rinsing water after metal etching. *Desalination,* 109, 187–193.

Yang, Z. F., Guha, A. K., and Sirkar, K. K. (1996) Novel membrane-based synergistic metal extraction and recovery processes. *Ind. Eng. Chem. Res.,* 35, 1383–1394.

Zaman, J., and Chakma, A. (1994), Review: inorganic membrane reactors, *J. Mem Sci.,* 92, 1.

Zaspalis, V., van Praag, W., van Ommen, J., Ross, J., and Burggraaf, A. (1991a), Reactions of methanol over alumina catalytically active membranes modified by silver, *Appl. Cat.,* 74, 235.

Zaspalis, V., Keizer, K., Ross, J., and Burggraaf, A. (1991b), Porous ceramic membranes in high temperature applications, *Key Eng. Mat.,* 61,62, 359.

Zaspalis, V., van Praag, W., Keizer, K., van Ommen, J., Ross, J., and Burggraaf, A. (1991c), Reactions of methanol over catalytically active alumina membranes, *Appl. Cat.,* 74, 205.

Zaspalis, V., van Praag, W., Keizer, K., van Ommen, J., Ross, J., and Burggraaf, A. (1991d), Reactor studies using vanadia modified titania and alumina catalytically active membranes for the reduction of nitrogen oxide with ammonia, *Appl. Cat.,* 74, 249.

Zhao, R., Govind, R., and Itoh, N. (1990), Studies on palladium membrane reactor for dehydrogenation reaction, *Sep. Sci. Technol.,* 25(13–15), 1473.

Ziaka, Z., Minet, R., and Tsotsis, T. (1993) A high temperature catalytic membrane reactor for propane dehydrogenation, *J. Mem. Sci.,* 77, 221.

Zolandz, R., Fleming, G. K. (1992) Gas permeation. In W. S. Winston Ho and K. K. Sirkar (Eds.), *Membrane Handbook,* Chaps. 2–6, Chapman and Hall, New York.

# 4

# Findings of the National Workshop on Process Waste Reduction via Separation Technologies and Separative Reactors

**AUTHOR**

**Peter P. Radecki**
National Center for Clean Industrial and Treatment Technologies,
Michigan Technological University, Houghton, Michigan

**WORKSHOP PRESENTERS**

**Earl Beaver**
Monsanto, St. Louis, Missouri

**Robert W. Carr**
University of Minnesota, St. Paul, Minnesota

**John C. Crittenden**
Michigan Technological University/CenCITT, Houghton, Michigan

**Edward L. Cussler**
University of Minnesota, St. Paul, Minnesota

**Russell F. Dunn**
Solutia, Inc.

**Darryl W. Hertz**
The M.W. Kellogg Company, Houston, Texas

**Jimmy L. Humphrey**
J.L. Humphrey & Associates, Austin, Texas

**WORKSHOP PRESENTERS (Continued)**

**Kerry Irons**
The Dow Chemical Company, Midland, Michigan

**Gregory Keeports**
Rohm & Haas

**George Keller II**
Consultant

**Kent Knaebel**
Adsorption Research Inc., Dublin, Ohio

**Richard Noble**
University of Colorado, Boulder, Colorado

**Pushpinder Puri**
Air Products and Chemicals, Inc., Allentown, Pennsylvania

**Peter P. Radecki**
Michigan Technological University/CenCITT, Houghton, Michigan

**Douglas M. Ruthven**
University of Maine, Orono, Maine

**Eric D. Sall**
Monsanto

**David R. Shonnard**
Michigan Technological University, Houghton, Michigan

**Kamalesh K. Sirkar**
New Jersey Institute of Technology, Newark, New Jersey

**Anna Lee Y. Tonkovich**
Pacific Northwest National Laboratory, Richland, Washington

**Jack Weaver**
Center for Waste Reduction Technologies, AIChE, New York, New York

**EDITORS**

**John C. Crittenden**
Michigan Technological University/CenCITT

**John L. Bulloch**
Michigan Technological University/CenCITT

## 4.1. Overview of Workshop

Eighty-nine industry, academic, and government representatives participated in a national workshop that was held in New Orleans on February 4–6, 1998 to discuss how adsorption, membranes, and separative reactor technologies may be used to reduce chemical process waste. The outcomes of the workshop include (1) documentation of perspectives from experts in these technology areas and (2) recommendations of candidate processes and process streams for application of the subject technologies to achieve process waste reduction. The workshop was a diverse and unusual gathering of technology experts, process engineers and treatment plant practitioners. This combination led to an interesting—sometimes challenging—dialogue in which each group worked to assimilate the perspectives and needs of the other groups to achieve the global objective of process waste reduction.

The workshop also served as a lead-in to a one-day roadmapping exercise that mapped strategies for further development of the three technologies. Roadmap results, which can be found in a separate document, *Vision 2020: 1998 Separations Roadmap,* published by the Center for Waste Reduction Technologies (CWRT, 1998), also documents the results of a follow-on workshop covering distillation, extraction, and crystallization. The roadmap identifies barriers to be overcome by the six separation technologies and outlines R&D needs for physical property data, new materials, process system improvements, and process modeling, in order for the technologies to address the challenges described in *Technology Vision 2020: The US Chemical Industry* (ACS et al, 1996).

The output from the workshop and the roadmap will enable researchers to focus scientific and engineering studies on "real world" needs which, if solved, would help the chemical industry reach the Vision 2020 objectives. This approach will lead to more efficient use of scarce research dollars than is achievable where the specific process focus of the research program is often not well defined.

Section 4.3 summarizes the workshop technical presentations while Section 4.4 catalogs a number of process streams that participants suggested would be good foci for membrane, adsorption and separative reactor research and technology demonstrations. Workshop sponsors and a list of participants can be found in Sections 4.2 and 4.5, respectively.

## 4.2. Workshop Sponsors

The workshop was sponsored by the Center for Waste Reduction Technologies (CWRT) which operates under the auspices of the American Institute of Chemical Engineers (AIChE). Co-sponsors were

- National Center for Clean Industrial and Treatment Technologies (CenCITT)
- U. S. Department of Energy—Office of Industrial Technologies (DOE-OIT)
- American Institute of Chemical Engineers—Separations Division
- Council for Chemical Research (CCR)
- American Chemical Society (ACS)—Industrial and Engineering Chemistry Division

In addition to identifying technical barriers and research needs, these organizations promote collaboration in research. Workshop and roadmap results can be used by these organizations as well as others, such as the National Institute for Standards and Technology, the National Science Foundation and the Environmental Protection Agency to help develop internal and external research programs.

## 4.3. Summaries of Technical Presentations

The workshop began with an overview of all the subject technologies and was followed by three technology sessions that focused on adsorption, membranes and separative reactors. Each technology session began with a plenary presentation that introduced the technology and the state-of-the-art, as well as suggestions of where the technology may be used to reduce process waste. Following each plenary presentation, a panel discussion was led by the monograph authors and others, and the technical barriers to technology implementation were discussed. Workshop participants then moved to breakout sessions where lists of process streams and research and development needs were identified to achieve waste reduction (see Section 4.4). Each session ended with summaries presented to the general assembly by the breakout session facilitators.

### 4.3.1. Introduction to the Center for Waste Reduction Technologies

*The workshop began with a presentation by Jack Weaver, Director of Technology Partnerships and Meetings for AIChE and the former Executive Director of CWRT. His presentation was entitled "Introduction and Opening Remarks—Review CWRT Mission and Needs of Industry."*

The CWRT mission is to benefit sponsors and society by identifying, developing and transferring *environmentally* beneficial technologies in a cost-effective and timely manner via leveraged resources. The following companies and organizations were members of CWRT at the time of the workshop:

3M
AM-RE Services
Argonne National Laboratory
Arthur D. Little
BFGoodrich
Celanese
CH2M HILL
Dow Chemical
Electrical Power Research Institute
Eastman Chemical
General Electric
ICI Americas
Kinetics Technology International
Los Alamos National Laboratory
M.W. Kellogg
Merck & Co., Inc.
Monsanto
C.W. Nofsinger
Owens Corning
Pacific Northwest National Laboratories
Rhone-Poulenc
Rohm and Haas
SmithKline Beecham
SRI International
Union Carbide
U.S. Department of Energy

Nonmember companies active in CWRT include Cytec, Eastman Kodak, Georgia Pacific, Bristol-Myers Squibb, Atlantic Richfield, Olin, SABIC (Saudi Arabia), and Procter & Gamble. CWRT also has a strategic alliance with the National Center for Clean Industrial and Treatment Technologies through which a variety of joint research, training, software development and other collaborations occur. This workshop and the project under which it was conducted are examples of the success of this alliance. In short, CWRT strives to promote and actively pursue networking, collaboration and technology transfer.

CWRT believes successful collaborations involve shared needs, some sense of urgency, the leadership of individual champions within the collaborating organizations, trust, and professional staff support. CWRT pursues collaborations in the following technical focus areas:

- Sustainability
- Source reduction via novel reaction technolgoies

- Separation technologies and in-process recycle
- Waste management
- Postprocess recycle/value recovery
- *In situ* remediation

Weaver provided a short introduction to the U.S. Chemical Industry's Technology Vision 2020 program. The Vision 2020 program prepared a document that outlines technology needs of the industry for it to remain competitive and vibrant in the year 2020 (ACS et al., 1996). Sponsoring organizations, such as AIChE, have taken on the responsibility of leading the development of these technologies, and the organizations are collaborating in coordination of the overall development efforts. Responsibilities thus far identified are

- CCR—*chemical synthesis, bioprocesses and biotechnology, materials technology, chemical measurement, computational technologies*
- AIChE—*manufacturing and operations*
- CCR and AIChE—*process science and engineering technology*
- Others—*supply chain management, information systems*

AIChE has also established a Vision 2020 Task Force to coordinate its efforts. Task force members and their institutional representations are

- Academia—Rex Reklaitis *(Purdue)*
- AIChE—Steve Weiner *(Battelle/PNNL)*
- Government—Bruce Cranford *(DOE)*
- Industry—Earl Beaver *(Monsanto)*
- Staff—Jack Weaver *(CWRT)*
- CCR Liaison—Paul Bryan *(UCC)*

Taken as a whole, the Vision 2020 program will provide the impetus to achieve the goals presented in the technology vision document, and will encourage a new level of multiorganizational cooperation which focuses on collaboration to compete on a world scale and reduce the environmental impact of the chemical industry.

### 4.3.2. Sustainability—The Future of Pollution Prevention

*This presentation was made by Earl Beaver, Director, Waste Elimination for Monsanto. His presentation was a visionary call-to-arms as all are challenged to make our manufacturing processes environmentally friendly, products recyclable and feedstocks renewable.*

In order for companies to become more sustainable, they will require ways to decouple economic growth from energy and natural resource consumption.

Realistically this will be difficult in today's marketplace. Accordingly, companies choosing to move toward sustainability are doing so by changing their product mixes. For example, Monsanto recently spun off its chemical division to Solutia. Most of the products of this division are made from nonrenewable feedstocks. The products and processes that remained at Monsanto will focus on new technologies and biological systems. In recognition of its commitment to sustainability and the social contract it perceives, Monsanto's promotional by-line is Food–Health–Hope™.

Today, accounting systems are very good at determining the capital and operating costs associated with production, but generally do not consider the inherent value of natural resources or social capital. For example, if we consider the impact of severe weather on the U.S. economy, which may be due to excess carbon dioxide emissions, we would realize that we are not paying enough for carbon dioxide emission and these should be regulated. In addition, one does not have to be concerned about global weather change and/or warming to realize the financial opportunities potentially available from a close scrutiny of carbon consumption.

Given the rate at which developing countries are growing, we must move fast if we are to avoid serious global environmental and environmentally driven socioeconomic problems. We need to rethink what "waste" represents. For example, before the Solutia spin off, less than 1% of total Monsanto wastes were reported in the Toxics Release Inventory. We should look upon process waste as a real indicator for inefficiency.

Barriers to sustainability include:

- The diversity of products
- Competitive attitudes
- Antitrust laws
- Lack of a critical mass of skills in companies to solve problems
- Lack of model streams for research studies
- Funding (this is described in a report from the President's Council on Science and Technology)
- Lack of understanding of the subject matter
- Lack of understanding of competencies needed
- Ineffectiveness of educational system—need for new reaction designs

Focusing a portion of our efforts in the following areas may lead to solutions:

- Advances in biological sciences, physics, psychology and information systems
- In accounting and modeling, substitute information for materials and energy*

* *Author's Note:* A provocative book proposing information as a fundamental quantity was written by George Johnson (Johnson, 1995).

- For example, write prescriptions based on weight and activity
- Look for common themes across industries
• Integrate research and development across the fields of chemical synthesis, manufacturing, information technology, information systems and biology
• Adapt separation technologies, including selective separations, for bio systems
• Integrate separation tools with life cycle concepts

### 4.3.3. An Overview of Emerging Adsorption, Membrane, and Separative Reactor Technologies for Process Pollution Prevention

*This presentation was made by Jimmy L. Humphrey, President of J. L. Humphrey and Associates. Dr. Humphrey is also the primary author of Chapter 1.*

In developing a strategy for technology selection, the following factors should be considered:

• Factors favoring selection of adsorption:
  - A high degree of purification is required.
  - Loaded bed(s) are more easily regenerated than partially loaded bed(s).
  - An adsorbent is available which is not susceptible to fouling or attack by feed components.
• Factors favoring selection of membrane processes:
  - Bulk rather than precise separations are sufficient.
  - Required processing rates are low to medium.
  - A membrane is available which is not subject to fouling by feed components.
  - Energy cost is a key consideration.
• Factors favoring selection of separative reactors:
  - A reversible reaction is involved: By removing product as it is formed, the reaction is driven to completion in the desired direction.
  - Reaction and separation temperatures (and other conditions) overlap.
  - Reaction is exothermic: For example, heat of reaction may be used to provide the heat for distillation.

The following should be considered in future development of these technologies for process applications and waste reduction:

• Materials research and development is a priority.
• Models to predict performance need improvement.
• Early economic analyses are needed to determine market potential and to facilitate decision-making in technology development investment.
• Technology transfer needs to be more efficient.

### 4.3.4. Introduction to Adsorption Technology

*This presentation was made by George Keller II, a consultant and recently retired Senior Corporate Fellow from Union Carbide. Keller's presentation provided an overview of a wide variety of emerging adsorption technologies.*

An adsorption process is a cyclic process in which a fluid is fractionated by (1) selectively adsorbing one or more compounds from the fluid onto a solid (adsorbent), and then (2) removing (regenerating, desorbing) the adsorbed compounds to restore the adsorbent to its state prior to step 1.

Examples of adsorption processes include

- Gas Bulk Separations
  - Air separations
  - Hydrogen purification
  - Normal paraffins/isoparaffins
  - Alcohol drying
- Gas Purifications
  - Gas drying
  - VOC removal from vent streams
  - Odors from air
  - Removal of sulfur compounds and other contaminants from various gas streams
- Liquid Bulk Separations
  - $p$-Xylene/other C10s
  - Glucose/fructose
  - Detergent-range olefins/paraffins
  - Citric acid/fermentation broth
- Liquid purifications
  - Removal of a wide range of relatively hydrophobic contaminants from water
  - Removal of water from organics
  - Odor and taste bodies from water
  - Decolorizing petroleum fractions, syrups, vegetable oils, etc.
  - Removal of various fermentation products from fermentor effluent
  - Examples of industrial adsorbents include:
- Hydrophilic (polar-compound selective)
  - A wide variety of zeolite molecular sieves (silica-alumina-based)
  - Silica gels
  - Corn grits
- Hydrophobic (non–polar-compound selective)
  - Activated carbons

- Carbon molecular sieves
- Silicalite (silica-based molecular sieve) and similar products
- Modified high-surface-area ion-exchange resins
• Other
  - A wide variety of irreversible adsorbents based on tenacious and specific bonding with various adsorbates
  - Biomass (biosorption or biofiltration)

Normal means of adsorbent regeneration are

- Pressure-swing adsorption (PSA)
  - Gas phase adsorbate partial pressure is reduced
  - Usually affected by total pressure reduction
  - Usually affected by mole fraction reduction in gas phase adsorbate
- Temperature-swing adsorption (TSA)
  - Reduced bonding tenacity between adsorbent and adsorbate
  - Desorption caused by raising system temperature
- Displacement-purge adsorption (DPA)
  - Desorption caused by displacing agent addition
  - Displacing agent adsorbs competitively with the adsorbate
- Irreversible adsorption (IA)
  - In a very few cases, the adsorbate–adsorbent bonds are very strong
  - Restoration of loaded adsorbent may require removal and chemical and/or thermal treatment

There are weaknesses associated with each of these regeneration methods:

- PSA: Normally only one pure product is possible, and the recovery of that product is less than quantitative.
- TSA: Cycle times are slow, leading to low productivities. In addition, the process is typically only applicable for removal of small amounts (usually less than about 5 wt %) of the feed.
- DPA: The process is very complex, usually involving one or more adsorption columns and two distillation columns.
- IA: The high cost of regeneration limits this process economically to removing about 100 kg or less per day of adsorbate.

In addition to regeneration challenges, adsorption has additional weaknesses:

- The adsorbent problem. There are too few different types to make all the separations we would like to do. Present-day methods of producing selectivity are based on
  - various forces such as van der Waals, hydrogen bonding, electrostatic, covalent, and pi bonding,

- separation by size (molecular sieving), and
- differences in intraparticle diffusion rates.
- The impracticality—at least until recently—of moving the bed from an adsorbing zone to a desorbing zone and back. The result: a process-complexity/investment problem.
- Removal of the heat of adsorption from the adsorbent and supply of the heat for desorption to the adsorbent

There are a variety of new adsorption systems, which allow for the adsorbent bed to be moved back and forth between the adsorbing and desorbing zones. These include the Polyad™ process, the rotary wheel adsorber and Calgon's fiber mat process. Polyad and the rotary wheel are described in Section 1.2.4. The fiber mat process involves an adsorbing flexible mat which is scrolled between two reels. Between the reels are the adsorption zone and a heated desorption zone.

The rotary wheel process allows cycle time between adsorption and desorption to be reduced, thus resulting in small bed volumes. This feature allows the designer to select from a wide variety of adsorbents that may have been cost-prohibitive in conventional process configurations. Leading companies in this technology are Seibu Giken, a Japanese company, and Durr, a German firm.

The rotary wheel can also be used as a concentrator. In solvent-containing processes, its use can result in more economical solvent destruction or recovery. Figure 4.1 depicts the in-process location for rotary concentrator installation. According to Ecopure®, rotary concentrators are generally applicable to processes with high volumetric flow rates, low pollutant (adsorbate) concentrations, and near-ambient conditions. Also according to Ecopure®, applications can be found in painting and coating processes, the semiconductor industry and in printing operations.

In considering future research and development in adsorption technologies, the following should be considered:

- The primary focus of adsorption processes will be in (a) cleanup of liquid and gaseous effluents, and (b) production and purification of raw materials and products.
- Additional possibilities for use are in complex separations of high-boiling or thermally unstable compounds which have geometric, polar or other differentiating interactions with adsorbents.
- Clearly most major improvements in adsorption in the last 10 years have been in process innovations and not in developing innovative adsorbents. These improvements include various moving-bed and simulated moving-bed technologies.
- Unfortunately, U.S. university and national-laboratory inputs to these innovations (and especially in the process-research area) have been miniscule. Most inputs have come from overseas and U.S. industry.

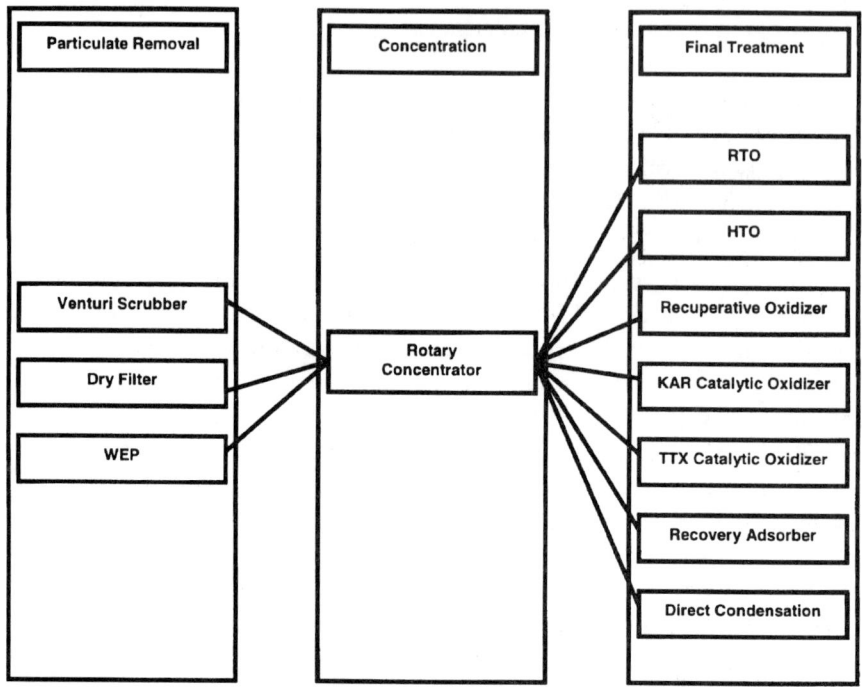

**Figure 4.1.** Rotary concentrator locations in solvent recovery processes.

## 4.3.5. Adsorption Technology Panel Presentations and Discussion

*Following the plenary presentation by Dr. Keller, four panelists presented 3–5-minute presentations on process and technology research needs associated with adsorption technologies. The panel was moderated by Erik D. Sall, a Research Engineering Consultant in the Chemical Sciences Department of Monsanto Corporate Research. The following subsections summarize the presentations.*

### 4.3.5.1. Adsorption's Role in Achieving In-Process Pollution Prevention

*This presentation was made by John C. Crittenden, Presidential Professor of Civil and Environmental Engineering and Director of CenCITT, Michigan Technological University.*

There is a practical hierarchy of preferences in pollution prevention approaches:

1. Source reduction (separative reactors)
2. In-process recycling (separation technologies)
3. On-site recycling (separation technologies)

## 4. Findings of the National Workshop on Process Waste Reduction

4. Off-site recycling (separation technologies)
5. Waste treatment to render the waste less hazardous (separation technologies)
6. Secure disposal

Adsorption's preferred role in pollution prevention is summarized as follows:

- Source reduction applications
  - Limited applications for nonreactive adsorption due to nature of the adsorption process
    ○ Purification of reactants and feedstocks yielding improved reaction selectivity and lower levels of process waste or harmful byproducts
  - Adsorptive reactors have potential for true "source reduction" in production processes.
- In-process or out-of-process recycle
  - Separation technologies are important for rendering the recycle stream pure enough for use.
  - Adsorption's largest niche is due to its ability to concentrate "dilute" streams.

Challenges to future adsorption development include

- Identification of areas where adsorption could be used in industry but currently is not due to technical limitations.
- Identification of technical limitations for application of adsorption in these areas.
- Identification of collaborative research teams and funding necessary to overcome these technical limitations.

### 4.3.5.2. Suggestions: New Areas of Waste Reduction via Adsorption
*This presentation was made by Kent Knaebel, President of Adsorption Research, Inc.*

The following conditions and attributes describe the state of the art in pressure- and temperature-swing adsorption processes:

- Pressure Swing Adsorption
  - oxygen from air (up to 95%)
  - nitrogen from air (up to 99%)
  - moisture from air
  - hydrogen from reformer off-gases
  - landfill gas ($CO_2/CH_4$)
  - synthesis gas ($CO/CO_2/H_2/CH_4$)
  - natural gas ($N_2/CH_4$)
  - argon recovery from ammonia syn-gas

- ozone enrichment
  - VOC vapor recovery
  - ammonia synthesis gas
- Temperature Swing Adsorption
  - solvents from air, for example:
    - methylene chloride
    - hexane
    - toluene
    - methyl ethyl ketone
  - $CO_2$ purification
  - ethanol drying
  - automotive evaporative fuel emissions
  - all use steam, flue gas, or hot $N_2$ for regeneration

There are also a variety of potential new applications and concepts for these areas:
- Pressure Swing Adsorption
  - natural gas: $CH_4/N_2/CO_2/H_2S/COS/NGL$
  - NOx/SOx from flue gas (attempted, but not commercialized)
  - olefin/paraffin ($C_2/C_2^=$, $C_4/C_4^=$, etc.)
  - ammonia recovery from synthesis byproduct
  - 99+% oxygen from air
  - ultra-high purity (>99.999%)
  - very high frequency operation (>10 Hz)
- Temperature Swing Adsorption
  - novel regeneration methods
    - electrical resistance (under development)
    - electrical induction
    - microwaves (tried in the 1980s)
  - semiconductor grade liquids
  - downsized units for small process vents, etc.

As one develops new adsorption technologies and attempts to bring them to the marketplace, a variety of barriers emerge which are related to communications. These include

- Education: adsorption principles are seldom taught to undergraduate students, and rarely to graduate students. Without widespread familiarity, the operation will remain under-utilized.
- Industrial users need to identify hurdles, either openly or in private, and universities need to produce more accessible (less esoteric) research. In many cases, one does not know what the other needs.
- Vendors are reluctant to reveal performance data or design parameters.

4. Findings of the National Workshop on Process Waste Reduction

There are also barriers associated with the technology itself:

- Modeling: there is a plethora of tools (ranging from "in-house" programs to ADSIM™). Problems exist in:
  - ease of use/training
  - standardization (model features, isotherm forms, etc.)
  - calibration (fitting coefficients) and verification
  - some common features are dubious (steady-state vs. cyclic)
- Adsorbents
  - A common database, so that alternate materials can be compared (being addressed for a few adsorbents with a few adsorbates).
  - More openness is needed.

Observations regarding adsorption research and development funding give some pause for consideration:

- Funding is the key to developing solutions to specific (as well as generic) problems, and creating simulation tools and databases.
- If "adsorption technology" had received only 1/2 the funding that "membrane technology" has received, where would the state-of-the-art be?
- The U.S. adsorption community has been very demure about pointing out the needs and possibilities. Meanwhile, greater effort is expended in Europe and Japan.

### 4.3.5.3. Perspectives for the Development of New Sorption Based Technologies

*This presentation was made by Douglas M. Ruthven, Professor and Chair of the Department of Chemical Engineering, University of Maine.*

Factors which hinder sorption process innovation include:

- Capital intensity of process industries
- Technical superiority of a new process not necessarily sufficient to ensure adoption
- Long development/implementation time scales (e.g. 30 years for PSA)

Future development should include the following themes:

- Recovery of valuable products from low-value (waste) streams
- Improved performance using combined separation processes
- Parallel passage contactors for high volume, low value-added applications (vapor phase)
- Development of small channel monolithic adsorbents

General problems associated with the recovery of valuable materials from waste streams include the frequent occurrence of multicomponent solutions, which are difficult to separate, and the fact that the valuable materials are usually present at low concentrations.

Table 4.1 outlines areas of progress one might expect within the next 1–5 years for several processes.

Process designers should consider synergistic relationships between separation processes. Current examples of synergistic coupling include:

- IFP/Chevron: Simulated Moving Bed Adsorber plus Crystallizer for production of high purity *p*-xylene (pX).
  - pX product purity specifications have risen from 99.5 to 99.8%. This purity is achievable by "Parex" but only at reduced throughput. Combination of initial separation by SMB with polishing by crystallization → improved purity at high recovery (lower cost).
  - Reduction of large recycle and associated distillation costs.
- Catalytic Dewaxing (Chevron): Size selective adsorption + catalysis.
  - SAPO-11 + Pt. Milder acid sites yield $C_5+$ liquids rather than lower value light gases, as in original Mobil process (HZSM-5).

**Table 4.1.** Adsorption Technology Development Areas over the Next 1–5 Years

| Process | Economic Driver | Critical Needs |
|---|---|---|
| Recovery of $C_3H_6$ (and $C_2H_4$) from cat-cracker off-gas | Demand for olefins exceeds supply. Steam cracking is very expensive. Large quantities of $C_3H_6$, $C_2H_4$ available but at low conc. and relatively low pressure | Highly selective adsorbent $Ca^{++}$ on $Al_2O_3$ (Yang); $Ag^+$ zeolite (Habgood). Novel process scheme for high throughput |
| Recovery of $H_2$ from low-BTU refinery gas | $H_2$ shortage in most refineries. Processing of heavier crudes. Elimination of undesirable hydrocarbons from fuels by hydrogenation. | Efficient adsorption or membrane process for large throughput, low pressure differential. (Selectivity is not a problem.) |
| Upgrading of low-BTU natural gas ($CO_2$, $N_2$ removal) | Increased demand for natural gas as a "clean fuel." Large reservoirs of sub-spec natural gas. | Highly selective adsorbent for $N_2$ relative to $CH_4$. Combined $CO_2 + N_2$ removal in one step. |
| VOC removal/recovery | Legislation. Economics of recovery vs. destruction of VOCs. | Low pressure differential (parallel passage or monolith in rotary wheel or other configuration) |

Future prospects for synergistic coupling of separation processes include:

- Zeolite Membranes (Adsorption & Permeation)
  - With or without direct reactor coupling
  - Selective removal of product to shift equilibrium
  - Debottlenecking of existing reactors/separators
  - Separation of differently branched hydrocarbon isomers
- Pressure Swing Adsorption + Membrane (e.g. $H_2$ recovery)
  - Integrated process improves energy efficiency (higher recovery at given purity and pressure differential)
- Rapid Cycle PSA (PSA + Monolith)
  - High throughput at low pressure differential
  - Requires parallel passage contactor with <100 micron gap (or equivalent monolith)

The following is a summary of breakthroughs needed for substantial adsorption technology improvements:

- Very thin coherent zeolite membranes
  - (~5mm)—crack free or effectively reparated
- Parallel passage contactor or monolith with uniform channels <100/mm.
- Highly selective adsorbents for olefin/paraffin
- Highly selective adsorbent for $N_2/CH_4$
- Methods for recovery and rejuvenation of spent catalysts and adsorbents

### 4.3.6. Design and Optimization of Waste Reduction by Adsorption with ADSIM™

*This presentation was made by Nick Hankins, an applications engineer with Aspen Technologies, Inc., and coauthored by Felix Jegede (who is also with Aspen). The presentation provided an introduction to Aspen's adsorption simulation modeling program, and described its uses in waste reduction applications.*

ADSIM™ is a flowsheet simulation package that enables process modeling and simulation for improved design and optimization of adsorption separation processes in gas and liquid streams. It has features that accommodate PSA, TSA, VSA, and reactive adsorption in gas-phase processes. In the liquid phase, it can handle bulk liquid adsorption, chromatography, and ion exchange. It can automatically generate flowsheets and column models, including mass, momentum, and energy transfer, for single and multiple components, and for systems with cyclic boundary conditions. ADSIM also contains a library of equipment models.

Challenges in waste reduction include

- Meeting environmental legislation on effluent quality and rate
- Minimizing capital and operating cost of process plants
- Flexibility to meet changing process demands

Specific to the design and operation of adsorption processes, technical challenges include

- Choice of adsorbent type(s) and layering
- Definition of operating cycle
- Size, arrangement, and number of beds
- Maintaining specs on effluent quality and rate
- Minimizing energy/fuel consumption per cycle
- Predicting response to feed disturbances
- Facilitating process control

ADSIM has demonstrated applications in a variety of process categories:

- Removal of VOCs from air and vent streams
- Removal of Organics from aqueous streams
- Removal of heavy metals by ion-exchange
- Recovery of CFC refrigerants
- Removal of $SO_2$ from flue gases
- $NO_x$ removal from wet nitric acid plant tail gas
- Methane recovery from fermentation gas of landfills/wastewater purification plants
- Nuclear waste management: radioactive Xe and Kr treated in charcoal delay systems

*Assuming the user has access to reliable adsorbent design data and detailed stream composition information,* ADSIM use in the design stage may provide benefits to the design process:

- Flexibility allowing rapid screening of process alternatives
- Reduced pilot plant effort
- Avoidance of under design and excessive over design
- Determination of the most economic and effective adsorbent
- Faster plant commissioning
- More reliable and flexible design

In the operation stage, ADSIM use can help the production engineer:

- Achieve optimum cyclic operation:
  - Maximize throughput and contaminant removal

- Minimize use of energy and raw materials
- Compensate for changes in:
  - Feed stock composition and throughput
  - Required effluent quality
  - Operating conditions
- Perform safety studies

In the future, adsorption will likely expand its applications, in part due to the following factors:

- Adsorption technology is set to occupy more a prominent position due to increasingly stringent environmental restrictions
- Improving prediction of multicomponent adsorption and mass transfer processes allows new rate-based processes using less energy
- Simulation of new, true moving bed and adsorbent wheel processes
- New generation of adsorbents available: biosorbents, irreversible adsorbents
- SMB PLUS now available: technology extended to small-scale separations
- Modeling "hybrid" systems—include catalytic process, e.g., modified activated carbon to adsorb $SO_2$ and reduce $NO_x$; distillation; membranes
- "Smart" control systems based on emerging computer technology (neural networks, artificial intelligence)

### 4.3.7. Process Integration for Pollution Prevention

*This presentation was made by Russell F. Dunn, Research Specialist of Solutia, Inc. and coauthored by B. K. Srinivas, Research and Development Engineer at General Electric Research and Development. An overview of process design integration methodologies via mass and energy exchange was presented in the context of process waste minimization. A listing of publications outlining mass and energy exchange methodologies was also provided and is reproduced here.*

Pollution prevention can best be achieved when one considers plant operations as a whole—that is, as an enterprise focus. From a technical perspective, this involves the compilation of individual process units, their plant-level utilities and whole site operations. From a business perspective, we observe that the larger the scope of the study, the more complex the design problem is, but there are greater pollution prevention opportunities and potential impact. Process integration design technology is a rapidly developing set of tools to maintain an enterprise focus in design.

Process integration design allows for

- The optimal allocation of mass and energy within a unit operation, process and/or site.

- Optimal allocation can be based on economic, environmental or other important objectives.
- Mass Integration + Energy Integration

The science of process integration involves:
- Identifying the optimal performance targets ahead of design
  - minimum operating cost, minimum waste generation, minimum water usage, maximum capacity, etc.
- Identifying the optimal strategy to attain the target
  - identifying the maximal attainable performance at no/minimal capital
    - redistribution of internal resources (raw material and energy)
    - optimizing the systems performance rather than that of individual units
  - identifying the maximal attainable performance with new technology
    - optimizing the choice and combination of technology
    - optimizing the applicable location for each technology
    - optimizing the performance load on each technology

Potential impacts of process integration for waste reduction include:

- Ability to deal with complex processes, integrated systems
- Techniques address important design issues
  - Reduce operating and capital costs
  - Debottleneck processes
  - Conserve raw materials
  - Reduce environmental emissions
  - Avoid capital costs for new processes
  - Evaluate new waste reduction technologies

Process integration design tools for separation technologies developed to date include:

- Synthesis of recycle/reuse systems
  - Mass exchange networks, MENs
  - Reactive mass exchange networks, REAMENs
  - Combined heat and reactive mass exchange networks, CHARMENs
  - Heat-induced separation networks, HISENs
  - Energy-induced separation networks, EISENs
  - Membrane networks/systems
    - Reverse osmosis networks, RONs
    - Pervaporation networks
    - Membrane-condensation hybrid networks
  - Distillation networks/trains, etc.

- Synthesis of source reduction systems
  - New chemistry and molecular/process structure
    o Environmentally acceptable reactions, EAR's
    o Reaction paths
    o Solvent–product systems
    o Reactor–separator systems
  - In-plant modifications
    o Waste interception and allocation networks, WINs
    o Heat-induced waste minimization networks, HIWAMINs

Industrial applications by Solutia of process integration resulted in the following:

- Site 1—Petrochemical Plant
  - 35% Decrease in site steam costs
  - 14% Reduction in site wastewater load
  - Debottlenecked process
  *Total Annual Savings: $ 5.0 MM/yr*

- Site 2—Food and Specialty Chemical Plant
  - 40% Decrease in site steam costs
  - 60% Reduction in site wastewater load
  - 9% Capacity increase
  *Total Annual Savings: $ 5.8 MM/yr*

- Site 3—Polymer Plant
  - 5% Decrease in site steam costs
  - 51% Reduction in site wastewater load
  - Debottlenecked site water supply
  *Total Annual Savings: $ 1.5 MM/yr*

Industrial applications by General Electric have also been successful:

- Site 1—Polymer Plant
  - 22% Decrease in site utility costs
  - 20% Reduction in site wastewater hydraulic load
  - Annual savings from materials recovery: $1.5 MM/yr.
  *Total Annual Savings: $ 4.0 MM/yr.*

- Site 2—Specialty Products Plant
  - 25% decrease in site utility costs
  - 15% Reduction in site wastewater hydraulic load
  - Annual savings from materials recovery: $2.0 MM/yr.
  *Total Annual Savings: $ 3.3 MM/yr.*

- Site 3—Polymer Plant
  - 10% Decrease in site utility costs
  - 10% Reduction in site wastewater hydraulic load
  - Increase in capacity by process debottlenecking: 4%

Total Annual Savings: $ 1.5 MM/yr.

Recent process integration publications include:

### Mass Exchange Networks (MEN)

Gupta, A. and Manousiouthakis, V., Waste reduction through multicomponent mass exchange network synthesis, *Comp. Chem. Eng.*, 18, S585, 1994.

Papalexandri, K. P., Floudas, C., and Pistikopoulos, E. N., Mass exchange networks for waste minimization: A simultaneous approach, *Trans. I. Chem. E.*, 72(Part a), 279, 1994.

Rossiter, A. P., Process integration and pollution prevention, *AIChE Symp. Ser.*, 90(303), 12, 1994.

Kiperstok, A., and Sharratt, P. N., On the optimization of mass-exchange networks for fixed removal of pollutants, *Trans. IChem E.*, 73B, 271, 1995.

Zhu, M. and El-Halwagi, M. M., Synthesis of flexible mass exchange networks, *Chem. Eng. Comm.*, 183, 193, 1995.

Dunn, R. F., and El-Halwagi M. M. "Design of Cost-Effective VOC Recovery Systems," Published by Tennessee Valley Authority Environmental Research Center/EPA Center for Waste Reduction, http://www.owr.ehnr.state.nc.us/ref/00034.htm, 1996, 1–82.

### Mass Exchange Networks with Regeneration (MEN/REGEN)

Garrison, G. W., Cooley B. L., and El-Halwagi, M. M., Synthesis of mass-exchange networks with multiple target mass separating agents, *Dev. Chem. Eng. Min. Proc.*, 3(1), 31, 1995.

### Reactive Mass Exchange Networks (REAMEN)

Srinivas, B. K., and El-Halwagi, M. M., Synthesis of reactive mass-exchange networks with general nonlinear equilibrium functions, *AIChE J.*, 40(3), 463–472, 1994.

Warren, A., Srinivas, B. K., and El-Halwagi, M. M., Optimal Design of waste reductions systems for coal-liquefaction plants, *J. Env. Eng..*, 121(10) 742, 1995.

### Combined Heat and Reactive Mass Exchange Networks (CHARMEN)

Huang, Y. L., and Edgar, T. F., Knowledge based design approach for the simultaneous minimization of waste generation and energy consumption in a petroleum refinery. In A. P. Rossiter, Ed., *Waste Minimization through Process Design*, McGraw Hill, New York, 1995. Pp. 181–196.

Srinivas, B. K., and El-Halwagi, M. M., Synthesis of Combined heat and reactive mass-exchange networks, *Chem. Eng. Sci.*, 49(13), 2059-2074, 1994.

### Heat-Induced Separation Networks (HISEN)

Dye, S. R., Berry, D. A., and Ng, K. M., Synthesis of crystallization-based separation systems, *AIChE Symp. Ser.*, 91(304), 238, 1995.

El-Halwagi, M. M., Srinivas, B. K., and Dunn, R. F., Synthesis of optimal heat-induced separation networks, *Chem. Eng. Sci.*, 50(1), 81–97, 1995.

### Energy-Induced Separation Networks (EISEN)

Dunn, R. F., Zhu, M., Srinivas, B. K., and El-Halwagi, M. M., Optimal design of energy-induced separation networks for VOC recovery, *AIChE J. Symp. Series*, 90(103), 74–85, 1994.

### Reverse Osmosis Networks (RON)

El-Halwagi, M. M., Optimal design of membrane-hybrid systems for waste reduction, *Sep. Sci. Tech.*, 28(1–3), 282 (1993).

### Pervaporation Networks (PERVAP)

Srinivas, B. K., and El-Halwagi, M. M., Optimal design of pervaporation systems for waste reduction, *Comp. Chem. Eng.*, 17(10), 957–970, 1993.

Wijmans, J. G., Kaschemekat, J., Davidson, J. E. and Baker, R. W., Treatment of organic contaminated waste water streams by pervaporation, *Env. Prog.*, 9(4), 262 (1990).

### Waste Interception and Allocation Networks (WIN)

El-Halwagi, M. M., Hamad, A. A., Garrison, G. W., Synthesis of waste interception and allocation networks, *AIChE J.*, 42(11), 3087 (1996).

El-Halwagi, M. M., and Dennis Spriggs, H., "An Integrated Approach to Cost and Energy Efficient Pollution Prevention." Paper presented at the 5th World Congress of Chemical Engineering Reprint, p. 344 (1996).

### Heat-Induced Waste Minimization Networks (HIWAMIN)

Dunn, R. F., and Srinivas, B. K., Synthesis of heat-induced waste minimization networks HIWAMINs, *Adv. Environ. Res.*, 1(3), 275-301.

### Membrane-Hybrid Systems

Hamad, A. A., Crabtree, E. W., Garrison, G. W., and El-Halwagi, M. M., "Optimal Design of Hybrid Separation Systems for In-Plant Waste Reduction," Paper presented at the 5th World Congress of Chemical Engineering Reprint, p. 453, (1996).

### 4.3.8. Membrane Basics

*This presentation was made by Edward L. Cussler, an Institute of Technology Professor at the University of Minnesota. Prior to this, he was professor of chemical engineering and materials science. He was also previously a director, vice president, and president of the American Institute of Chemical Engineers from 1989 through 1995. This presenta-*

tion began with a brief overview of mass transfer rate (flux) expressions for basic membrane systems, both with and without pores. This material is covered in Chapter 3, and is therefore not repeated here. In addition, Dr. Cussler recently published an overview (Cussler, 1997)

The current membrane market is modest. Medical applications account for $600 million in gross sales, reverse osmosis/ultrafiltration about $1 billion, and gas separations, $100 million.

Keys to feasible systems include:

- Keep membrane material costs below $10/m^2$
- Membrane thickness need be no thinner than 0.1 $\mu$m for polymer membranes; making them thinner usually results in flux being controlled by boundary layer effects
- Area to volume ratio of about 3000 $m^2/m^3$

Frequent design considerations are that membrane systems are usually only one stage. That is, the nature of the technology does not lend itself to recycle loops or multiple series stages. Membranes are susceptible to fouling.

Part of the complexity of modeling membrane systems is the diversity of driving forces under which they operate. For example, glass membranes are typically used for gas separations. In such systems, diffusion through the membrane is the controlling feature. Rubber membranes are often used for vapors. For these systems, solubility is normally the controlling feature.

The following new directions are recommended for future research:

- Membrane chemistry
  - Olefin/alkane separations
  - Vapors
- Module design to reduce fouling
  - As a function of flow
  - As a function of backflush
- Untried separations
  - Target applications where a minor component is permeable (selective separations)

### 4.3.9. Membrane Technology Panel Presentations and Discussion

*Gregory Keeports, Manager, Hazard Assessment and Environmental Engineering, Rohm & Haas, and Chair of the CWRT Advisory Board moderated this panel presentation.*

### 4.3.9.1. Membrane Technologies: Applications—Present and Future

*This presentation was made by Richard W. Baker, President of Membrane Technology & Research, Inc.*

Rubbery membranes are solubility selective. They can be used to separate condensible from noncondensible compounds, such as separating propylene from nitrogen. Even though nitrogen is smaller, propylene preferentially passes because it is more soluble in the membrane. Solubility selective membrane applications include

- HCFC from $N_2$ in refrigeration plant vents
- VCM from $N_2$ from PVC polymerization
- Olefins from $N_2$ polyolefin resin purge bins and distillation column vents

Figure 4.2 provides a schematic for use of a membrane for monomer recycle in polyolefin manufacturing.

New industrial applications in the future may include the following:
- Utilizing existing membranes
  - $Cl_2/N_2$ separation in chlor alkali plant tail gas
  - $H_2/C_{3+}$ hydrocarbons in refinery fuel gas
  - $C_2H_4/N_2$, air in ethylene oxide reactor vents (oxygen reactor vents)—can expect these in 2–3 years
- Utilizing new membranes
  - Vapor/vapor separations
    o Propylene/propane
    o Ethylene/ethane

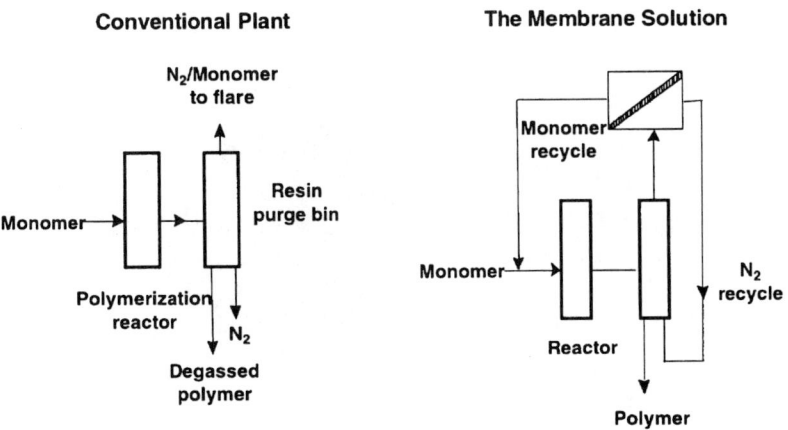

**Figure 4.2.** Membrane alternative for monomer recovery in polyolefin plants.

### 4.3.9.2. Applications of Membrane Technologies for Process Waste Reductions in Gaseous Streams

*This presentation was made by Pushpinder S. Puri. In his current position in the Corporate Science and Technology Center of Air Products and Chemicals, Inc. (APCI), he is responsible for the global leadership of gas separation membrane technology, and management of APCI's polymer technology program.*

Membranes have a variety of features which should insure their continued market growth:

- Easy to operate—"it's like operating a pipe"
- Versatility—ability to handle varied feed compositions and concentrations
- Modular—easy to turn-up and turn-down, linear scale-up
- Wide operating ranges—temperature, pressure, environments
- Commercial availability—high selectivity and productivity
- Chemical compatibility
- Low cost

The membrane market continues to grow. Figure 4.3 provides statistics on the current gaseous system membrane market. In addition, membrane materials research has made some impressive strides. For example, membranes only 300–400 Å thick are being used under 150 atm pressure.

Specific applications that should be considered for future development efforts include

- Recovery and recycle of organic vapors from air/gas streams
  - Airborne VOC recovery—fuel transfer, dry cleaning, etc.
  - Residual solvents and monomer recovery—paints, coatings, adhesives, etc.

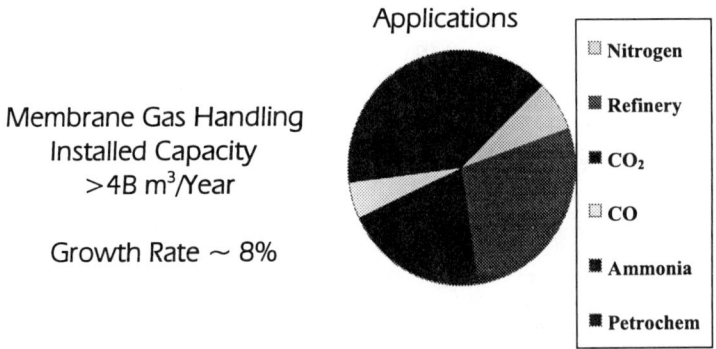

**Figure 4.3.** Gaseous streams membrane installed capacity and applications.

- Acid gas recovery from fuel and combustion gases
  - $CO_2$, $SO_x$, and $NO_x$ recovery from stack/exhaust gases
  - Natural gas upgrading
- Recovery and recycle of perfluorocompounds (PFCs) and hydrofluorocompounds (HFCs) in semiconductor industry

Research and development challenges include:

- Need membrane materials beyond current commercial polymers
  - Mixed-matrix, inorganic–organic nanocomposites, molecular sieves, etc.
- Exploitation of novel membrane transport mechanisms
  - Membrane as gas–liquid, gas–solid contactors (membrane absorbers/adsorbers)
  - Size selective membranes (sieving membranes)
  - Sorption enhanced separations (SSF membranes)
- Hybrid Processes
  - Membranes as a novel tool in the toolbox to develop economic hybrid process schemes

### 4.3.9.3. Environmental/Health Implications

*This presentation was made by David R. Shonnard, assistant professor in the Department of Chemical Engineering, Michigan Technological University.*

As companies seek to reduce the environmental impact of their products and processes, the environmental research community needs to respond with methods to quantify the relative impact of competing design options. Example environmental impacts include global warming, ozone depletion, smog formation, human carcinogenic potential, human noncarcinogenic hazard potential, and biotic hazards. Toward this end, there are a number of research opportunities to improve the utilization of environmental and health data:

- Research foci needed to improve identification of chemicals of concern
  - environmental impacts
  - economic value
- Research emphases needed to improve waste streams characterization within processes
  - stream properties
  - foster industrial collaborations
- New membrane materials and modules needed to expand in-process recycle applications

Example applications described in Chapter 3, which clearly address environmental impacts include

- Ozone depletion—CFC-12 recovery from distillation column vents
- Global warming—$CO_2$ recovery from fossil fuel combustion
- Smog formation—VOC recovery from aqueous waste streams
- Human health—heavy metal recovery from aqueous waste streams

### 4.3.9.4. Process Waste Reduction via Membranes

*This presentation was made by Kamalesh K. Sirkar. He is a Distinguished professor in the Department of Chemical Engineering at the New Jersey Institute of Technology, where he also holds the sponsored chair in membrane separations. This presentation provides a fairly detailed list of research and development needs specific to waste reduction.*

The following are strategies to reduce process waste:

- Apply existing membrane separation technologies
- Apply a hybrid of membrane and other separation technologies
- Develop new membrane separation technologies
- Develop new separation concepts using existing membrane devices

Pollutant-specific needs for membrane separation technology development include:

- Gaseous: solvent vapors, $C_2$–$C_5$ Hydrocarbons from $N_2$/air/waste gas
  - Compact, cheaper polymeric hollow fiber vapor permeation devices for
    - High capacity (>2000 cfm) process exhaust streams
    - Low capacity (2–20 cfm) process applications (high concentration)
  - Higher vapor permeation membrane selectivity (especially for $C_2$–$C_5$ hydrocarbons)
  - Membrane-based absorbers for high purification followed by membrane strippers for absorbent regeneration
  - Cyclic membrane devices for high purification (replace carbon beds)
- Gaseous: $CO_2$
  - For low pressure feed (flue gas, etc.), cheap, high-capacity amine-based membrane absorbers and regenerators—avoid membrane wetting
  - High $CO_2/CH_4$ selectivity membrane for high pressure natural gas streams: plasticization-resistant ceramic-polymer composite
  - Fuel cell (gasoline reforming-based), syngas ($CO_2$–$H_2$–CO)
- Gaseous: $SO_2$–$NO_x$ (flue gas)
  - Cheap high capacity crossflow membrane absorbers (bottleneck: desorber in series)
- Gaseous: Oxidizing agents/reactants
  - Chemically stable porous/nonporous membrane absorber–reactor for oxidizing agents and other reactants or species ($Cl_2$, $Br_2$, $O_3$, etc.)

- Liquids: Organic pollutants in water
  - Develop more selective pervaporation membranes for polar volatile solutes (alcohols, organic acids)
  - Develop cheap, crossflow-based fouling-resistant hollow fiber modules for pervaporation
  - Develop membrane-based strippers over higher temperature ranges
  - Develop membrane solvent extraction applications
- Develop membrane systems coupled with reactions for biological or chemical degradation
- Liquids: Oil–water Separation
  - Develop oil-selective membrane devices
- Liquids: Salts in Water
  - More applications of bipolar membrane technology: convert salt into the corresponding acid and the base
  - Adopt continuous deionization and avoid use of chemicals needed for ion exchange bed regeneration
  - Recover waste heat and develop efficient membrane, device and process for membrane distillation to recover and recycle water
- Liquids: Acids/alkalis in water
  - Apply diffusion dialysis to recover from metal-contaminated streams
  - For nonvolatile acids/alkalis, recover and recycle water by membrane distillation
  - Develop stable facilitated transport hollow fiber membrane to recover and concentrate acids in strip stream as salts
- Liquids: Heavy metals in water
  - Remove and recover via functionalized microfiltration membranes—develop membranes and systems
  - Develop stable supported liquid membranes and modules to remove and concentrate heavy metals for recovery
  - Recover, concentrate and recycle heavy metals by membrane solvent extraction
  - Remove acids via diffusion dialysis
- Liquids: Organic solutes/homogeneous catalysts in organic solvent
  - Develop solvent-resistant nanofiltration membranes to retain 250–1000 mol. wt. solutes
  - Develop solvent-resistant reverse osmosis membranes to retain 50–250 mol. wt. solutes
  - Use nanofiltration membranes to retain homogeneous catalysts
  - Develop selective pervaporation membranes for organic–organic separation over a range of temperatures

Research and development needs may also be organized by the various categories of membrane technologies:

- Reverse osmosis
  - Need organic solvent-resistant membranes/modules
  - Need oxidation resistant membranes/modules
- Nanofiltration
  - Need organic solvent-resistant sharp cut-off membranes, modules: In-process streams for pollution prevention and productivity improvement
- Microfiltration: gas streams
  - Need improved particulate-fouling resistant membranes for high temperature gas streams with/without catalysts (e.g., removal of $NO_x$, VOCs)
- Supported liquid membranes (SLMS)
  - Develop stable SLM and a device for removing, recovering and concentrating heavy metals, organic acids, bases, etc. from dilute waste and process aqueous streams
  - Develop stable SLM and a device for removing and recovering $CO_2$, $H_2S$, VOCs from $N_2$/air
- Bipolar membrane technology
  - Apply bipolar membranes in electrodialysis to convert salts in waste streams to corresponding acid and base
- Pervaporation
  - Organic solvent resistant, high temperature—stable membranes and modules for organic/organic separation
  - Need highly selective membranes for removing/concentrating polar organic solutes from water
  - Cheaper and compact modules for organic dehydration and VOC removal from water
- Vapor permeation
  - Compact, cheaper, more selective vapor permeation membranes
  - Efficient device for all-vapor streams
- Membrane gas absorption
  - Develop membrane and contactor device for nonpore wetting absorption
  - Apply membrane absorbers to absorption of $CO_2$, VOCs, $C_1$–$C_5$, $H_2O$ from natural gas, $SO_2$–$NO_x$ removal from flue gas
- Membrane solvent extraction
  - Develop asymmetric hydrophilic membranes and module for back extraction
  - Apply to many aqueous and organic streams
- Membrane distillation
  - Develop efficient direct contact membrane distillation device for water recovery

In addition to the above specific R&D opportunities, the membrane technology development community should challenge itself to address these questions:

- How can one achieve high purification in membrane gas separation systems?
- How can one minimize the volume of concentrate or reject in membrane separation technologies?

### 4.3.10. Introduction to Separative Reactor Technology

*This presentation was made by Anna Lee Tonkovich, Senior Research Engineer at Pacific Northwest National Laboratory.*

Separative reactor technology achieves process synergy through combination of unit operations. Given good process selection, separative reactors hold the potential for process waste reduction, or pollution prevention, by achieving higher per-pass product yields and purity, producing fewer by-products and consuming less energy. Target processes include oxidation and dehydrogenation reactions. Existing oxidation processes typically have low yields, which accounts for more than 20 billion lb/year $CO_2$ emissions in the commodity chemical sector. Existing dehydrogenation process technologies typically have low yields, which accounts for large separation and recycle loads. Separative reactors have the potential to address these and other processes, achieving waste reduction through improved product selectivity and reduced energy usage.

New industrial applications have demonstrated:

- Syn gas yields exceeding 90%
- Oxidative coupling yields above 30%
- Methanol synthesis
- Olefin production from alkanes
- Preferential production of monoalkylates
- Shift toward the desired reaction pathway in phthalic anhydride production

Separative reactors are best described as a developing technology area. Demonstrated industrial applications are few. The state of the art in separative reactor development is summarized below:

- Adsorptive reactors
  - Simulated countercurrent moving-bed chromatographic reactors
    - 20+ years of R&D; no commercial reactors
    - Dehydrogenation, oxidation, esterification
  - Pressure swing reactors
    - 10 years of R&D; no commercial reactors
    - Isomerization, dehydrogenation, oxidation

o Isomerization, dehydrogenation, oxidation
  – Gas–solid–solid trickle flow reactor
   o 10 years of limited R&D activity
   o Methanol synthesis
 • Membrane reactors
  – Organic membranes
   o 30+ years of R&D and extensive commercial use
   o Operating regime: up to 200 C (limits reactive applications)
   o Many materials options
  – Inorganic membranes
   o 10 years of R&D; limited commercial use
   o Operating regime: up to 1000 C (extends reactive applications)
   o Few materials options
  – Perovskite membranes for oxidation reactions
  – Metal alloy membranes for dehydrogenations
  – Advanced molecular sieve membranes

Existing processes, which are candidates for possible replacement by adsorptive separative reactors, typically involve equilibrium-limited reactions and low per-pass yields. Examples include dehydrogenation, oxidation, methanol synthesis, esterification reactions and reverse water gas shift reactions. Figure 4.4 provides a schematic of a multiple bed adsorptive reactor. Multiple bed systems can be used for processes requiring both adsorption and catalyst. They can also be used to achieve cyclic operation within steady-state plants.

A variety of processes are candidates for membrane reactor application:

 • Dense membrane application: oxidation reactions (ex: Syn gas production)
  – Air separation prior to oxidation reaction
  – Membrane role
   o passive or active $O_2$ transport to reaction zone
   o may create different reaction surfaces

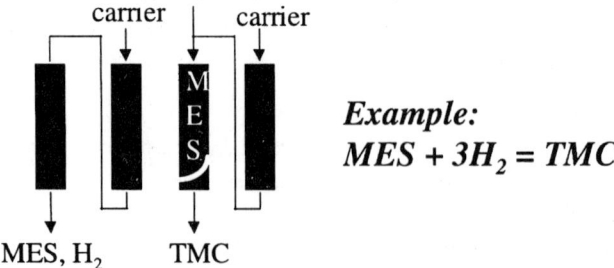

**Figure 4.4.** Multiple bed adsorptive reactor schematic.

- Materials
    - perovskites
    - zirconia
  - Pollution prevention opportunity
    - eliminate downstream product–$N_2$ separation
    - increase selectivity = reduce by-product formation
- Dense membrane application: dehydrogenation reactions (ex: i-butene)
  - $H_2$ separation during reaction to shift equilibrium
  - Membrane role
    - atomic hydrogen transport
    - surface recombination
  - Materials
    - Palladium
    - Palladium-alloys
  - Pollution prevention opportunity
    - eliminate downstream product- $H_2$ separation
    - increase selectivity = reduce by-product formation
- Porous membrane applications: oxidation, alkylation, dehydrogenation
  - Parallel or series-parallel reaction pathway
  - Materials: advanced molecular sieve membranes
    - 40 Å commercially available
    - mesoporous membranes
      - § organic templating agent to create 20 Å pores
      - § macroporous substrates
    - zeolite membranes
      - § thin films with ~5Å pores
      - § macroporous substrates
  - Pollution prevention opportunities
    - favor desired reaction through controlled reactant addition
    - improve selectivity by separating product
  - Membrane reactors pose a number of challenges for implementation:
- Temperature incompatibility
  - Separation selectivity increases as temperature drops
  - Reaction may require high temperature
- Process rates
  - Equal product formation and separation rates
  - Slow step dominates process
- Materials
  - Selectivity
  - Inert for reaction OR catalytic for desired reaction

- Stable under reaction conditions (gradients)
- Reactor design issues
  - Thermal integration (heat removal or addition)
  - Catalyst location and design
  - Sealing
  - Complex processes
- Process economics benchmarking
  - Single vs. multiple unit operations?
  - Higher capital vs. lower operating costs?
  - Different downstream separation loads
  - Fewer by-products

Future research needs include:

- Temperature issues
  - Separations at high temperatures
    - selective inorganic sorbent and membrane materials
    - separations based on zeolites, molten salts, oxides, ceramics
    - composite inorganic–organic materials to raise operating temperatures by 100–200°C (ex: polyphosphazenes)
  - Reactions at low temperatures
    - alternative reaction initiation schemes, i.e., nonthermal activation (plasma, microwave, sonic)
    - novel low temperature reaction pathways, i.e., biocatalysis
    - redirect focus on condensed phase reaction pathways
- Process rates
  - Rigorous simulation capability to set targets for materials development and process designs
- Next generation of materials
  - Membranes
    - Mesoporous materials (pore diameters <20 Å)
    - Oxide membranes with molecular sieve capability for gases (pore diameters between 5 and 10 Å)
    - Zeolite membranes that are active for desired reaction and passivated for undesired pathways
    - Coatings to alter surface properties
  - Sorbents
- Process simplification
- Economic evaluation
  - Cost studies
  - Cost comparisons

o lower operating costs vs higher capital costs
  o different product purity may require alternate downstream processing
  – Assessment of risks for implementation
  o safety
  o retrofit of existing process
  o total process replacement
  o added capacity for growth chemicals

We are probably 3–5 years away from seeing the first commercial separative reactors. With new materials and designs, we hope to see robust systems in use before the year 2020. A research and development investment strategy should include the following objectives:

- *Near term*—develop existing concepts
  – Economic evaluation
  – Process simplification and scalability
  – Rigorous integrated process development
- *Mid to Long term*—seed technology "pipeline"
  – Raise technology robustness and overcome limitations
  – Achieve major pollution prevention targets

### 4.3.11. Separative Reactor Technology Panel Presentations and Discussion

*This panel was moderated by Kerry Irons, a Global Core Technical Leader in the Engineering Sciences/Market Development Department of the Dow Chemical Company. He is responsible for leveraging Core Technologies R&D capabilities into Dow Manufacturing's technology implementation work processes. He also is the global coordinator for Dow's internal and external environmental technology process research.*

#### 4.3.11.1 Adsorption-Based Separative Reactor Technology

*This presentation was made by Robert W. Carr, professor, Department of Chemical Engineering and Materials Science at the University of Minnesota. Much of this presentation focussed on simulated countercurrent moving bed chromatographic reactors, which is covered in Chapter 2, and is not repeated here.*

Technical and research issues for further development of adsorptive reactors include:

- Reactors
  – Optimal design (configuration)
  – Heat integration in high temperature reactions

- Catalysts
  - High selectivity
  - Low temperature (compatibility with adsorbents)
- Adsorbents
  - Adsorbent selection
  - New adsorbents
  - Surface heterogeneity
  - Purging of strongly adsorbed substances

### 4.3.11.2. Membrane-Based Separative Reactor Technology

*This presentation was made by Richard Noble, professor of chemical engineering at the University of Colorado in Boulder and Co-Director of the NSF I/U CRC Center for Separations Using Thin Films. Dr. Noble introduced a series of catalytic membrane reactor configurations and reaction mechanisms involved in membrane reactor processes. The reader is directed to Chapter 3 of this book and Chapter 14 in* Membrane Separations Technology: Principles and Applications *(Falconer et al., 1995)*

Recent advances in materials include the development of solid oxide and zeolite membranes; in applications, use in selective oxidation reactions. Technical barriers that need overcoming include:

- Materials
  - Nanoporous materials
  - Controlled pore size and porosity
- Operations
  - Temperature and pressure range
  - Lifetime
  - Chemical resistance
- Applications
  - Controlled permeation of reactants and products
  - Uniform access to catalyst
  - Selective oxidations
  - Dehydrogenation
  - Isomerization

### 4.3.11.3. Synthesis of an Integrated Separative Reactor Process

*This presentation was made by Jeffrey J. Siirola, a research fellow in the Chemical Process Research Laboratory of Eastman Chemical Company. It describes a separative reactor alternative for methyl acetate production.*

Figure 4.5 depicts a conventional acetate flowsheet. Figure 4.6 is a flowsheet for the original methyl acetate process used in this case study. Figures 4.7 and 4.8

4. Findings of the National Workshop on Process Waste Reduction 295

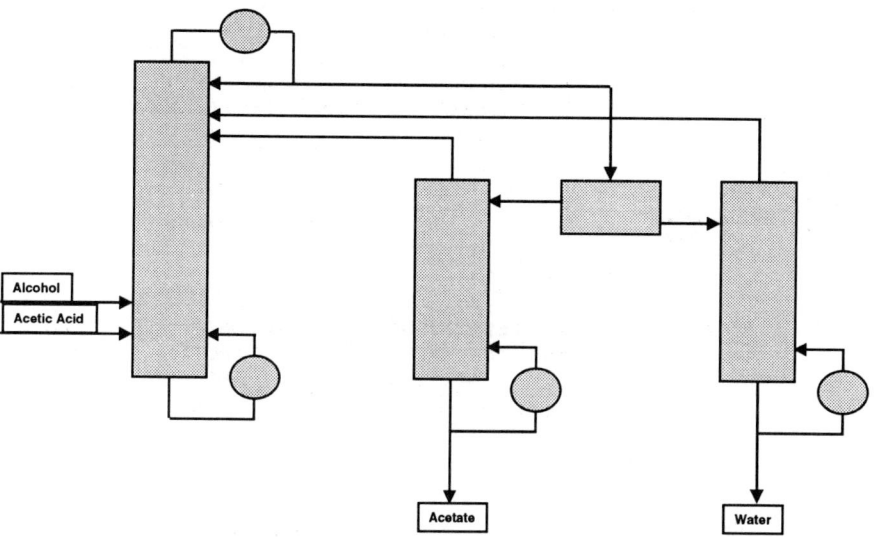

**Figure 4.5.** Conventional acetate process flowsheet.

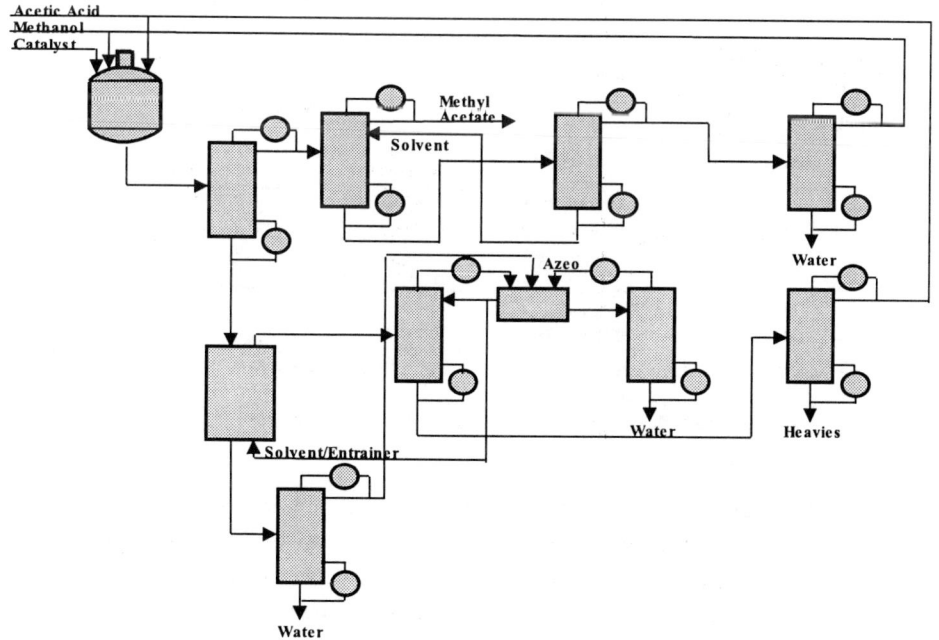

**Figure 4.6.** Original methyl acetate flowsheet.

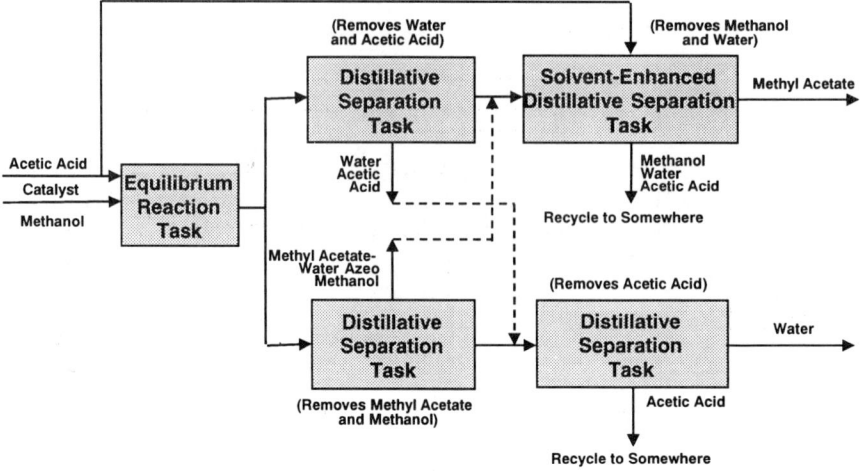

**Figure 4.7.** Task Approach to methyl acetate process synthesis.

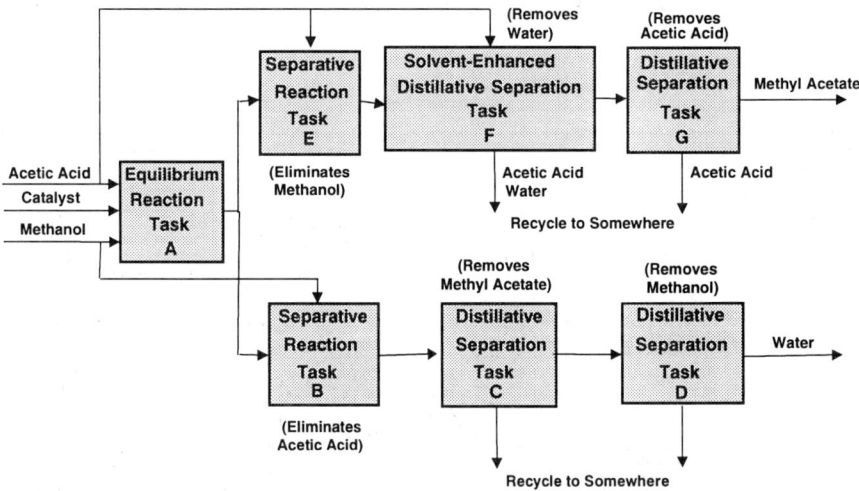

**Figure 4.8.** Additional reaction tasks to alter separation problems.

provide task flows for methyl acetate synthesis and the additional reaction tasks necessary to alter process separation problems, respectively. Finally, Figure 4.9 shows the configuration of the integrated methyl acetate separative reactor process. Both capital and operating costs are cut 80% from the original flowsheet (Figure 4.6). One will also note that the number of major pieces of equipment is cut from 28 to three.

4. Findings of the National Workshop on Process Waste Reduction    297

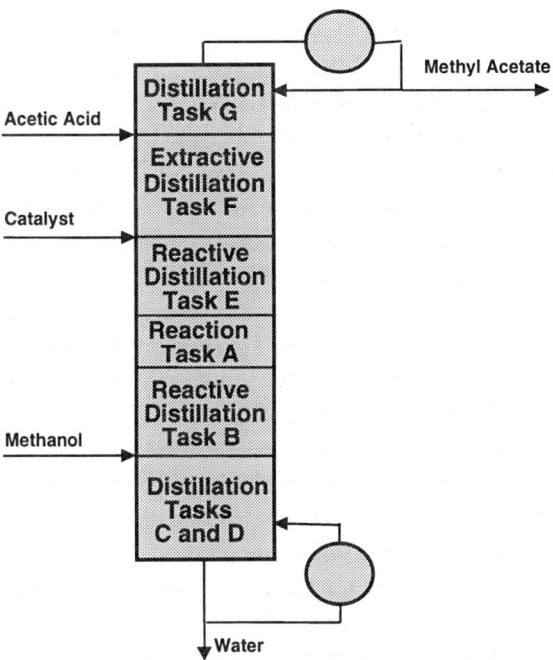

**Figure 4.9.** Integrated methyl acetate separative reactor process.

### 4.3.12. Conclusions of CWRT Separation and Separative Reactor Monograph

*This presentation was made by Peter P. Radecki, executive director, corporate relations for Michigan Technological University, and Corporate Liaison for CenCITT.*

The monograph (this book) is part of a CWRT long-term objective of developing technologies and tools useful for promoting waste minimization practices in industry. Other components of the project include the described workshop, technology fairs and roadmapping exercises. Organizations who have interest in developing separation technology investment strategies and who are partners in this program include the U.S. Department of Energy Office of Industrial Technologies (OIT) and the National Center for Clean Industrial and Treatment Technologies (CenCITT). This monograph, which focuses primarily on adsorption and membrane separation technologies, also contains information on the emerging science of separative reactors using membranes, adsorption, and reactive distillation.

The combination of the monograph and this workshop is an experiment. Traditionally, technology researchers and developers approach their work through fundamental characterization and exploitation of the driving forces of the technologies. They start with the driving forces, develop the technology and then seek applications. Here, however, technology experts are presented with a desired endpoint—cost-effective waste reduction—and are challenged to (1) characterize the roles that their technology areas could play in achieving that endpoint, and (2) propose target streams for research and development objectives.

In the Monsanto Million Dollar Challenge, a process stream that needed "fixing" was described and researchers the world over were asked to propose R&D programs to "fix" it. That is, an industry-relevant design objective was presented and technology researchers were asked to adapt their skills to meet the objective. The project under which this monograph is being written effectively treats the areas of adsorption, membrane and separative reactor research in much the same way. That is, workshop participants identified many process streams that need fixing in order to achieve the design objective of process waste reduction. Research funding programs in the future, including those of the sponsoring organizations, can combine the technology state-of-the-art characterizations and the process streams in this monograph to devise process waste reduction funding programs. Research centers can use the findings to devise their technology development strategies. The end result could be many industry-relevant process waste reduction "million dollar challenges" being addressed. Making this happen will require the leadership of many organizations to continue the process and to collaborate. This monograph is only the first step in the process.

As indicated earlier, this project is an experiment. If successful, we will develop a culture of cross-sector collaboration in R&D. The roadmapping exercise, which followed immediately after the workshop, is potentially a first step in development of a design objective oriented research agenda. The challenge will be to maintain the focus on problem solving.

## 4.4. Processes and Process Stream Recommendations

Breakout session discussions at this workshop resulted in identification of over 40 target processes and process streams for research involving adsorption, membrane and separative reactor technologies. The processes and streams are categorized in two tables: Table 4.2 shows candidate processes and streams for in-process recycle, feed purification and feed preprocessing, and Table 4.3 shows candidate processes and streams for reaction by-product elimination and posttreatment recycle. These tables include many process or stream attributes, and however, time did not permit all attributes to be defined. It is hoped that others may complete the stream attributes as part of future research projects.

**Table 4.2.** Candidate Processes and Process Streams for In-Process Recycle, Feedstream Purification and Feed Preprocessing

| Process Stream Description | Chemical or Production Process Name | Applicable Separation Breakthrough Area | Typical Contaminant(s) Conc. (% or ppm) | Batch or Continuous? | Process Stream Matrix Description | Flowrate Range | Separation Efficiency Desired (%) | Recovered Stream Purity Required (% or ppm) | Applicable Temp. Range | Applicable Press. Range | Other Separation Technologies to Consider | Remarks |
|---|---|---|---|---|---|---|---|---|---|---|---|---|
| Chiral Molecules, Feed Stream | Agricultural and specialty chemical manufacturing | Adsorbent with required selectivity yet capable of bulk separation | Bulk separation 50% v/v | B & C | Organic solvent | $1 \times 10^6$ lb/yr | 0.99 | 0.99 | Near ambient | Near ambient | Distillation, chromatography, crystallization | |
| Polyolefin Purge Bin Vent Streams | Polyolefin production | Process and materials, cost, selectivity | 10% monomer, 1% heavy organics | C | Nitrogen is remaining matrix | $1 \times 10^7$ lb/yr monomer | 0.9 | 0.9 | Ambient | Ambient | Membrane, refrigeration/condensation | Selectivity needed, high throughputs |
| Ethanol, acid and sugar | Ethanol production from glucose and sulfuric acid | Adsorbent and process | 10% acid, 10% sugar | C | Water, xylose, natural organics | $1 \times 10^6$ gal/yr poss. | >95% | >95% | Ambient | Ambient | Chromatography, neutralization | Cost must be very low to compete with gasoline for fuel, eliminate TDS |
| Propylene Glycol from Fermentation | Fermentation | Selective adsorbent that won't be fouled in complex matrix | < 5% | B & C | Separate from water, organic acids, cells, inorganic nutrients | $1 \times 10^6$ lb/yr | >95% | >95% | Ambient | Ambient | Coarse filtration followed by distillation | Matrix complexation |

299

| Process Stream Description | Chemical or Production Process Name | Applicable Separation Breakthrough Area | Typical Contaminant(s) Conc. (% or ppm) | Batch or Continuous? | Process Stream Matrix Description | Flow-rate Range | Separation Efficiency Desired (%) | Recovered Stream Purity Required (% or ppm) | Applicable Temp. Range | Applicable Press. Range | Other Separation Technologies to Consider | Remarks |
|---|---|---|---|---|---|---|---|---|---|---|---|---|
| Ammonia recovery from off gas | Metal processing | Selective adsorbent | 0.03% Ammonia | B & C | Air stream w/ammonia, carbon dioxide, water vapor | $5 \times 10^6$ lb/yr of $NH_3$ | As much as possible | As much as possible | Ambient | Ambient | Venting to air, acid scrubbing | Humidity negatively impacts selectivity for ammonia |
| $CO_2$ and $H_2O$ Separation | Steel making | Selective adsorbent to remove $CO_2$ and $H_2O$ from $H_2$, CO | 20–40% $CO_2$ | C | $H_2S$, CO, $H_2$, $NH_3$, water, $CO_2$ | Large flow-rates | >90% | >90% | Ambient to 350°F | 100 to 500 psi | Direct reduction process | Currently used only as combustion gas |
| Acetone Separation from Air | Solution spinning of fiber, transport and storage | Selectively remove acetone at low concentrations | <100 ppmv | C | Air, acetone, humidity | $50 \times 10^6$ lb/yr acetone to purge warehouses | Any | Any | Ambient | Ambient | Biofiltration | Adsorption used up to this point to remove 99% of acetone in fiber process |
| Silane feed stream purification | Silicone polymer production | Selectivity to separate di-methyl silane from mono- and tri-methyl silanes | ~95% di-methyl silane and 5% mono- and tri- evenly distributed | C | Mixed silanes | $5 \times 10^8$ lb/yr | 0.99 | 0.9999 | Ambient to 200°C | Ambient | Currently expensive distillation w/ rel. vol. ~1.03; extractive distillation also considered | Liquid stream polymerizes in presence of water; need water-free adsorbent |

| Process Stream Description | Chemical or Production Process Name | Applicable Separation Break-through Area | Typical Contaminant(s) Conc. (% or ppm) | Batch or Continuous? | Process Stream Matrix Description | Flow-rate Range | Separation Efficiency Desired (%) | Recovered Stream Purity Required (% or ppm) | Applicable Temp. Range | Applicable Press. Range | Other Separation Technologies to Consider | Remarks |
|---|---|---|---|---|---|---|---|---|---|---|---|---|
| Metal Finishing Rinse Streams | Cu, Ni, Pb, etc. metal finishing | Adsorbent material development, process improvement | .01–1 % | C | Water, organic, other metals | Vari- able | 0.95 | | Ambient | Ambient | Current method is hydroxide precipitation | Not economic to separate now |
| Tail gas cleanup | | | ~20% $C_{i2}$ | | | | | | | | | |
| Remove nitrogen, carbon dioxide, and hydrogen sulfide from natural gas | Natural Gas Purifica- tion | Adsorbent selectivity | Remove $N_2$ to <3%; $H_2S$ from ~1% to 4 ppmv w/o reducing $CO_2$ to this level | C | $N_2$, $CO_2$, and $H_2S$ in natural gas | Vari- able | | | Ambient | High Press- ure | Membranes, absorption | |
| Fuel gas purifica- tion | | | Remove RCRA compds. (e.g. isocyanates, aromatics) present at <1% | C | | | | | | | | |

301

| Process Stream Description | Chemical or Production Process Name | Applicable Separation Breakthrough Area | Typical Contaminant(s) Conc. (% or ppm) | Batch or Continuous? | Process Stream Matrix Description | Flow-rate Range | Separation Efficiency Desired (%) | Recovered Stream Purity Required (% or ppm) | Applicable Temp. Range | Applicable Press. Range | Other Separation Technologies to Consider | Remarks |
|---|---|---|---|---|---|---|---|---|---|---|---|---|
| Inorganic/ corrosive gases | | Need resistant materials with selectivity for picking up corrosive gases from mixtures | | | | | | | | | | |
| Pharmaceutical chemical separation | Pharmaceuticals | Solvent resistant membranes. separative reactors for selective addition of reactants eliminating the separation after reaction | | | Recover 250–700 dalton compounds from small solvent molecules and other organics | | | | | | | Small scale so small econ. driver |
| Removal of product or reaction inhibitors so they don't build up in the bioreactor | Bioreactors | Selective materials and novel process | | | | | | | | | | |

302

**Table 4.3.** Candidate Processes and Process Streams for Reaction By-Product Elimination and Posttreatment Recycle

| Process Stream Description | Chemical or Production Process Name | Applicable Separation Breakthrough Area | Typical Contaminant(s) Conc. (% or ppm) | Batch or Continuous? | Process Stream Matrix Description | Flowrate Range | Separation Efficiency Desired (%) | Recovered Stream Purity Required (% or ppm) | Applicable Temp. Range | Applicable Press. Range | Other Separation Technologies to Consider | Remarks |
|---|---|---|---|---|---|---|---|---|---|---|---|---|
| $H_2$ sep. from C1, C2, C3 | Refineries | Process design | 30–60% $H_2$, 10–20% C3+ | C | ? | 10–50 MM SCF day | 50% | 90% $H_2$ | 50°C | 100 psi | Cryogenic, membrane, PSA | |
| Low MW organics in water (ethylene and propylene glycol, acetic acid) | De-icer, e.g. | Adsorbent membrane | 2% | Semi-B | | ca. 50 gpm | 90% | 50% in product/ 90% vol reduction | Ambient to 100°C | Ambient | RO, distillation, bio? | |
| Less than 50 ppm VOC's in vapor | Many | | <10–50 ppm | Cyclic B | VOC in air | 10–100k SCFM | 90%+ | Less than ppm | Ambient to 50°C | 1–30 bar | Adsorption (TSA) | |
| Oxygenated VOC's (ketones) | Coatings | Adsorbents or process | 2–5k ppm | Cyclic B | VOC in air | 10–100k SCFM | 90%+ | 90%+ | 150–250°C | Ambient | TSA or cryogenic absorption | |
| HCl and other | Chemical mfg. | reversible recovery | 500 ppm to 10% | C | in air with other contaminants | 10–1000 SCFM | 90%+ | >50% | 50°C | 1–10 bar | Ad/absorption | |

| Process Stream Description | Chemical or Production Process Name | Applicable Separation Breakthrough Area | Typical Contaminant(s) Conc. (% or ppm) | Batch or Continuous? | Process Stream Matrix Description | Flowrate Range | Separation Efficiency Desired (%) | Recovered Stream Purity Required (% or ppm) | Applicable Temp. Range | Applicable Press. Range | Other Separation Technologies to Consider | Remarks |
|---|---|---|---|---|---|---|---|---|---|---|---|---|
| Olefin/paraffin | PP prod'n. | Membrane or PSA | 20% paraffin in olefin | C | $H_2$, higher HC's | 500 SCFM | 50%+ | ppm | Ambient to 50°C | 10 bar | Adsorption (PSA)- UOP OLEX process, membrane, distillation | |
| CS2 | Acrylonitrile | ? | 1000 ppm | C | Air | 1–10k SCFM | 90%+ | 90% volume reduction | Ambient | 1–2 bar | Biofiltration | |
| Small volume mixed waste | Specialty chemical production | ? | < 5–10% | B | Mixed organics in water | less than 1000 gpd | 90%+ | | | Ambient | SCWO, WAO, membranes | |
| PFC $CHF_3$, $CF_4$ formed, $C_3F_8$; Five components Electronic off gas streams from electronic component production, flow and conc. is variable. | PF Compound, $CF_4, C_2F_6, NF_3$, $CHF_3, SF_6, N_2$ | Economics. Adsorption capacity is very low. High selectivity for these nonpolar compounds with low polarizability. Obtain adsorbents with high selectivity | 0 to 1%; average is .1 % PFC | C | $N_2$ | 100 to 300 or 10K to 30K L/min, Flow is variable. | Recover 95 % of the PFCs and separate each | > 99.9 pure for each PFC | Ambient | 750 torr, ambient | Membrane distillation, cryogenic adsorption | PFC gas is expense and needs to be removed for environmental reasons. |

| Process Stream Description | Chemical or Production Process Name | Applicable Separation Breakthrough Area | Typical Contaminant(s) Conc. (% or ppm) | Batch or Continuous? | Process Stream Matrix Description | Flowrate Range | Separation Efficiency Desired (%) | Recovered Stream Purity Required (% or ppm) | Applicable Temp. Range | Applicable Press. Range | Other Separation Technologies to Consider | Remarks |
|---|---|---|---|---|---|---|---|---|---|---|---|---|
| Fermentation | Removal of penicillin, removal of color bodies, hormones, etc. | Need high selectivity in similar chemical separations and low cost/lb of separation | ~1% | B, semi-C | | | | | | | | |
| Removal of color bodies | Sugars, glycerine (leading to cosmetics, soaps, etc.) | | PPM or lower | B | | | | | | | | |
| Removal of potential color bodies | Pharmaceuticals | | PPM or lower | B | | | | | | | | |
| Removal of color bodies | Specialty monomers | | PPM or lower | B or C | | | | | | | | |
| Oxidative coupling of methane | | Removal of $O_2$, $H_2$ | High % | C | | | | | | | | |
| Oxidation of $E_tH_2$ to acetaldehyde | | | | C | | | | | | | | |

| Process Stream Description | Chemical or Production Process Name | Applicable Separation Breakthrough Area | Typical Contaminant(s) Conc. (% or ppm) | Batch or Continuous? | Process Stream Matrix Description | Flowrate Range | Separation Efficiency Desired (%) | Recovered Stream Purity Required (% or ppm) | Applicable Temp. Range | Applicable Press. Range | Other Separation Technologies to Consider | Remarks |
|---|---|---|---|---|---|---|---|---|---|---|---|---|
| Adipic acid precursors to cyclohexanone | | | | C | | | | | | | | |
| Odor body removal | Agricultural products, organic manufacturing (e.g., sulfur based chemistry) | | | B | | | | | | | | |
| Avoiding crystallizations/ isolations | Specialty chemicals, paraxylene, agricultural chemicals | | | B or C | | | | | | | | |
| Recovery of linear alpha olefins from branched chain olefins | | | Low % | | | | | | | | | |
| Chlorination – sequester chlorinated materials at just the right point in processes | | | | | | | | | | | | |

| Process Stream Description | Chemical or Production Process Name | Applicable Separation Breakthrough Area | Typical Contaminant(s) Conc. (% or ppm) | Batch or Continuous? | Process Stream Matrix Description | Flowrate Range | Separation Efficiency Desired (%) | Recovered Stream Purity Required (% or ppm) | Applicable Temp. Range | Applicable Press. Range | Other Separation Technologies to Consider | Remarks |
|---|---|---|---|---|---|---|---|---|---|---|---|---|
| | Alkylation | | | | | | | | | | | |
| Pre-purifications to remove oxygenates from otherwise chlorinated solvents | | | | | | | | | | | | |
| Manufacture of EDTA—nitrilotriacetic acid by-product control by removal of ammonia | | | | | | | | | | | | |
| Fine chemical synthesis routes have many competing reactions needing selective removals of by-products | | | | | | | | | | | | |

| Process Stream Description | Chemical or Production Process Name | Applicable Separation Breakthrough Area | Typical Contaminant(s) Conc. (% or ppm) | Batch or Continuous? | Process Stream Matrix Description | Flowrate Range | Separation Efficiency Desired (%) | Recovered Stream Purity Required (% or ppm) | Applicable Temp. Range | Applicable Press. Range | Other Separation Technologies to Consider | Remarks |
|---|---|---|---|---|---|---|---|---|---|---|---|---|
| Chiral adsorbates for racemates (similar to chromatography processes) | | | | | | | | | | | | |
| Stoichiometric catalysis | | | | | | | | | | | | |
| Selective removal of nuclear by-products (removal of radioactive material from phosphates) | | | | | | | | | | | | |
| Use adsorbents to replace solvents | | | | | | | | | | | | |
| Bound reagents/controlled release via adsorb/desorb | | | | | | | | | | | | |

## 4.5. List of Workshop Participants

A diverse group of eighty-nine participated in the workshop, representing twenty-eight companies, eleven government agencies and laboratories, and eight universities. Table 4.4 lists the participants.

**Table 4.4.** Workshop Participants

| Name | Organization |
| --- | --- |
| Adler, Stephen | CWRT |
| Amarnath, Ammi | EPRI |
| Baker, Richard | Membrane Technology Research |
| Beaver, Earl | Monsanto |
| Bodnaruk, Dan | DOE/Chicago |
| Bontha, Jagan | Pacific Northwest Nat. Lab |
| Brosey, Bill | Oak Ridge National Lab. |
| Bryan, Paul F. | Union Carbide |
| Bulloch, John | CenCITT-MTU |
| Carr, Robert W. | Univ. of Minn. |
| Cichy, Paul | Rohm & Haas |
| Constable, David | SmithKline Beecham |
| Cranford, Bruce | AIChE |
| Crittenden, John | CenCITT-MTU |
| Cussler, Ed | University of Minn. |
| Datta, Rathin | Argonne Nat. Lab |
| Dobbs, Gregory M. | United Technologies |
| Doody, Dennis | Bristol-Myers Squibb |
| Dunn, Russell | Solutia |
| Dye, Rob | Los Alamos Nat. Lab. |
| Faulkner, Doug | U.S. DOE |
| Flammino, Tony | Merck |
| Frank, Jim | Argonne Nat. Lab |
| Gold, Harris | Foster-Miller |
| Hankins, Nick | Aspen Technology |

| Name | Organization |
|---|---|
| Harten, Teresa | U.S. EPA-ORD-NRMRL |
| Hassan, Neguib | Westinghouse |
| Helling, Rich | Dow Chemical |
| Hertz, Darryl | M.W. Kellogg Co. |
| Hsiung, Thomas | Air Products & Chemicals |
| Humphrey, Jimmy L. | J.L. Humphrey & Associate |
| Irons, Kerry | Dow Chemical |
| Jarvinen, Gordon | Los Alamos Nat. Lab. |
| Jurgensen, John A. | DuPont Dacron |
| Kaempf, Douglas | U.S. DOE |
| Keeports, Greg | Rohm & Haas |
| Keller, George E. | Consultant |
| Kelley, Steve | NREL |
| Killat, George | Dow Chemical |
| King, Kathy | Rohm & Haas |
| Knaebel, Kent | Adsorption Research, Inc. |
| Krause, Ted | Argonne Nat. Lab |
| Li, David | Air Liquide |
| Marr, Dave | Akzo Nobel |
| Meyer, Don | Nofsinger, Inc. |
| Miller, Robert N. | Air Products & Chemicals |
| Mills, Ken | Norton Chemical Proc. Prod. |
| Muhlebach, George | CWRT |
| Munns, Thomas | National Research Council |
| Myers, Alan L. | Univ. of Penn. |
| Noble, Richard | Univ. of Colorado |
| Ong, Say Kee | Iowa State Univ. |
| Paulson, Jim | U.S. DOE, Chicago |
| Pellegrino, John | NIST |
| Pennington, Tim | ARCO Chemical |
| Pereira, Candido | Argonne Nat. Lab |
| Peterson, Eric | Lockheed Martin-Idaho |

## 4. Findings of the National Workshop on Process Waste Reduction

| Name | Organization |
|---|---|
| Ponciroli, Dana | CWRT |
| Puri, Pushpinder S. | Air Products & Chemicals |
| Radecki, Pete | CenCITT-MTU |
| Reynolds, John | Lawrence Livermore Nat. Lab. |
| Robinson, Sharon | Oak Ridge National Lab. |
| Rogers, Joseph | CWRT |
| Roundhill, Max | Texas Tech. Univ. |
| Ruthven, Douglas | University of Maine |
| Sall, Erik | Monsanto |
| Salvo, Joseph J. | GE Corp., R&D |
| Sauer, Nancy | Los Alamos Nat. Lab. |
| Schmidt, Tom | Oak Ridge National Lab. |
| Schucker, Robert C. | Exxon R&D Labs |
| Seale, Jerry | Eastman Chemical |
| Shirley, Arthur | BOC Gases R&D |
| Shonnard, David | Michigan Tech. Univ. |
| Siirola, Jeff | Eastman Chemical |
| Sirkar, Kamelesh | NJIT |
| Smith, Barbara F. | Los Alamos Nat. Lab. |
| Smith, Ron | SRI Consulting |
| Stone, Mark | INEEL/LMITCO |
| Swink, Denise | DOE/OIT |
| Thompson, Tyler | Dow Chemical |
| Tonkovich, Lee | Pacific Northwest Nat. Lab |
| Tran, Thanh | Argonne Nat. Lab |
| Vandergrift, George F. | Argonne Nat. Lab |
| Watson, Jack | Oak Ridge National Lab. |
| Watz, Jill | Lawrence Livermore Nat. Lab. |
| Weiner, Steven C. | Pacific Northwest Nat. Lab |
| Woinsky, Sam | EPRI, Chem., Petro., & N.Gas Center |
| Wooley, Robert | NREL |
| Zhang, Jenny | ARCO Chemical |

## References

ACS, AIChE, CMA, CCR and SOCMA (1996), "Technology Vision 2020—The U.S. Chemical Industry." Report available from American Chemical Society, Dept. of Government Relations and Science Policy, Washington, DC.

CWRT (1998), *Vision 2020: 1998 Separations Roadmap,* CWRT/AIChE, New York.

Cussler, E. L. (1997), *Diffusion Mass Transfer in Fluid Systems,* 2nd ed., Cambridge University Press, New York, Chapter 17.

Falconer, J. L., et al. (1995), Catalytic membrane reactor. In *Membrane Separation Technology: Principles and Applications*, Chapter 14, Elsevier, New York.

Johnson, G. (1995), *Fire in the Mind—Science, Faith and the Search for Order*, Vintage Books, New York.

# Index

## A

Absorption
  carbon dioxide, 182
  gas, membrane-based, current processes, 159–161
Absorption reactors
  process modifications, 2
  separative reactor processes, 25–26
Acetic acid, esterification of, simulated countercurrent moving bed chromatographic reactor, 104–105
Activated alumina adsorbents, described, 51–52
ADSIM simulation, adsorption technologies, waste reduction, 275–277
Adsorbents, 48–54
  activated alumina, 51–52
  carbon, 48–49
  characteristics of, 5
  clay, 52–53
  examples of, 5–6
  polymeric and ion exchange resins, 51
  research needs, 3–4, 75–76
  selection of, 53–54
  silica gel, 52
  suppliers of, 86–88
  zeolites, 49–50
Adsorption, carbon dioxide, 182–183
Adsorption-based separative reactor processes, 293–294
Adsorption equilibrium
  adsorption process design, 69–70
  adsorption technologies, 38–43
  simulated countercurrent moving bed chromatographic reactor, 98–101
  single-component, engineering design information, 77–78

Adsorption process design, 54–74
  configurations, 55–69
    moving bed adsorbers, 63–66
    nonregenerative, 55–56
    pressure drop, 66–69
    regenerative, 56–63
  cost efficiency, 71–72
  equilibrium and mass transfer parameters, 69–70
  mechanisms, 54–55
  pollution prevention, 72–74
Adsorption processes, 4–17
  adsorbents, 5–6
  advantages/disadvantages, 11
  applications summary, 1–4
  bibliography, 128–129
  configurations, 7–11
    bed movements, 10
    fixed beds, 7–8
    moving/fluid beds, 8–9
    rotary wheels, 9–10
    simulated moving beds, 10, 11
  cost efficiency, 12–15
  factors favoring, 11
  fundamentals of, 4–5
  future directions, 15–17, 266, 273–275
  implementation progress, 74–79
    engineering design information, 77–79
    materials development, 75–76
  nonreactive uses, 79–86. *See also* Nonreactive adsorption processes
  pollution prevention, 12
  regeneration cycles, 6–7
  suppliers of, 86–88
Adsorption reactors
  process modifications, 2
  separative reactor processes, 26

313

Adsorption technologies, 33–54
  adsorbents, 48–54. *See also* Adsorbents
  adsorption equilibrium, 38–43
  bibliography, 128–129
  current processes, 35–36
  fundamentals of, 36–38
  heat of adsorption, 43–48
    kinetics, 44–46
    molecular simulation, 46–48
  overview, 34–35, 267–270
  pollution prevention role, 270–271
  waste reduction
    design and optimization, 275–277
    overview, 271–273
Adsorptive chemical reactors, 90–129
  countercurrent moving bed chromatographic reactor, 94–96
  gas–solid–solid trickle flow reactor, 120–122
  natural gas utilization, 108–120
    methanol from synthesis gas, 114
    oxidative coupling of methane, 109–112
    partial oxidation of methane to methanol, 112–113
  overview, 90–92
  pressure swing reactor, 114–120
  reaction chromatography, 92–93
  rotating cylindrical annulus chromatographic reactor, 93–94
  simulated countercurrent moving bed chromatographic reactor, 97–108
    equilibrium stage model, 98–101
    esterification of acetic acid, 104–105
    multiple column configuration, 101–104
    reactor dynamics, 105–108
  temperature swing reactor, 122–124
Airborne organics, recovery of, process modifications, 3
Air drying, nonreactive adsorption processes, 82–83
Aqueous streams
  metal ion recovery from, 199–212. *See also* Metal ion recovery from aqueous waste streams
  pervaporation/aqueous streams, 212, 215–219. *See also* Pervaporation/aqueous streams

**B**

Biological scrubbing, separative reactor processes, 26
Biostripping, separative reactor processes, 26
Bioventing, separative reactor processes, 25–26

By-product streams, recovery from, process modifications, 3

**C**

Carbon adsorbents, described, 48–49
Carbon-based membranes, membrane technology, 170–171
Carbon dioxide, sources of, greenhouse gases, 181–183. *See also* Greenhouse gases
Catalyst recovery, process modifications, 2
Center for Waste Reduction Technologies, 262–264. *See also* Separative reactor processes
Ceramic membranes
  greenhouse gases, 184–185
  membrane technology, 168–170
Chromatographic reactor
  adsorptive chemical reactors, 92–93
  countercurrent moving bed, adsorptive chemical reactors, 94–96
  rotating cylindrical annulus, adsorptive chemical reactors, 93–94
  simulated countercurrent moving bed, adsorptive chemical reactors, 97–108
Clay adsorbents, described, 52–53
Climate change. *See* Greenhouse gases
Close-boiling isomers, moving bed adsorbers, 65
Continuous flow contactor, improvements, 77
Cost efficiency
  adsorption processes, 12–15, 66–69, 71–72
  dehydrogenation reactors, 236
  hydrocarbon-selective oxidation, 232
  membranes in separative reactors, 243–248
  metal ion recovery, 210, 212
  methane reforming reactions, 242–243
  natural gas BTU adjustments, 84
  pervaporation/aqueous streams, 216–219
  solvent vapor recovery from gas streams, 193–198
Countercurrent moving bed chromatographic reactor
  adsorptive chemical reactors, 94–96
  simulated, adsorptive chemical reactors, 97–108
Cracked gas drying, nonreactive adsorption processes, 82
Cryogenic process, carbon dioxide, 183

**D**

Dehydrogenation, separative reactor processes, 27–29

Dehydrogenation reactors, 234–237
Diafiltration, current processes, 143, 144
Dialysis, current processes, 145–146
Diffusion dialysis, electrodialysis (ED), 148–149
Displacement purge adsorption (DPA), regeneration cycles, 7
Displacement-purge cycle, regenerative adsorption processes, fixed bed systems, 58
Distillation vents, solvent vapor recovery from gas streams, 195–196
Donnan dialysis, electrodialysis (ED), 149–150

### E

Economics. *See* Cost efficiency
Electrodesorption, adsorption processes, 16
Electrodialysis (ED)
  current processes, 146–150
  membrane processes, 22
Electromagnetic desorption, adsorption processes, 16
Emulsion liquid membranes (ELM), current processes, 150–152
Equilibrium. *See* Adsorption equilibrium
Esterification, of acetic acid, simulated countercurrent moving bed chromatographic reactor, 104–105
Expensive materials, moving bed adsorbers, 66

### F

Fixed beds
  adsorption processes, 7–8
  regenerative adsorption processes, 57–63
Flat membrane modules, membrane technology, 175
Functionalized MF membranes, metal ion recovery, 207–210

### G

Gas absorption, membrane-based, current processes, 159–161
Gaseous streams
  solvent vapor recovery from, 192–198. *See also* Solvent vapor recovery from gas streams
  waste reduction, membrane technology, 284–285
Gases/bulk separations, nonreactive adsorption processes, 81–82
Gases/purifications separations, nonreactive adsorption processes, 82–83
Gas separation, membrane processes, 19–22
Gas separation membrane polymers, membrane technology, 161–163
Gas–solid–solid trickle flow reactor, adsorptive chemical reactors, 120–122
Glass, membrane technology, 168
Global climate change. *See* Greenhouse gases
Greenhouse gases, 180–191
  carbon dioxide sources, 181–183
  commercial applications, 187–191
    natural gas streams, 189–190
    power generation industry, 187–189
    synthesis gases, 190–191
  overview, 180–181
  technology, 183–187
    ceramic membranes, 184–185
    membrane gas-liquid contactor, 185–187
    selective gas separation membranes, 183–184

### H

Heat of adsorption, adsorption technologies, 43–48. *See also* Adsorption technologies
Hollow-fiber membrane modules, membrane technology, 175–177
Hydrocarbon-selective oxidation, 227–233
  considerations, 230–233
  generally, 227–228
  technology, 228–230
Hydrogen separation, nonreactive adsorption processes, 85–86
Hydrogen sulfide, separative reactor processes, 29

### I

Inert-purge cycle, regenerative adsorption processes, fixed bed systems, 57–58
Inorganic materials (membrane technology), 164–172
  carbon-based, 170–171
  ceramics, 168–170
  generally, 164–168
  glass, 168
  membranes in separative reactors, 224
  zeolites, 168
Integrated separative reactor processes, 294–297
Ion-exchange membranes, electrodialysis (ED), 146–150
Ion exchange resins, described, 51

### K

Kinetics, adsorption technologies, 44–46

## L

Leak-off streams, recovery from, process modifications, 3
Ligand enhanced membrane processing, metal ion recovery, 207
Liquid separation membrane polymers, membrane technology, 163–164
Low-grade refinery gases, hydrogen separation, nonreactive adsorption processes, 85–86

## M

Mass transfer parameters, adsorption process design, 69–70
Membrane-based gas absorption, current processes, 159–161
Membrane-based separative reactor processes, 294
Membrane-based solvent extraction, current processes, 152–154
Membrane-based stripping, current processes, 155–156
Membrane gas permeation, current processes, 156–158
Membrane processes, 17–24
  advantages/disadvantages, 18–19
  applications, 180–219
    metal ion recovery from aqueous waste streams, 199–212. *See also* Metal ion recovery from aqueous waste streams
    pervaporation/aqueous streams, 212, 215–219. *See also* Pervaporation/aqueous streams
    pollution prevention, 19–22
    solvent vapor recovery from gas streams, 192–198. *See also* Solvent vapor recovery from gas streams
    summary of, 1–4
  factors favoring, 19
  fundamentals of, 17–18
  future directions, 22–24, 266
  separative reactors using, 220–258. *See also* Membranes in separative reactors
Membrane reactors
  process modifications, 2
  separative reactor processes, 26–29
Membranes in separative reactors, 220–258
  dehydrogenation reactors, 234–237
  hydrocarbon-selective oxidation, 227–233
    considerations, 230–232
    generally, 227–228
    technology, 228–230
    vendors and contacts, 232
  implementation considerations, 243–248
  methane reforming reactions, 237–243
    considerations, 242–243
    generally, 237–238
    technology, 238–242
  overview, 220–226
Membrane technology, 132–180
  applications, 283–285
  current processes, 136–161
    dialysis, 145–146
    electrodialysis (ED), 146–150
    emulsion liquid membranes (ELM), 150–152
    generally, 136–138
    membrane-based gas absorption, 159–161
    membrane-based solvent extraction, 152–154
    membrane-based stripping, 155–156
    membrane gas permeation, 156–158
    microfiltration (MF), 143–145
    nanofiltration (NF), 142–143
    pervaporation, 154–155
    reverse osmosis (RO), 138, 142
    summary table, 139–141
    ultrafiltration (UF), 143, 144
    vapor permeation, 158–159
  environmental/health implications, 285–286
  materials, 161–172
    inorganic, 164–172. *See also* Inorganic materials
    polymeric, 161–164
  modules
    flat membrane modules, 175
    hollow-fiber and tubular membrane modules, 175–177
    membrane modules, 172–174
  overview, 132–136, 281–282
  systems, 172–173, 177–180
Metal ion recovery from aqueous waste streams, 199–212
  background, 199–200
  considerations, 210, 212
  membrane selection and applications, 200–204
  overview, 199
  technology, 204–210
    functionalized MF membranes, 207–210
    ligand enhanced membrane processing, 207
    membrane contactors, 210
    modules and membranes, 204–207
    system selection, 210, 211
  vendors and contacts, 212, 214

Methane
  oxidative coupling of, adsorptive chemical reactors, 109–112
  oxidative coupling to methanol, adsorptive chemical reactors, 112–113
Methane reforming reactions, 237–243
  considerations, 242–243
  generally, 237–238
  technology, 238–242
Methanol
  oxidative coupling to methane, adsorptive chemical reactors, 112–113
  from synthesis gas, adsorptive chemical reactors, 114
Microfiltration (MF)
  current processes, 143–145
  membrane processes, 21
Microwave desorption, adsorption processes, 16
Molecular simulation
  adsorption technologies, 46–48
  engineering design information, 78
Monolith contactor
  adsorption process design, 67, 69
  improvements, 76
Mosaic resins, adsorption processes, 16
Moving bed adsorbers, adsorption process design, 63–66
Moving/fluid beds, adsorption processes, 8–9
Multicomponent adsorption modeling, engineering design information, 78
Multiple column configuration, simulated countercurrent moving bed chromatographic reactor, 101–104

## N

Nanofiber adsorption technology, 75
Nanofiltration (NF), current processes, 142–143
National Workshop on Process Waste Reduction, recommendations of, 298–308. *See also* Separative reactor processes
Natural gas
  adsorption processes, 15–16
  adsorptive chemical reactors, 108–120
    methanol from synthesis gas, 114
    oxidative coupling of methane, 109–112
    partial oxidation of methane to methanol, 112–113
  BTU adjustments in, 83–85
  improvements, 75
Natural gas drying, nonreactive adsorption processes, 82

Natural gas processing, nonreactive adsorption processes, 82
Natural gas streams, greenhouse gases, 189–190
Natural gas sweetening, nonreactive adsorption processes, 82
Nitrogen production, nonreactive adsorption processes, 81
Nonreactive adsorption processes, 79–86
  adsorption separation examples, 80–81
  BTU adjustments in natural gas, 83–85
  gases/bulk separations, 81–82
  gases/purifications, 82–83
  hydrogen separation, 85–86
Nonreactive membrane processes applications, 180–191, 180–219. *See also* Greenhouse gases; Nonreactive membrane processes applications
Nonregenerative adsorption processes, design, 55–56
$n$-Paraffin separations, nonreactive adsorption processes, 81

## O

Organics, airborne, recovery of, process modifications, 3
Oxidative reaction. *See also* Hydrocarbon-selective oxidation
  hydrocarbon-selective oxidation, 229–230
  of methane, adsorptive chemical reactors, 109–112
  of methane to methanol, adsorptive chemical reactors, 112–113
  partial, methane reforming reactions, 239–242
Oxygen production, nonreactive adsorption processes, 82

## P

Packed bed problems, adsorption process design, 66–69
Parallel passage contactor
  adsorption process design, 67, 69
  improvements, 76
Parallel reactions, membranes in separative reactors, 221–222
Partial oxidative reaction, methane reforming reactions, 239–242
Pervaporation
  current processes, 154–155
  membrane processes, 21

Pervaporation/aqueous streams, 212, 215–219
  considerations, 216–219
  technology, 212, 215–216
Polarity, differing, moving bed adsorbers, 65
Pollution reduction and prevention
  adsorption processes, 12, 72–74
  adsorption technologies, 270–271
  membrane processes, 19–22
  natural gas BTU adjustments, 84
  process integration for, 277–281
  process modifications, 2
  separative reactor processes, 264–266
Polymeric materials
  adsorbents, described, 51
  membranes in separative reactors, 224
  membrane technology, 161–164
Polyolefin polymerization vents, solvent vapor recovery from gas streams, 194–195
Power generation industry, greenhouse gases, 187–189
Pressure drop, adsorption process design, 66–69
Pressure swing adsorption (PSA)
  regeneration cycles, 7
  regenerative adsorption processes, fixed bed systems, 60–63
Pressure swing reactor, adsorptive chemical reactors, 114–120
Process integration, pollution prevention, 277–281
Process waste, reduction of, membrane technology, 286–289
Purge streams, elimination of, process modifications, 2

## R

Radial flow contactor, improvements, 76
Reaction chromatography, adsorptive chemical reactors, 92–93
Reactive distillation, separative reactor processes, 24–25
Refinery gases, low-grade, hydrogen separation, nonreactive adsorption processes, 85–86
Regeneration cycles, adsorption processes, 6–7
Regenerative adsorption processes, design, 56–63
Research & development
  adsorbents, 3–4, 75–76
  adsorption processes, 15–17, 76–79
  membrane processes, 22–24
  separative reactor processes, 29

Reverse osmosis (RO)
  current processes, 138, 142
  membrane processes, 19–21
Rotary wheel adsorbers
  adsorption processes, 9–10
  improvements, 77
Rotating cylindrical annulus chromatographic reactor, adsorptive chemical reactors, 93–94

## S

Safety considerations, membranes in separative reactors, 243–244
Selective gas separation membranes, greenhouse gases, 183–184
Separative reactor processes, 24–29, 259–312
  absorption reactors, 25–26
  adsorption-based, 293–294
  adsorption reactors, 26
  applications summary, 1–4
  factors favoring, 24
  fundamentals of, 24
  future directions, 29, 266
  integrated, 294–297
  membrane-based, 294
  membrane processes and, 220–258. See also Membranes in separative reactors
  membrane reactors, 26–29
  reactive distillation, 24–25
  technical presentations, 262–298
    adsorption technology, 267–277
    future trends, 266
    membrane technology, 281–289
    process integration, 277–281
    separative reactor technology, 289–293
    sustainability, 264–266
  technical presentations, Center for Waste Reduction Technologies, 262–264
  workshop overview, 261
  workshop participants, 259–260, 309–311
  workshop sponsors, 261–262
Series-parallel reactions, membranes in separative reactors, 222
Series reactions, membranes in separative reactors, 222
Silica gel adsorbents, described, 52
Simulated countercurrent moving bed chromatographic reactor, adsorptive chemical reactors, 97–108
Simulated moving beds, adsorption processes, 10, 11
Simulation models, adsorption processes, 17

Single-component adsorption equilibrium, engineering design information, 77–78
Solvents, elimination of, process modifications, 2
Solvent vapor recovery from gas streams, 192–198
  considerations, 193–198
  technology, 192–193
Steam reforming (water-gas shifting) reactions, methane reforming reactions, 238–239
Sustainability, separative reactor processes, 264–266
Synthesis gases
  greenhouse gases, 190–191
  production of, hydrocarbon-selective oxidation, 230

## T

Temperature swing adsorption, regenerative adsorption processes, fixed bed systems, 58–60
Temperature swing reactor, adsorptive chemical reactors, 122–124
Thermal swing adsorption (TSA), regeneration cycles, 6
Thermodynamically limited reactions, membranes in separative reactors, 221
Transient adsorption process modeling, engineering design information, 78–79
Tubular membrane modules, membrane technology, 175–177

## U

Ultrafiltration (UF)
  current processes, 143
  membrane processes, 21
Unsaturated hydrocarbon separations, nonreactive adsorption processes, 81

## V

Vapor permeation, current processes, 158–159
Very-high-boiling materials, moving bed adsorbers, 65
Volatile organic compounds (VOCs)
  adsorption processes, 12–15, 73–74
  pervaporation/aqueous streams, 212, 215–219. *See also* Pervaporation/aqueous streams
  solvent vapor recovery from gas streams, 192–198
  vapor permeation, 158–159

## W

Waste reduction
  adsorption technologies
    design and optimization, 275–277
    overview, 271–273
  gaseous streams, membrane technology, 284–285
  process waste, membrane technology, 286–289
Waste streams, aqueous, metal ion recovery from, 199–212. *See also* metal ion recovery from aqueous waste streams
Wastewater, recovery of, process modifications, 3
Water-gas shifting (steam reforming) reactions, methane reforming reactions, 238–239
Water-splitting technology, electrodialysis (ED), 148

## Z

Zeolites
  described, 49–50
  membrane technology, 168